AF360731

Impact of Natural Hazards on Forest Ecosystems and Their Surrounding Landscape under Climate Change

Impact of Natural Hazards on Forest Ecosystems and Their Surrounding Landscape under Climate Change

Editors

Jaroslav Vido
Paulína Nalevanková

MDPI • Basel • Beijing • Wuhan • Barcelona • Belgrade • Manchester • Tokyo • Cluj • Tianjin

Editors
Jaroslav Vido
Technical University in Zvolen
Zvolen

Paulína Nalevanková
Technical University in Zvolen
Zvolen

Editorial Office
MDPI
St. Alban-Anlage 66
4052 Basel, Switzerland

This is a reprint of articles from the Special Issue published online in the open access journal *Water* (ISSN 2073-4441) (available at: https://www.mdpi.com/journal/water/special_issues/Natural_Hazards_Impact_Forest_Ecosystems#).

For citation purposes, cite each article independently as indicated on the article page online and as indicated below:

LastName, A.A.; LastName, B.B.; LastName, C.C. Article Title. *Journal Name* **Year**, *Volume Number*, Page Range.

ISBN 978-3-0365-2004-9 (Hbk)
ISBN 978-3-0365-2005-6 (PDF)

Cover image courtesy of Jaroslav Vido.

Contents

About the Editors

Jaroslav Vido (Dr.) Deals with research of the climate change related natural hazards and relevant interactions in ecosystems and landscape. He works at the Faculty of Forestry of the Technical University in Zvolen (Slovakia).

Paulína Nalevanková (Dr.) Deals with research of the forest tree species physiological interactions on extreme meteorological and climate conditions. He works at the Faculty of Forestry of the Technical University in Zvolen (Slovakia).

Both editors carry out their research activities through the research project VEGA No. 1/0370/18 "Assessing the vulnerability of selected natural and disturbed ecosystems to hydrometeorological extremes".

Editorial

Impact of Natural Hazards on Forest Ecosystems and Their Surrounding Landscape under Climate Change

Jaroslav Vido [1,2,*] **and Paulína Nalevanková** [1,2]

[1] Department of Natural Environment, Faculty of Forestry, Technical University in Zvolen, T.G. Masaryka 24, 960 01 Zvolen, Slovakia

[2] Oikos NGO, Environmental Laboratory, Na karasíny 247/21, 971 01 Prievidza, Slovakia; admin@oikos.space

* Correspondence: vido@tuzvo.sk

1. Introduction

In the last decades, the increasing frequency of natural hazards has impacted forest ecosystems and their surroundings. It is because of climate change that the dynamics of the ecosystem structure, feedbacks, and relationships are changing. These structural changes are too complicated and complex to be entirely, or at least satisfactorily, explained. However, it is possible to explain at least some of these interconnections. Water is the primary transport medium for energy and material fluxes in ecosystems, and therefore, it is a common denominator of the complex interconnections between their partial components.

Consequently, we paid attention to water as the primary agent driving the impact of natural hazards in forest ecosystems and their surroundings. Water scarcity causes drought, and its surplus causes flood, respectively. Additionally, it is also necessary to understand temporal distribution patterns of water in a warmer climate and ecophysiological consequences in forest structures. Thus, we decided to prepare a Special Issue in which contributors tried to explain some water-related examples of natural hazard impacts on the forest and the surrounding ecosystem.

The Special Issue we introduce consists of 11 original research papers [1–11] divided into three groups based on research interests, namely, 1. hydrological and atmospheric aspects of natural hazards in changing climate and environment, 2. ecophysiological and ecological water-related impacts of natural hazards, and 3. methodological approaches in natural hazard evaluation.

2. Overview of the Special Issue Contributions

2.1. Hydrological and Atmospheric Aspects of Natural Hazards in Changing Climate and Environment

Long-term drought trend analyses provided by Vido and Nalevanková [1] implied that the drought risk evaluated from the meteorological point of view depends mostly on altitude in the area of the Central Carpathian Mountains. An interesting fact is that the beech ecosystems are predominantly located in drought-prone areas (under 1000 m above sea level). On the other hand, trend analyses of the individual months indicate an increasing trend toward wetter conditions in winter. Nevertheless, the authors discussed that it is necessary to investigate how increasing winter temperature changes the snow regime, which could negatively impact river discharge in the spring season.

Response to these challenges is partially discussed in the article by Danáčová et al. [2]. The authors deal with estimating the effect of deforestation on runoff in small mountainous basins in Slovakia. In addition to evidence that rising temperature increase river discharge in mountain creeks in early spring, they also found proof that deforestation is increasing water discharge in the area, especially in the summer months during thunderstorms or torrential rains. The most dramatic result of this study is the impact of climate change and river basin deforestation. In the Boca River basin, the estimated modeled floods increased by 59%, and in the Ipoltica River basin by 172% in the case of the 100-year return period.

Citation: Vido, J.; Nalevanková, P. Impact of Natural Hazards on Forest Ecosystems and Their Surrounding Landscape under Climate Change. *Water* **2021**, *13*, 979. https://doi.org/10.3390/w13070979

Received: 19 March 2021
Accepted: 31 March 2021
Published: 2 April 2021

Another contribution of Fleischer et al. [3] deals with carbon balance and streamflow at a small catchment scale 10 years after the severe natural disturbance (Windthrow of 2004 followed by a forest fire and bark beetle outbreak) in the Tatra Mts, Slovakia. Authors studied carbon fluxes and streamflow 10 years after the forest destruction in three small catchments, which differ in size, land cover, disturbance type, and post-disturbance management. Interestingly, 10 years after the windstorm of 2004, most of the windthrow sites acted as carbon sinks (from -341 ± 92.1 up to -463 ± 178 gC m^{-2} y^{-1}). In contrast, forest stands strongly infested by bark beetles regenerated much slowly and on average emitted 495 ± 176 gC m^{-2} year^{-1}. Moreover, 10 years after the forest destruction, the annual carbon balance in studied catchments was almost neutral in the least disturbed catchment.

Authors found that different post-disturbance management has not influenced the carbon balance yet. Interesting findings are that streamflow characteristics did not indicate significant changes in the hydrological cycle.

Long-term temporal precipitation quality changes in Slovak mountain forests were studied by Mind'aš et al. [4]. Authors found significant declining trends for almost all evaluated chemical components (S-SO4, N-NH4, N-NO3, Ca, Mg, and K), which can significantly affect element cycles in mountain forest ecosystems. The evaluated 41-year-period (1987 to 2018) is characterized by significant changes in the precipitation regime in Slovakia. The obtained results indicate possible directions in which the quantity and quality of precipitation in the mountainous areas of Slovakia will develop with ongoing climate change.

The last paper in this section ("Hydrological and atmospheric aspects of natural hazards in changing climate and environment") deals with the density of seasonal snow in the mountainous environment of five Slovak ski centers (Mikloš et al.) [5]. Climate change increases the role of artificial snow due to winter mountain tourism. These problems have many positive but also negative ecological consequences. The paper by Mikloš et al. brings new insights to snowpack development processes in a manipulated mountainous environment through examinations of temporal and spatial variability in snow densities and an investigation into the development of natural and ski piste snow densities over the winter season.

2.2. Ecophysiological and Ecological Water-Related Impacts of Natural Hazards

An exciting topic about the influence of warmer and drier environmental conditions on species-specific stem circumference dynamics in submontane forests has been introduced by Leštianska et al. [6]. The study's motivation was to understand better the species-specific effects of weather conditions on tree growth because this could lead to better future forest management. The results showed that studied species (Abies alba and Larix decidua) could cope with changing environmental conditions. However, the long-term increase in air temperature and more frequent heat waves, coupled with more intense and prolonged drought episodes, could affect species' ability to respond to environmental changes.

Lukasová et al. [7] studied the autumn phenological responses of European beech to summer drought and heat waves. The results showed that the meteorological drought in the warmer climate led to earlier leaf coloring of European beech. That could be in the future decades a significant ecophysiological impact on whole beech areal in the middle altitudes.

Another example of the ecophysiological impact of various meteorological components and water stress on European beech was investigated by Nalevanková et al. [8]. The study of sap flow monitoring with related environmental factors confirmed that soil water deficit leads to a radical limitation of stand transpiration and significantly affects the relationship between transpiration and environmental drivers. Additionally, it was demonstrated that a time lag exists between the course of transpiration and environmental factors on a diurnal basis. An application of the time lags within the analysis increased the strength of the association between transpiration and the variables. However, due to the occurrence and duration of soil water stress, the dependence of transpiration on the environmental variables became weaker and the time lags were prolonged.

Ecological drought impacts demonstrated by the Carabus population in lowland oak hornbeam forest have been studied by Šiška et al. [9]. The study was carried out during meteorologically two different years—2017 and 2018. Authors found that drought negatively influenced population abundance, and the effect of drought is likely to be expressed with a two-year delay.

The last study on the ecological or ecophysiological impacts of natural hazards was proposed by Kubov et al. [10]. This study deals with the drought impact amplified by the physiological influence of hemiparasitic yellow mistletoe on the oak ecosystem. The study showed how the cumulative effect of biotic hazard represented as hemiparasitic shrub and drought as abiotic natural hazard could worsen the forest ecosystems' ecosystem stability.

2.3. Methodological Approaches in Natural Hazard Evaluation

A methodological contribution to water-related natural hazard impact on forest ecosystems is represented study of Středová et al. [11]. Forest ecosystems and their surroundings faced a variety of natural hazards. That includes abiotic hazards as floods, drought episodes, torrential rains or windstorms, and biotic hazards represented by pathogen outbreaks or changing ecosystem structures. Due to this, the authors decide to define crucial abiotic stressors affecting central European forest ecosystems and, concerning their possible simultaneous effect, develop a universal method of multi-hazard evaluation. That could be helpful in future natural hazard management in forest ecosystems.

We hope that the Special Issue combines many viewpoints on water-related natural hazards impacts on forest ecosystems. Looking at the Special Issue's content, we see that it is not possible to cover this topic with one or even more editions. Further and systematic research is needed. We wish you a pleasant reading.

Funding: This research was funded by VEGA research projects funded by the Science Grant Agency of the Ministry of Education, Science, Research and Sport of the Slovak Republic No. 1/0370/18.

Acknowledgments: We would like to acknowledge all authors who contributed to the Special Issue for their valuable research. We would like to acknowledge also the technical support team of the MDPI who greatly helped with this Special Issue, with special thanks to Editors in Water Editorial Office.

Conflicts of Interest: The authors declare no conflict of interest.

References

1. Vido, J.; Nalevanková, P. Drought in the Upper Hron Region (Slovakia) between the Years 1984–2014. *Water* **2020**, *12*, 2887. [CrossRef]
2. Danáčová, M.; Földes, G.; Labat, M.M.; Kohnová, S.; Hlavčová, K. Estimating the Effect of Deforestation on Runoff in Small Mountainous Basins in Slovakia. *Water* **2020**, *12*, 3113. [CrossRef]
3. Fleischer, P., Jr.; Holko, L.; Celer, S.; Čekovská, L.; Rozkošný, J.; Škoda, P.; Olejár, L.; Fleischer, P. Carbon Balance and Streamflow at a Small Catchment Scale 10 Years after the Severe Natural Disturbance in the Tatra Mts, Slovakia. *Water* **2020**, *12*, 2917. [CrossRef]
4. Minďaš, J.; Hanzelová, M.; Škvareninová, J.; Škvarenina, J.; Ďurský, J.; Tóthová, S. Long-Term Temporal Changes of Precipitation Quality in Slovak Mountain Forests. *Water* **2020**, *12*, 2920. [CrossRef]
5. Mikloš, M.; Skvarenina, J.; Jančo, M.; Skvareninova, J. Density of Seasonal Snow in the Mountainous Environment of Five Slovak Ski Centers. *Water* **2020**, *12*, 3563. [CrossRef]
6. Leštianska, A.; Fleischer, P., Jr.; Merganičová, K.; Fleischer, P.; Střelcová, K. Influence of Warmer and Drier Environmental Conditions on Species-Specific Stem Circumference Dynamics and Water Status of Conifers in Submontane Zone of Central Slovakia. *Water* **2020**, *12*, 2945. [CrossRef]
7. Lukasová, V.; Vido, J.; Škvareninová, J.; Bičárová, S.; Hlavatá, H.; Borsányi, P.; Škvarenina, J. Autumn Phenological Response of European Beech to Summer Drought and Heat. *Water* **2020**, *12*, 2610. [CrossRef]
8. Nalevanková, P.; Sitková, Z.; Kučera, J.; Střelcová, K. Impact of Water Deficit on Seasonal and Diurnal Dynamics of European Beech Transpiration and Time-Lag Effect between Stand Transpiration and Environmental Drivers. *Water* **2020**, *12*, 3437. [CrossRef]
9. Šiška, B.; Eliašová, M.; Kollár, J. Carabus Population Response to Drought in Lowland Oak Hornbeam Forest. *Water* **2020**, *12*, 3284. [CrossRef]

10. Kubov, M.; Fleischer, P., Jr.; Rozkošný, J.; Kurjak, D.; Konôpková, A.; Galko, J.; Húdoková, H.; Lalík, M.; Rell, S.; Pittner, J.; et al. Drought or Severe Drought? Hemiparasitic Yellow Mistletoe (*Loranthus europaeus*) Amplifies Drought Stress in Sessile Oak Trees (*Quercus petraea*) by Altering Water Status and Physiological Responses. *Water* **2020**, *12*, 2985. [CrossRef]
11. Středová, H.; Fukalová, P.; Chuchma, F.; Středa, T. A Complex Method for Estimation of Multiple Abiotic Hazards in Forest Ecosystems. *Water* **2020**, *12*, 2872. [CrossRef]

 *water*

Article

Drought in the Upper Hron Region (Slovakia) between the Years 1984–2014

Jaroslav Vido [1,2,*] and Paulína Nalevanková [1,2]

[1] Department of Natural Environment, Faculty of Forestry, Technical University in Zvolen, 960 01 Zvolen, Slovakia; nalevankova.paulina@gmail.com

[2] Oikos NGO., Environmental laboratory, Na Karasíny 247/21, 971 01 Prievidza, Slovakia

* Correspondence: vido@tuzvo.sk; Tel.: +421-45-5206-215

Received: 15 September 2020; Accepted: 13 October 2020; Published: 16 October 2020

Abstract: Climate change causes an increase in the frequency and severity of weather extremes. One of the most relevant severe and damaging phenomena in Europe is drought. However, a difference in the spatial frequency of the occurrence and drought trends is evident between southern and northern Europe. Central Europe and particularly the West Carpathian region form a transitional zone, and drought patterns are complicated because of the geomorphologically complicated landscape. Since almost half of the Slovak state territory is represented by such natural landscape, it is necessary to investigate regional drought specifics. Therefore, we decided to analyze drought occurrence and trends using the SPI (Standardised Precipitation Index) and the SPEI (Standardised Precipitation Evapotranspiration Index) at available climatological stations of the Slovak Hydrometeorological Institute (SHMI) in the upper Hron region within the 1984–2014 period. We found that (1) drought incidence decreased with increasing altitude, (2) increasing air temperature increased the difference in drought trends between lowlands and mountains during the studied period, and (3) abrupt changes in time series of drought indices, that could indicate some signals of changing atmospheric circulation patterns, were not revealed. Finally, we constructed a simplified map of drought risk as an explanation resource for local decision-makers.

Keywords: drought; Slovakia; Hron river; trend analyses; altitude; climate change

---

1. Introduction

Ongoing climate change causes an increase in weather extremes, especially droughts and floods [1–4]. There are rising trends in forest and wildfires in southern Europe [5], more frequent occurrence of extreme flood and storm situations in northwestern Europe [6], and droughts in central and South Europe [6,7]. In the context of drought [6,8,9] argue that there is an evident difference between drought trends in southern and northern Europe. While southern Europe is experiencing an increasing incidence of extreme drought episodes, the trend is the opposite in northern Europe. Following this statement, Alfieri [10] argued that, based on climate change scenarios, the difference between rainfall trends and, therefore, drought frequency between northern Europe and southern Europe will continue to grow. The primary driver of worsening drought trends and their frequency in southern Europe is rising air temperature due to climate change (and therefore evapotranspiration) [10–12]. Stagge et al. [6] analyzed the influence of precipitation and air temperature on the spatial patterns of drought occurrence in Europe. They found that droughts are less frequent in northern Europe due to higher rainfall, while in the south, droughts are more frequent and extreme due to higher air temperatures and less rainfall.

Although primary driving factors of drought are usually well understood and discussed on continental scales, the situation could be more complicated on a regional and sub-regional scale [13]. For instance, Vido et al. [14] reported increasing drought frequency in the Podunajská nížina valley in Slovakia between 1966 and 2013. However, Škvarenina et al. [15] found the opposite drought trend based on meteorological stations located only 60 km away. The reason for this is the complicated geomorphological structure of Slovakia, where the mountains of the western Carpathians with numerous geological depressions—the Inner Carpathian valleys—meet the Pannonian Plain in the south [16]. That creates specific climatic conditions that are invisible on the continental scale, but cause significant local differences in precipitation and air temperature regimes on a regional scale [15–21]. Historically, Zlatnik [22] stated significant climatic differences in Slovakia and also formulated the so-called "climatic line of Slovakia" based on biogeographical observations. The line divided the area of Slovakia into two areas (North and South). The area to the north of this line was determined as relatively wetter and colder (influence of the Baltic Sea climate) than the southern one, which is drier and warmer due to the influence of the Pannonian climate [13,17]. Nevertheless, the study of Vilček et al. [23] on the thermal continentality of Slovakia suggested that this distinction of climatic zones was more strongly related to the concentration of high mountains in northern Slovakia than to the direct influence of the sea or continental climate. Also, Zeleňáková [24] highlighted the significant effect of altitude on rainfall totals in Slovakia. That implies that altitude is a significant driver of climatic conditions in Slovakia that influence drought occurrence. That was previously implied by [13,17].

In such a heterogeneous geographical region exist relevant assumptions of climate change's specific influence on drought evolution and occurrence. That was implied by [24] and historically [25] in the context of the spatial and temporal precipitation distribution over Slovakia.

However, more detailed research of drought trends and occurrence on the scale of the inner Carpathian basins and valleys, which are typical of almost half of Slovakia's territory, has not yet been carried out.

Therefore, we decided to analyze drought occurrence and trends using the SPI (Standardised Precipitation Index) and the SPEI (Standardised Precipitation Evapotranspiration Index) at available climatological stations of the Slovak Hydrometeorological Institute (SHMI) in the upper Hron region that corresponds with the Upper Hron river basin (partial basin of the Hron river) within the period 1984–2014. The region is a typical representative of the inner Carpathian basins and valleys [16]. To achieve the goal, we formulated the following particular aims:

1. Find out the trends of SPI and SPEI along the studied area's altitudinal gradient.
2. Detect possible abrupt changes in temporal trends of the SPI and SPEI.
3. Analyze the trends of the above indices in individual months.
4. Try to spatially identify drought-prone areas based on the SPI and SPEI time series' temporal evolution.

2. Materials and Methods

2.1. Study Area

The Upper Hron region corresponds to the Upper Hron river basin located in central Slovakia. The river basin studied in the presented paper is a partial basin of the river Hron (Gravelius' stream order "2"). The Upper Hron river source (also the source of the whole river Hron) is located beneath the Kráľová Hoľa peak in Low Tatra Mts. near the village Telgárt in the East of the basin. The altitude of the river source is 980 m a.s.l. (metres above sea level). The outlet is located on the West of the basin near Zvolen city at an altitude of 273 m a.s.l. Generally, the river basin has an East–West elongated shape. Principal tributaries of the river are the Slatina river (left tributary), Starohorský creek (right tributary), Čierny Hron river (left tributary), and Rohožná river (left tributary). Our partial basin area is 2846.9 km^2, which represents 52% of the whole Hron river basin (5453 km^2). The mainstream (Hron river) is 128 km long in the upper Hron river basin (river km 150–278).

The river basin is surrounded by Nízke Tatry Mts., Starohorské vrchy Mts., and Veľká Fatra Mts. to the North, Kremnické vrchy Mts. to the West, Štiavnické vrchy Mts and Javorie Mts to the South, and Veporské vrchy Mts. and Spišsko-Gemerský karst Mts. to the South-East (Figure 1). In the center of the basin is a well-preserved caldera of the Neogenic stratovolcano—Poľana Mts. The highest point of the river basin is Ďumbier peak (2025 m a.s.l.) located in the North of the Nízke Tatra Mountains.

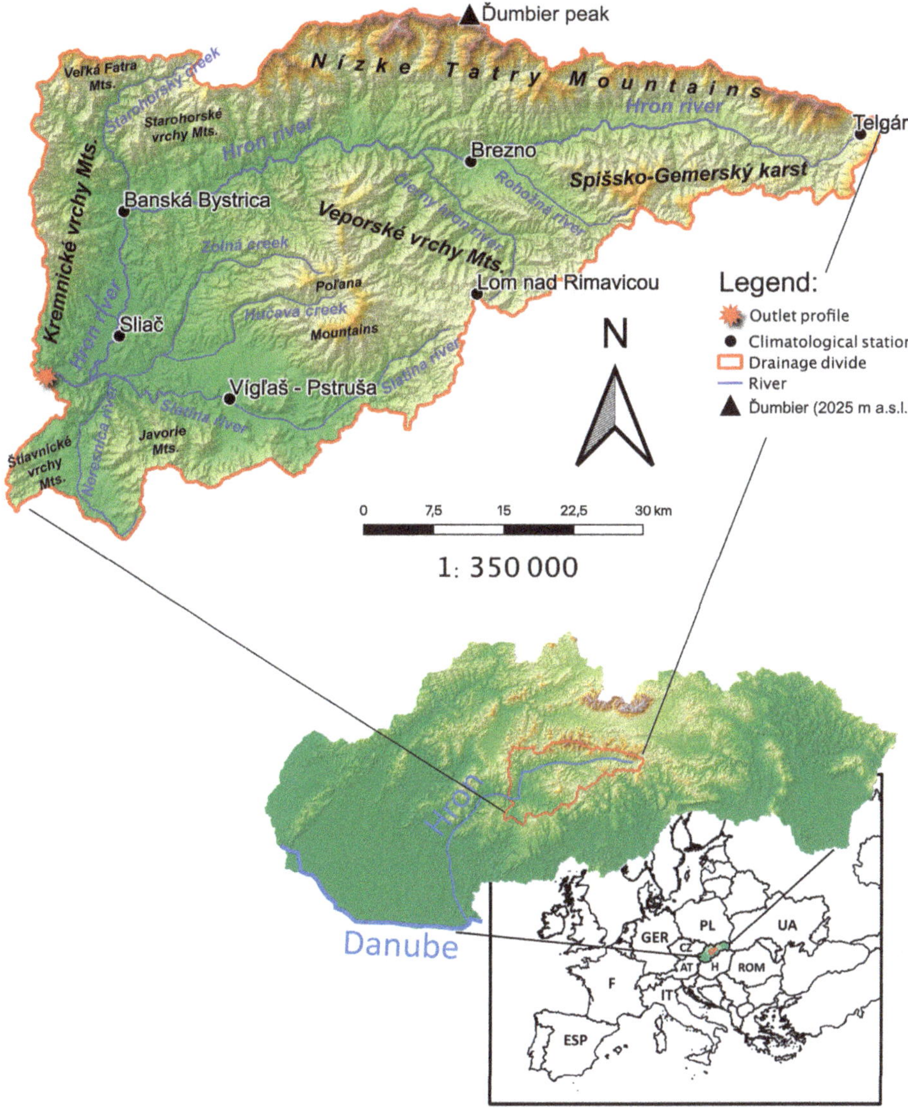

Figure 1. The river basin of the Upper Hron river and its localization within Slovakia and Europe.

The studied partial river basin is a highly forested area. Of 2846.9 km^2 of the area, 1847 km^2 is forested (65%). Forests are located mainly in surrounding mountainous areas and the central part of the basin around Poľana and Veporské vrchy Mts.

2.2. Climate of the Area

The climate conditions of the studied area differ due to its terrain variability. In general, we can divide the climatic conditions into the climate of intra-Carpathian basins and valleys and the mountain climate. Further climatic characteristics were adopted from the Climate Atlas of the Slovak Republic [26] based on the reference period 1961–2010. The basin and valley climate is divided into a warm, slightly humid climate with a mild winter (Zvolenská kotlina valley—southwest of the area) and a slightly warm, humid, highland climate (high basins in the north-east of the area). The mountain areas vary between a cold mountain climate (Nízke Tatry Mountains and Veľká Fatra Mts.) to moderately cold (other mountains in the studied area). The mean annual temperature in the area ranges from −0.5 °C in the area of the highest mountains (Nízke Tatry Mountains—north of the area) to +8.5 °C in the Zvolenská kotlina valley (southwest of the area). The July (the hottest month) average temperature ranges from +8.2 °C at the highest mountain (Nízke Tatry Mountains—the north) to +18.8 °C in the Zvolenská kotlina valley (southwest of the area). The coldest month (January) mean temperature varies between −2.8 °C in the Zvolenská kotlina valley to −8.2 °C in Nízke Tatry Mountains. The highest average annual precipitation totals are recorded in the highest mountain areas of Nízke Tatry Mountains (1209 mm), and the lowest in the Zvolenská kotlina valley (southwest), with total rainfall up to 630 mm [27].

2.3. Climate Data

To achieve the goals of the study, we used climatological data, monthly mean air temperature [°C], and monthly precipitation totals [mm] from six climatological stations situated within the selected river basin of the Upper Hron river (Table 1). Our study period was 1984–2014. In Table 1 are also calculated monthly means of air temperature as well as monthly means of precipitation totals. The localization of the stations used is depicted in Figure 1. Data used in our investigation were obtained from the Slovak Hydrometeorological Institute (SHMI). According to the World Meteorological Organization's internal and international standards, preliminary data processes (i.e., quality checking, homogenization, and preparation for the end-user) were carried out by the Slovak Hydrometeorological Institute [28]. Climatological stations are spatially well distributed within the study area and along the altitudinal gradient (range from 313 m a.s.l. to 1018 m a.s.l.).

Table 1. Climatological stations of the Slovak Hydrometeorological Institute (SHMI) used in the study with mean monthly air temperature and precipitation totals based on the period 1984–2014.

Station Name		Jan.	Feb.	Mar.	Apr.	May	Jun.	Jul.	Aug.	Sep.	Oct.	Nov.	Dec.
Sliač	WMO index 11903;	Altitude 313 m a.s.l.;				Latitude [φ] 48°38′33″;				Longitude [λ] 19°08′31″			
	Precipitation [mm]	44.7	41.6	42.5	47.7	65.2	83.5	72.6	67.5	54.7	54.0	63.8	56.0
	Temperature [°C]	−3.3	−1.2	3.3	9.1	14.0	17.2	18.9	18.1	13.7	8.6	3.4	−1.9
Vígľaš-	WMO index 11904;	Altitude 368 m a.s.l.;				Latitude [φ] 48°32′39″;				Longitude [λ] 19°19′19″			
Pstruša	Precipitation [mm]	31.5	30.9	31.6	46.1	70.1	85.4	73.1	62.1	50.8	47.3	53.6	44.2
	Temperature [°C]	−3.4	−1.2	3.2	8.7	13.5	16.5	18.1	17.6	13.4	8.3	3.3	−1.9
Banská	WMO index 11898;	Altitude 427 m a.s.l.;				Latitude [φ] 48°44′01″;				Longitude [λ] 19°07′01″			
Bystrica	Precipitation [mm]	55.9	51.4	52.4	55.9	82.1	89.3	81.4	71.0	62.2	66.0	80.4	71.1
	Temperature [°C]	−2.5	−0.6	3.3	9.1	13.8	16.8	18.4	17.8	13.5	8.6	3.6	−1.3
Brezno	WMO index 11917;	Altitude 487 m a.s.l.;				Latitude [φ] 48°48′06″;				Longitude [λ] 19°38′14″			
	Precipitation [mm]	41.7	39.6	45.3	51.4	88.4	98.7	96.1	81.0	59.5	56.8	56.5	50.1
	Temperature [°C]	−3.7	−2.0	2.3	8.3	13.3	16.4	18.2	17.3	12.6	7.7	2.7	−2.5
Telgárt	WMO index 11938;	Altitude 901 m a.s.l.;				Latitude [φ] 48°50′55″;				Longitude [λ] 20°11′21″			
	Precipitation [mm]	33.6	40.6	41.9	63.1	113	121	92.6	87.1	60.6	63.3	70.7	42.2
	Temperature [°C]	−5.3	−4.0	−0.5	4.5	9.6	12.6	14.3	13.6	10.2	5.7	0.3	−4.0
Lom nad	WMO index 11910;	Altitude 1018 m a.s.l.;				Latitude [φ] 48°39′38″;				Longitude [λ] 19°39′57″			
Rimavicou	Precipitation [mm]	48.7	56.4	51.8	64.2	100	124	99.5	96.4	64.6	74.8	87.2	66.8
	Temperature [°C]	−5.4	−4.1	−0.6	4.3	9.7	12.6	14.5	13.9	10.3	5.6	0.3	−4.0

2.4. Drought Indices

We used two widely used drought indices to achieve the goals of the paper; Standardized Precipitation Index (SPI) and Standardized Precipitation Evapotranspiration Index (SPEI). The reason was to investigate how drought patterns differ, assuming the influence only of precipitation (SPI) and the balance between precipitation and evapotranspiration (SPEI). For our purpose, we used two time scales of the indices; SPI and SPEI for one month and SPI and SPEI for twelve months. Our selection is based on a previous investigation pointing out that the one-month scale refers to short-term drought fluctuations (meteorological drought) [27,28]. In contrast, the indices for the 12-month scale are used when assessing long-term (cumulated) drought episodes with a severe impact on ecosystems and the socio-economic structure [12,29–31].

These indices, SPEI and SPI, for one and twelve months were therefore used to show long-term trends within the studied period.

However, for drought trend investigation in individual months within the studied period, SPI and SPEI for only one month (which are not cumulative compared to 12-month indices) were used.

2.4.1. Standardised Precipitation Index (SPI)

The Standardized Precipitation Index [32] is a drought index calculated based on the probability of the occurrence of a certain amount of precipitation in a given period. The calculation requires a long-term monthly precipitation database with 30 years or more of data. The probability distribution function is derived from the long-term record by fitting a gamma function to the data. The cumulative distribution is then transformed using equal probability to a normal distribution with a mean of zero and a standard deviation of one, so the SPI values are really in standard deviations [33]. Full mathematical descriptions of the principles and calculation of the SPI are given in [33]. Positive SPI values indicate greater than median precipitation, while negative SPI values indicate less than median precipitation. The magnitude of departure from zero represents the probability of occurrence so that decisions can be made based on this SPI value. Thus, SPI values of less than −1.0 occur 16 times in 100 years, SPI of less than −2.0 occurs two to three times in 100 years, and an SPI of less than −3.0 occurs once in approximately 200 years. The SPI can be calculated for a variety of time scales. This feature allows the SPI to monitor short-term water supplies (such as soil moisture) and longer-term water resources such as groundwater supplies or lake levels [34]. Cumulated SPI values may be therefore used to analyze drought severity. The principle of this accumulation is that if we have sequences of monthly sums of precipitation and we want to calculate the SPI values for, e.g., three-month periods (SPI for three months), then the first element of a new sequence is the sum of the first three months, the second element is formed by summing precipitation in the 2nd, 3rd, and 4th months, and the next is a sum of the 3rd, 4th, and 5th months, and so on [35]. The same logic applies for all time scales. The SPI for a specific month is then calculated from this new time series as follows:

$$SPI = \frac{x_i - \overline{x}}{\sigma} \tag{1}$$

where x_i is the precipitation of the selected period during the year i, $\overline{x}$ is the long-term mean precipitation, and σ is the standard deviation for the selected period [36].

2.4.2. Standardised Precipitation Evapotranspiration Index (SPEI)

The principle of the SPEI calculation is based on the standardized precipitation index (SPI) [32], which evaluates the deviations of precipitation from the long-term normal at different time scales (usually from 1 to 24 months). The SPI has long been used for drought monitoring in several countries in the world [37]. One of the SPI limitations is that it does not include the passive components of the hydrological regime (i.e., evapotranspiration). Vicente-Serrano et al. [38] used both precipitation and potential evapotranspiration (PET) to generate the SPEI values that include the deviation of the whole climatic balance (P-PET) from the normal (i.e., positive values represent a positive balance and vice versa). Following the methodology of Vicente-Serrano et al. [38], a drought episode starts (similar to SPI methodology) when a negative value of the index appears and lasts until the first positive value. However, the index must reach or exceed −1 for at least one month during the specific episode. The calculation of the potential evapotranspiration in SPEI is based on the equation of Thornthwaite [39]. Calculation of SPEI requires a time series of at least thirty years of monthly average air temperatures and monthly precipitation totals from each station.

2.5. Trend Analyses

We applied two non-parametric methods for trend analyses:

(i)   Mann–Kendall trend test (MK) and
(ii)   Cumulative sum of Rank Difference test (CRD).

The Mann–Kendall test [40] is a standard non-parametric test for trend detection [41]. However, as stated by Onyutha [42], the Mann-Kendall test is a purely statistical method, so there are no detailed insights about the specifics of the studied trends. We adopted the CRD test due to the need to detect abrupt changes in the temporal trend of the SPIs and SPEIs. This need arose because some authors implied that significantly changed patterns of atmospheric circulation occurred in the late eighties and early nineties, which influenced rain patterns over central Europe [39,40]. Therefore, we used this relatively new method to analyze whether some abrupt changes in the time series of the SPIs and SPEIs were evident.

This method combines both statistical and graphical approaches in trend analyses. A detailed description of the method and computation procedures is presented by Onyutha [42].

The combination of the CRD test with the MK test can bring deep insights into trend behaviour (i.e., cyclical anomalies and abrupt changes in trends).

In our case, we used CRD plots of the SPIs and SPEIs. Computation of the CRD was carried out using the CRD-NAIM_v.3 tool, which was downloaded together with its user manual via the link: https://sites.google.com/site/conyutha/tools-to-download (accessed on 31 July 2020).

Since the computations were based on monthly scale, CRD parameters in the tool were as follows:

- significance level set to 5%
- time scale representing a moving average of 60 (60 months = 5 years)
- the initial block set to 10
- number of Monte Carlo runs for resampling set to 1000 (default setting)

The MK test was used as a general indicator of a trend toward aridity or humidity; CRD graphical outputs were used as an additional indicator of changes in trend directions throughout the studied period. For the MK test, the significance level default was set to $\alpha = 0.05$.

For trend analyses aimed at investigating trends in particular months within the studied period, only the MK test was used.

2.6. Spatial Identification of the Drought-Prone Area

To spatially identify potential drought-prone areas, we used station-based analyses of the SPI and SPEI for twelve months. The reason was to identify areas that have a higher potential to be endangered by severe droughts.

The QGIS geographical information system (release 3.12.1, Bucuresti) was used. The spatial delimitation of drought-prone areas was limited by the altitude corresponding to the climatological station's altitude, which showed a significant trend (towards wetter conditions) of both analyzed drought indices (SPI and SPEI). This process was carried out using the Raster Calculator of the QGIS, set to delimit the area with altitudes corresponding to altitudes equal to and higher than the specific climatological station altitude (stations with a prevailing number of significant trends toward wetter conditions).

3. Results

3.1. Trends of SPI and SPEI within the Period 1984–2014

3.1.1. Trends of the SPI within the Studied Period

Trend analyses of the SPI for one month showed no significant and very slight trends (R^2 between 0.0013 to 0.0018) toward wetter conditions for four climatological stations situated at low altitudes (Sliač, Vígľaš-Pstruša, Banská Bystrica, and Brezno) (Figure 2a–d). Exceptions were the two highest stations, which recorded a significantly increasing trend toward wetter conditions (Telgárt and Lom nad Rimavicou) (Figure 2e,f and Table 2).

However, these two trends had a slight slope (R^2 = 0.0129 and 0.0106). The alternation of drought episodes with wetter (precipitation-rich) periods was relatively regular and frequent throughout the study period. However, the last five years in the time series (2009–2014) are an exception. During this period, drought episodes and wet episodes lasted relatively longer than usual. The secondary relatively wetter episode was recorded between the years 1994–1996.

These facts are much more pronounced based on the twelve-month SPI (Figure 3). Based on the twelve-month SPI, the dry and long-lasting drought period of 1985–1993 becomes evident. This drought period is evident for all studied stations, but interestingly most pronounced for the highest station Lom nad Rimavicou situated in the southeast of the studied region.

Because the twelve-month index shows accumulated precipitation patterns, drought and wet episodes were much more pronounced. This feature has an influence on the trend slope and its significance. All stations showed a significantly rising trend towards wetter conditions when applying SPI for twelve months (Table 2 and Figure 3).

Considering altitude as a climatic driving factor (as mentioned in the introduction of the article), we see that with increasing altitude a decreased number of drought episodes during the studied period. On the other hand, it is interesting that the highest located station Lom nad Rimavicou (Figure 3f) recorded less pronounced wet episodes (maximum magnitude of the SPI for 12 months was +1) between the years 1994–2000 compared to stations located at lower altitudes (Figure 3a–e).

Table 2. Trend indicators of time series 1984–2014 for studied stations.

Station Name	SPI 1m	SPI 12m	SPEI 1m	SPEI 12m
Sliač	–	▲	–	–
p-value	0.249	0.000	0.989	0.71
Vígľaš-Pstruša	–	▲	–	–
p-value	0.145	<0.0001	0.946	0.862
Banská Bystrica	–	▲	–	–
p-value	0.109	<0.0001	0.752	0.083
Brezno	–	▲	–	–
p-value	0.183	<0.0001	0.823	0.217
Telgárt	▲	▲	–	▲
p-value	0.021	<0.0001	0.161	<0.0001
Lom nad Rimavicou	▲	▲	–	▲
p-value	0.023	<0.0001	0.198	<0.0001

▲ Significant rising trend toward wetter conditions (significance level α = 0.05), – No trend recorded.

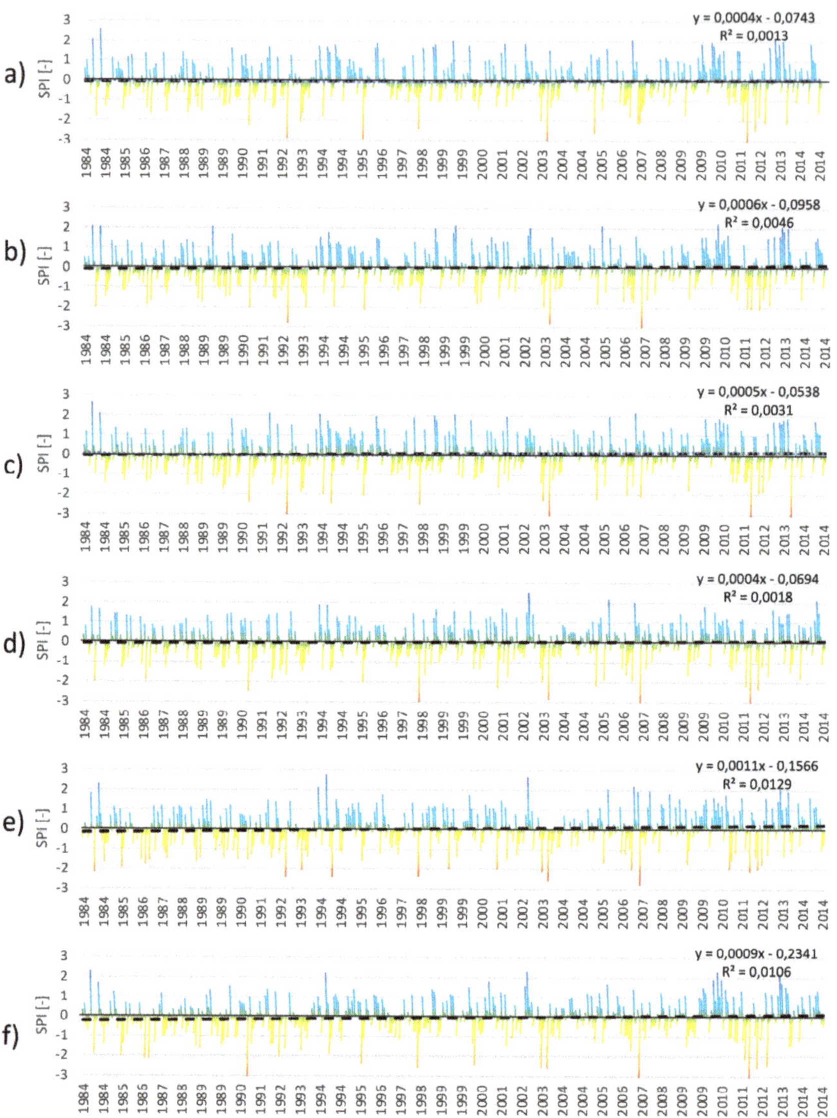

Figure 2. Temporal course of one-month SPI with the linear trend of the time series. Letter (**a**) refers to climatological station Sliač, (**b**) station Vígľaš-Pstruša, (**c**) station Banská Bystrica, (**d**) station Brezno, (**e**) station Telgárt, (**f**) station Lom nad Rimavicou. The dashed line represents the linear trend line.

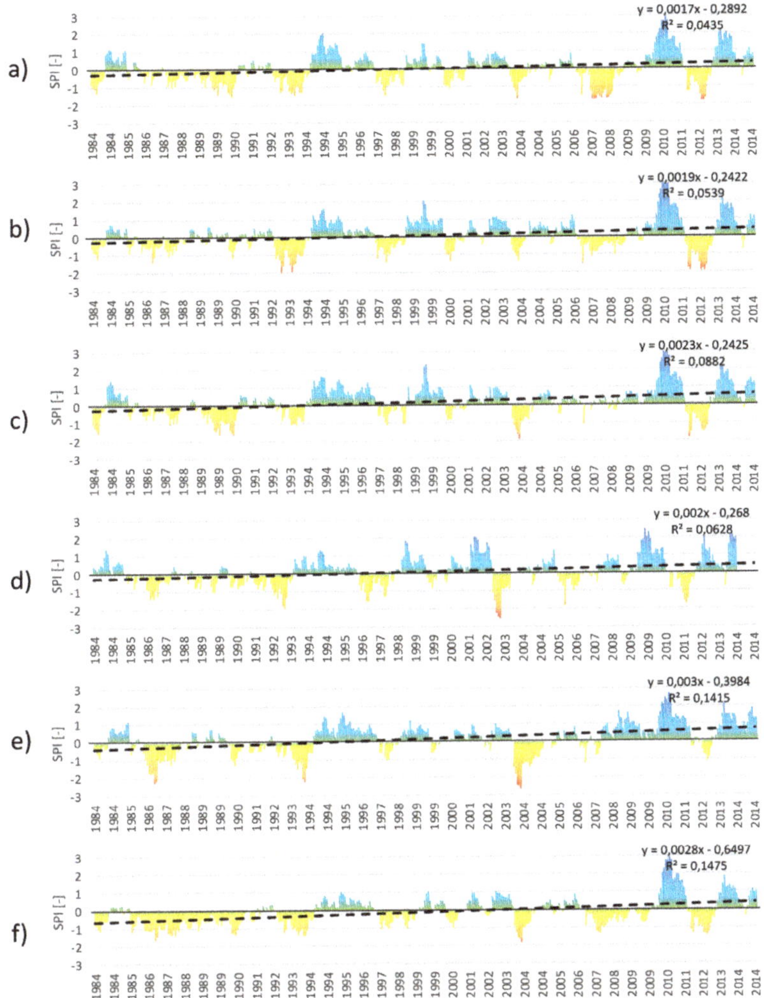

Figure 3. Temporal course of a twelve-month SPI with the linear trend of the time series. Letter (**a**) refers to climatological station Sliač, (**b**) station Vígľaš-Pstruša, (**c**) station Banská Bystrica, (**d**) station Brezno, (**e**) station Telgárt, (**f**) station Lom nad Rimavicou. The dashed line represents the linear trend line.

3.1.2. Trends of the SPEI within the Studied Period

Previous results have taken into account only precipitation (SPIs). However, in drought analyses, it is necessary to also take into account evapotranspiration (mainly driven by air temperature) to see the complex influence of these parameters on temporal trend evolution due to climate change involved in temperature increases. Therefore, we also analyzed the time series and linear trends of the SPEI for one and SPEI for twelve months.

It is evident that by incorporating evapotranspiration, the previously detected severe and long-lasting drought episodes in the late eighties and early nineties and drought episodes of 2003, 2007–2009, and 2011–2012 were more pronounced.

Based on trend analyses of the SPEIs for one month (Figure 4), no trends were recorded within the studied period. Slight insignificant humid trends were recorded for the highest located stations,

Telgárt and Lom nad Rimavicou (both stations R^2 = 0.0044) (Figure 4e,f). However, in comparison to the SPI for one month, it is evident that air temperature (evapotranspiration) influenced the flattening of all the trends (Table 2). SPEI for 12 months highlighted these trends (Figure 5). SPEI for 12 months showed a significant trend toward wetter conditions for the two highest stations, Telgárt and Lom nad Rimavicou (R^2 = 0.0562 and 0.0453) (Table 2 and Figure 5e,f). Nevertheless, for the remaining stations, SPEIs for twelve months showed no trend except for Banská Bystrica (Figure 5c). However, the MK test revealed its trend as insignificant (Table 2). Thus, comparing the SPI and SPEI analysis results, we can state that evapotranspiration (driven by air temperature) has changed the temporal evolution of the drought trends (Table 2). When we compare SPI and SPEI, it is evident that the driver of the divergent drought trend evolution is a continuous rising of air temperature. However, at higher altitudes, in comparison to lower altitudes, we observe lower influence of rising temperature on evapotranspiration, and therefore, the trends of the SPI and SPEI remain unchanged at higher altitudes.

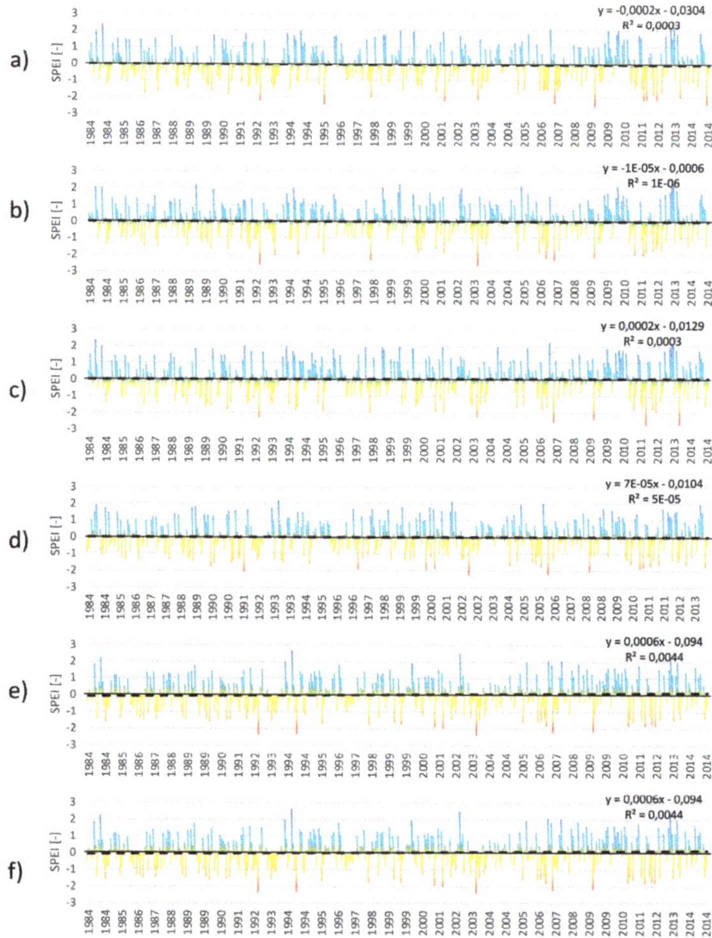

Figure 4. Temporal course of a one-month SPEI with the linear trend of the time series. Letter (**a**) refers to climatological station Sliač, (**b**) station Vígľaš-Pstruša, (**c**) station Banská Bystrica, (**d**) station Brezno, (**e**) station Telgárt, (**f**) station Lom nad Rimavicou. The dashed line represents the linear trend line.

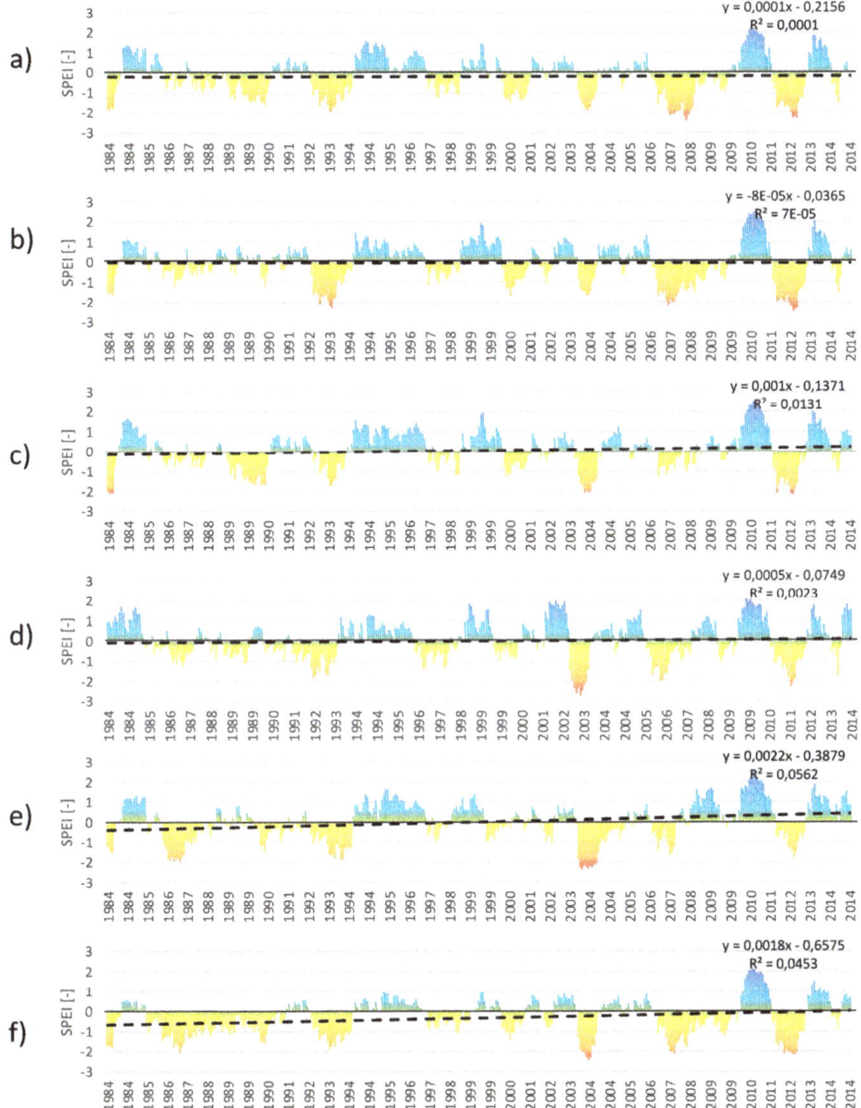

Figure 5. Temporal course of a twelve-month SPEI with the linear trend of the time series. Letter (**a**) refers to climatological station Sliač, (**b**) station Vígľaš-Pstruša, (**c**) station Banská Bystrica, (**d**) station Brezno, (**e**) station Telgárt, (**f**) station Lom nad Rimavicou. The dashed line represents the linear trend line.

3.2. Detection of Abrupt Changes in Temporal Trends of SPI and SPEI within the Period 1984–2014

CRD plots constructed to detect abrupt changes and possible sub-trends in time series within the studied period confirmed the general trends of the SPI's temporal evolution. No abrupt changes in general rising trends of SPI for one and twelve months were detected within the studied period (Figure 6). However, the CRD plot of SPI for one month at the lowest station Sliač (Figure 6a) implies that trend direction toward wetter conditions is almost indistinct. This result also corresponds with an insignificant MK test and low R^2 (0.0003) for this station mentioned in the previous Section 3.1.2. That, however, applies only to SPI for one month. When assessing SPI for 12 months, the CRD plot for all the stations showed a positive trend direction (Figure 6 right), which implies that no abrupt change in the rain regime was observed in the studied period, and we see increasing (even though insignificant) or no changing trend in precipitation within the studied period.

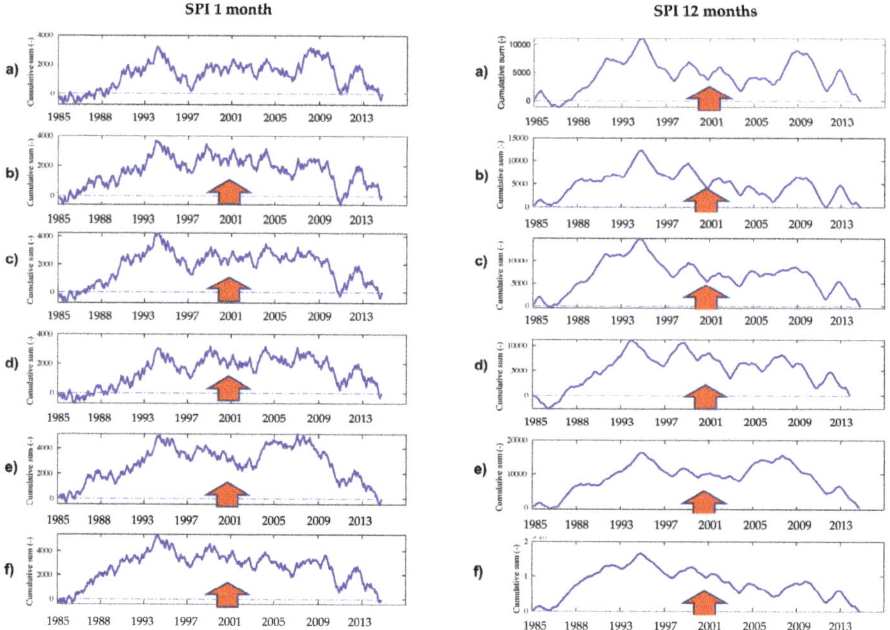

Figure 6. CRD plot of the SPI for one month (**left**) and twelve months (**right**) within the studied period 1984–2014. Letter (**a**) refers to climatological station Sliač, (**b**) station Vígľaš-Pstruša, (**c**) station Banská Bystrica, (**d**) station Brezno, (**e**) station Telgárt, (**f**) station Lom nad Rimavicou. The red arrow indicates a positive trend direction.

However, CRD plots constructed for calculated SPEIs for one and twelve months showed how air temperature (evapotranspiration) influences the trend direction (Figure 7). As stated in Section 3.1.2, SPEI indicates that rising temperature during the studied period also changed drought trends from humid to indistinct trends (except for the highest meteorological stations that retained humid trends). CRD plots also confirmed this previous result. Although abrupt changes in time series were not detected similarly as by CRD for SPIs, CRD for SPEIs recorded pronounced indistinct trends for all stations with the exception of the two highest meteorological stations (Telgárt and Lom nad Rimavicou) where clear trends toward wetter conditions were retained (positive trend directions). This result showed that abrupt changes during the studied period were not observed, but rising temperatures significantly influenced the slope of the SPEI trends compared to SPI. This finding, therefore, supports the results described in Section 3.1.

SPEI 1 month   SPEI 12 months

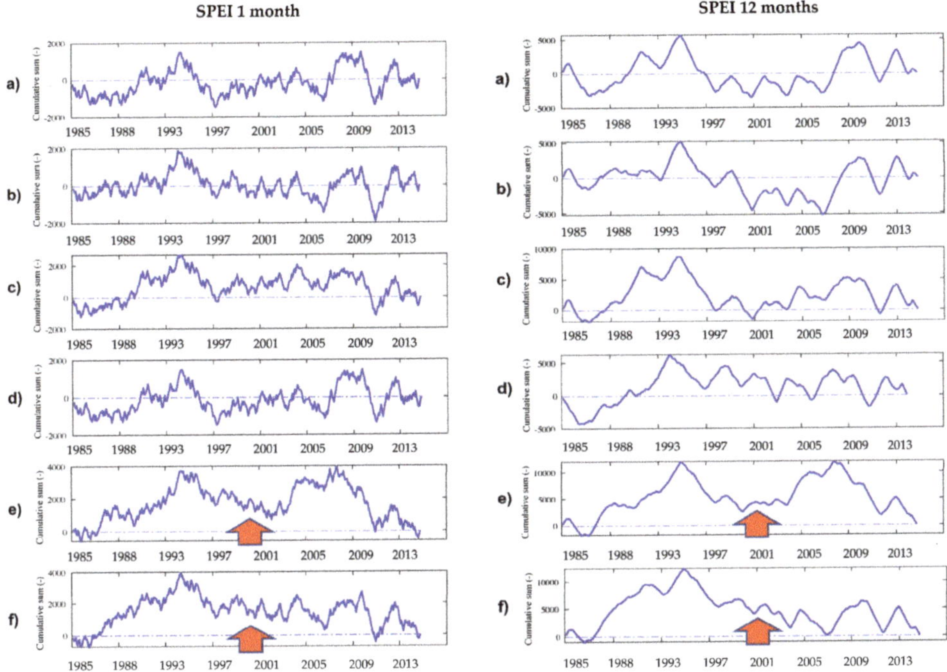

Figure 7. CRD plot of the SPEI for one month (**left**) and twelve months (**right**) within the studied period 1984–2014. Letter (**a**) refers to climatological station Sliač, (**b**) station Vígľaš-Pstruša, (**c**) station Banská Bystrica, (**d**) station Brezno, (**e**) station Telgárt, (**f**) station Lom nad Rimavicou. The red arrow indicates a positive trend direction.

3.3. Trend Analyses of SPI and SPEI for Individual Months in the Period 1984–2014

Trend analyses of the SPIs and SPEIs showed significant (tested by the MK test) rising trends toward wetter conditions in July. Only at stations Sliač and Brezno were these trends insignificant. The second most frequent occurrence of significant trends toward wetter conditions was recorded in January.

The remaining monthly trends at all the stations were insignificant. Interesting trends, although insignificant, were recorded in April and May. All the stations recorded decreasing trends toward drier conditions based on both indices (SPI and SPEI). In December, all stations recorded an increasing trend toward wetter conditions, but these trends were insignificant. An interesting summary fact is that significant trends (July and January) were recorded along the whole altitudinal gradient. The summary of the trend analyses is presented in Table 3.

Table 3. Trend indicators of time series 1984–2014 for individual months at the studied stations.

SPI (Standardised Precipitation Index)												
Station Name	**Jan.**	**Feb.**	**Mar.**	**Apr.**	**May**	**Jun.**	**Jul.**	**Aug.**	**Sep.**	**Oct.**	**Nov.**	**Dec.**
Sliač	▲	–	–	▼	▼	▲	▲	–	▼	–	▼	▲
p-value	0.083	0.773	0.946	0.812	0.341	0.333	0.011	0.812	0.760	0.986	0.658	0.496
Vígľaš-Pstruša	▲	▲	–	▼	▼	–	▲	–	–	▲	–	▲
p-value	0.074	0.234	0.932	0.227	0.454	0.865	0.006	0.671	0.812	0.386	0.367	0.395
Banská Bystrica	▲	▲	▲	▼	▼	–	▲	–	–	–	▼	▲
p-value	0.004	0.683	0.518	0.529	0.292	0.540	0.004	0.671	1.000	0.878	0.734	0.465
Brezno	▲	–	–	▼	▼	–	▲	▲	–	▲	▼	▲
p-value	0.043	1.000	0.946	0.276	0.529	0.905	0.025	0.598	1.000	0.646	0.598	0.367
Telgárt	▲	–	–	▼	▼	▲	▲	▲	▲	▲	–	▲
p-value	0.110	0.799	0.812	0.434	0.405	0.118	0.004	0.359	0.825	0.367	0.852	0.385
Lom nad Rimavicou	▲	▲	▲	▼	▼	–	▲	–	▲	▲	–	▲
p-value	0.036	0.773	0.308	0.496	0.316	1.000	0.024	0.878	0.734	0.563	1.000	0.316
SPEI (Standardised Precipitation Evapotranspiration Index)												
Station Name	**Jan.**	**Feb.**	**Mar.**	**Apr.**	**May**	**Jun.**	**Jul.**	**Aug.**	**Sep.**	**Oct.**	**Nov.**	**Dec.**
Sliač	▲	–	–	▼	▼	▲	▲	▼	▼	–	▼	▲
p-value	0.096	0.919	0.812	0.234	0.292	0.892	0.069	0.465	0.622	0.973	0.191	0.454
Vígľaš-Pstruša	▲	▲	▲	▼	▼	–	▲	–	–	–	–	▲
p-value	0.004	0.683	0.518	0.529	0.292	0.540	0.004	0.671	1.000	0.878	0.734	0.465
Banská Bystrica	▲	▲	–	▼	▼	▼	▲	▼	▼	–	▼	▲
p-value	0.004	0.598	0.892	0.234	0.227	0.507	0.040	0.518	0.825	0.919	0.865	0.529
Brezno	▲	▲	▼	▼	▼	▼	▲	▲	▲	–	▼	▲
p-value	0.034	0.886	0.844	0.058	0.643	0.682	0.087	0.872	0.592	0.901	0.225	0.605
Telgárt	▲	–	▲	▼	▼	▲	▲	▲	–	▲	▼	▲
p-value	0.110	0.773	0.878	0.341	0.350	0.350	0.014	0.658	0.959	0.444	0.760	0.385
Lom nad Rimavicou	▲	▲	▲	▼	▼	▼	▲	–	–	▲	–	▲
p-value	0.036	0.721	0.563	0.158	0.341	0.575	0.045	0.658	0.747	0.696	0.721	0.316

▲ trend toward wetter conditions, ▼ trend toward drier conditions, – no trend, grey shaded cells imply the statistical significance of the trend (significance level $\alpha = 0.05$).

3.4. Spatial Identification of the Drought-Prone Areas

Attempts to identify drought-prone areas were based on results (summarized in Table 2) that the highest frequency of trends toward humid conditions was recorded at the stations Telgárt (901 m a.s.l.) and Lom nad Rimavicou (1013 m a.s.l.). We used this information to construct a map that is spatially divided along with the altitude of the station Telgárt (901 m a.s.l.). For the remaining lower located stations, only in one case (SPI for 12 months) was the trend recorded as significant toward wetter conditions. Also, significant trends toward the humid condition of the SPEI for 12 months were detected only for the highest stations Telgárt and Lom nad Rimavicou. Therefore, since the SPEI for 12 months detects severe droughts more precisely than the SPI for twelve months due to accounting of evapotranspiration, the argument for delimitation of drought-prone areas below altitudes of 901 m a.s.l. is even more credible. Besides, significant trends of the SPEI for 12 months toward humid conditions at stations Telgárt and Lom nad Rimavicou imply lowering the frequency of severe drought occurrence within the studied period at altitudes higher than 901 m a.s.l. Based on this assumption, a map of the relative drought risk (Figure 8) was constructed. The map showed that relatively lower drought risk areas (trends toward humid conditions based on the SPI and SPEI for 12 months) are located in mountainous, mainly forested areas. These areas are also source areas of the Hron river (East of the Telgárt settlement) and all main tributaries. Hence, this implies relatively good prospects concerning the hydrological drought. On the other hand, in the south of the studied area, drought-prone areas were identified, where Štiavnické vrchy Mts. and Javorie Mts. are located. Therefore, in these locations and the Neresnica river basin (including the river's source area), drought could become a severe problem if the evolution of drought trends will remain as occurred in the studied period.

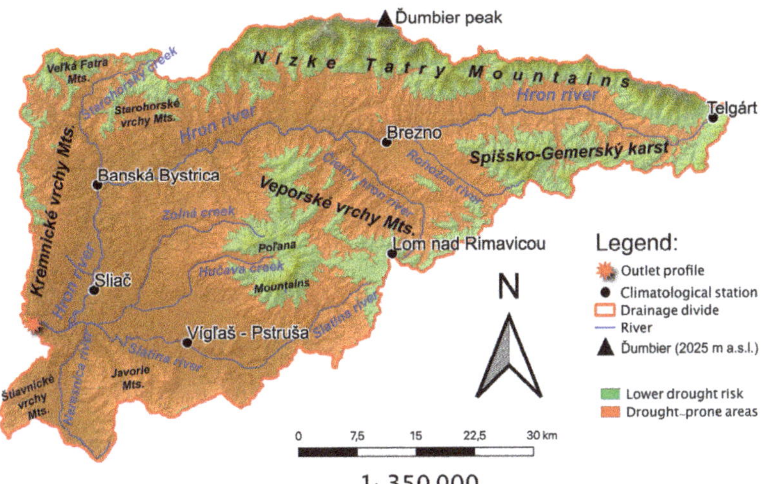

Figure 8. Spatial identification of the areas with lower drought risk and relatively drought-prone areas based on the drought indices' temporal evolution.

4. Discussion

Observed generally rising trends in the SPIs toward humid conditions are likely connected to climate change involved in increasing the absolute humidity [20]. These trends are positively correlated with rising altitude within the area. This pattern was confirmed by results that showed rising R^2 as well as the rising significance of the SPI trends with rising altitude (Figures 4 and 5). However, these results were indistinct based on the SPI for one month. Assessing the SPI for twelve months indicated significant trends toward humid conditions. Another possible interpretation could be that severe drought episodes became less frequent in the given area during the studied period due to significant humid trends of the SPIs for twelve months, which represent severe drought with cumulative drought impacts, as stated in [30]. However, it is necessary to consider that the SPI takes into account only precipitation.

Therefore, we also utilized the SPEI, which also takes potential evapotranspiration into account. The influence of evapotranspiration flattened generally rising (humid) trends, especially at lower and middle altitudes (Figures 4 and 5), and humid trends were revealed only for the highest located stations (Telgárt and Lom nad Rimavicou). That implies that rising temperatures (based on SPEI recalculated to evapotranspiration) within the studied period significantly influenced drought trend evolution. That has been confirmed particularly by the SPEI for twelve months. Results demonstrated the influence of the potential evapotranspiration (compared to the SPI) on drought episode extension and magnitude. That is evident in drought episodes of the early nineties, during the pan-European drought of the 2003, 2007, and 2011/2012 drought episodes. Based on this, it is evident that rising evapotranspiration, possibly linked with rising temperatures due to ongoing climate change [18,43], strongly influenced (and we argue that will continue to influence) drought patterns in the studied period. However, with rising altitudes, the evaporative demand of the atmosphere is lower [44]. That was distinct in retaining humid trends at the highest altitudes (Table 2). This feature somehow implies a relatively lower drought risk at higher altitudes. However, we are discussing standard weather patterns, and under this assumption, we do not consider severe pan-continental drought situations linked with intense anticyclonal situations or with tropical air advection [45]. There is no doubt that these situations influence mountainous areas the same as low altitudes. That was confirmed by [4,14,30], and we confirmed that in Figure 5. Besides, in such a geographically heterogeneous region, drought occurs (especially hydrological droughts), also related to winter snow regime and spring air temperatures, mainly in mountainous areas. In this context, our results showed significantly increasing humid trends in January and insignificantly increasing trends in December. That could imply a better snow regime in mountainous areas. However, [46,47] argue that because of the rising air temperature in mountain regions during winter, these trends are instead linked with increasing water discharge from the river basins. That has a paradoxically negative influence on drought regimes in the cold part of the year and early spring. However, these drought aspects are beyond the scope of our investigation. We leave these aspects for further studies.

The most pronounced pattern that applies to almost all stations with two exceptions was a significant increasing trend of both indices in July. This is possibly linked to the increase of atmospheric convective activity in the hottest month [48]. Although this could imply that this could have a possible effect related to lowering drought risk in the region, some authors [48–50] argue that this convective precipitation situation in summer has a rather torrential character that leads to fast water discharge and flash floods and an insignificant influence on increasing soil and groundwater supplies. We argue that this has to be taken into account when considering the results of our investigation. Another result we would like to address is insignificant trends toward drier conditions in April and May.

Interestingly, all the stations had decreasing trends in these two months. Although these trends were insignificant, we would like to highlight this general trend, since April and May are the most crucial months for agricultural and forestry activities in the landscape [51]. Similar results were indicated in [14,24]. We argue that this fact should be an objective for further investigation of the highest priority.

CRD plots detected no abrupt changes in time series of the SPIs and SPEIs. Although there were some signals of sub-trends, based on the CRD methodology [42], we cannot reliably state that these signs are relevant. Constructed CRD plots, therefore, clearly confirmed analyses of linear trends tested in Section 3.1. In comparison to the results of previous research, it is interesting that CRD analyses of the SPEI for north of the Danube lowland (Požitavie region—50 km west of our study area) revealed a distinct negative trend direction [14], opposite to our findings. Based on this, it seems that climatic conditions in the inner Carpathian region in the context of drought and drought evolution in terms of ongoing climate change will have a very different course. That should be, however, the subject of further research interest.

Our results finally led to the construction of the drought risk map based on drought trends along the altitudinal gradient, since we found that drought trends correlated with altitude within the studied region. Our hypothesis is based on the assumption that at stations that showed no trends based on the 12-month SPEI, there exists a strong assumption that continual rising air temperatures in the coming decades will increase the frequency of droughts and thus lead to arid trends. Our hypothesis is supported by previous results [14], which clearly showed how increasing air temperature leads to an intensification of drought effects and its higher frequency due to increased evapotranspiration. We understand that this map simplifies some geographical aspects which could have a local influence on the drought patterns. However, the general overview of the drought risk spatial distribution over the region is clearly described.

Based on this prerequisite, we identified as drought-prone areas all altitudes up to 901 m a.s.l. (elevation of the Telgárt climatological station). The map is depicted as the most drought-prone agricultural lands situated in valleys of the Hron river and the Slatina river [51]. Forest ecosystems characterized as mixed forest [52] also fall to the drought-prone areas. In these forests, there are widely abundant European spruce, relatively drought-sensitive species [53,54]. That has to be taken into consideration in forest management of the area. Another fact that we suggest as an objective for further investigation is hydrological analyses focused primarily on the basin of the Neresnica river located in the south of the studied area. Besides other rivers in the area, only this river has a source area solely in the identified drought-prone area.

We argue that such a map, although simple, was missing until now and as the first attempt for regional-based mitigation and adaptation measures by local authorities, is sufficient. We understand the need to improve the map in a future investigation in terms of the incorporation of all possible geographical–climatological aspects.

Author Contributions: Conceptualization, J.V.; methodology, J.V.; software, J.V.; validation, J.V.; formal analysis, J.V.; investigation, J.V. and P.N.; data curation, J.V.; writing—original draft preparation, J.V. and P.N.; writing—review and editing, J.V. and P.N.; visualization, J.V. and P.N.; supervision, J.V. All authors have read and agreed to the published version of the manuscript.

Funding: This research was funded by VEGA research projects funded by the Science Grant Agency of the Ministry of Education, Science, Research and Sport of the Slovak Republic No. 1/0370/18 and by the Slovak Research and Development Agency under the contract No. APVV-18-0347, APVV-19-0340, and APVV-19-0183.

Conflicts of Interest: The authors declare no conflict of interest.

References

1. Lindner, M.; Fitzgerald, J.B.; Zimmermann, N.E.; Reyer, C.; Delzon, S.; Van Der Maaten, E.; Schelhaas, M.-J.; Lasch, P.; Eggers, J.; Van Der Maaten-Theunissen, M.; et al. Climate change and European forests: What do we know, what are the uncertainties, and what are the implications for forest management? *J. Environ. Manag.* **2014**, *146*, 69–83. [CrossRef]

2. Andrade, C.; Leite, S.M.; Santos, J.A. Temperature extremes in Europe: Overview of their driving atmospheric patterns. *Nat. Hazards Earth Syst. Sci.* **2012**, *12*, 1671–1691. [CrossRef]

3. Blauhut, V.; Stahl, K.; Stagge, J.H.; Tallaksen, L.M.; De Stefano, L.; Vogt, J.V. Estimating drought risk across Europe from reported drought impacts, drought indices, and vulnerability factors. *Hydrol. Earth Syst. Sci.* **2016**, *20*, 2779–2800. [CrossRef]

4. Gobiet, A.; Kotlarski, S.; Beniston, M.; Heinrich, G.; Rajczak, J.; Stoffel, M. 21st century climate change in the European Alps—A review. *Sci. Total. Environ.* **2014**, *493*, 1138–1151. [CrossRef]

5. Ruffault, J.; Martin-StPaul, N.K.; Duffet, C.; Goge, F.; Mouillot, F. Projecting future drought in Mediterranean forests: Bias correction of climate models matters! *Theor. Appl. Clim.* **2013**, *117*, 113–122. [CrossRef]

6. Stagge, J.H.; Kingston, D.G.; Tallaksen, L.M.; Hannah, D.M. Observed drought indices show increasing divergence across Europe. *Sci. Rep.* **2017**, *7*, 14045. [CrossRef] [PubMed]

7. Vicente-Serrano, S.M.; Lopez-Moreno, J.-I.; Beguería, S.; Lorenzo-Lacruz, J.; Sanchez-Lorenzo, A.; García-Ruiz, J.M.; Azorin-Molina, C.; Morán-Tejeda, E.; Revuelto, J.; Trigo, R.; et al. Evidence of increasing drought severity caused by temperature rise in southern Europe. *Environ. Res. Lett.* **2014**, *9*, 044001. [CrossRef]

8. Trnka, M.; Balek, J.; Štěpánek, P.; Zahradníček, P.; Možný, M.; Eitzinger, J.; Žalud, Z.; Formayer, H.; Turňa, M.; Nejedlík, P.; et al. Drought trends over part of Central Europe between 1961 and 2014. *Clim. Res.* **2016**, *70*, 143–160. [CrossRef]

9. Spinoni, J.; Naumann, G.; Vogt, J. Spatio-temporal seasonal drought patterns in Europe from 1950 to 2015. *EGU Gen. Assem.* **2016**, *18*, 12268.

10. Alfieri, L.; Burek, P.; Feyen, L.; Forzieri, G. Global warming increases the frequency of river floods in Europe. *Hydrol. Earth Syst. Sci.* **2015**, *19*, 2247–2260. [CrossRef]

11. Dai, A. Increasing drought under global warming in observations and models. *Nat. Clim. Chang.* **2012**, *3*, 52–58. [CrossRef]

12. Twardosz, R.; Cezak, U.K. Thermal anomalies in the Mediterranean and in Asia Minor (1951–2010). *Int. J. Glob. Warm.* **2019**, *18*, 304. [CrossRef]

13. Vicente-Serrano, S.M.; García-Herrera, R.; Barriopedro, D.; Azorin-Molina, C.; López-Moreno, J.I.; Martín-Hernández, N.; Tomás-Burguera, M.; Gimeno, L.; Nieto, R. The Westerly Index as complementary indicator of the North Atlantic oscillation in explaining drought variability across Europe. *Clim. Dyn.* **2015**, *47*, 845–863. [CrossRef]

14. Vido, J.; Nalevanková, P.; Valach, J.; Šustek, Z.; Tadesse, T. Drought Analyses of the Horné Požitavie Region (Slovakia) in the Period 1966–2013. *Adv. Meteorol.* **2019**, *2019*, 1–10. [CrossRef]

15. Škvarenina, J.; Tomlain, J.; Hrvol', J.; Škvareninová, J.; Nejedlík, P. Progress in dryness and wetness parameters in altitudinal vegetation stages of West Carpathians: Time-series analysis 1951–2007. *Idojaras* **2009**, *113*, 47–54.

16. Miklós, L.; Hrnčiarová, T. *Atlas Krajiny Slovenskej Republiky [Landscape Atlas of the Slovak Republic]*; Ministry of Environment of the Slovak Republic, Slovak Agency of Environment: Basnská Bystrica, Slovak, 2002.

17. Zeleňáková, M.; Purcz, P.; Blišťan, P.; Vranayová, Z.; Hlavatá, H.; Diaconu, D.C.; Portela, M.M. Trends in Precipitation and Temperatures in Eastern Slovakia (1962–2014). *Water* **2018**, *10*, 727. [CrossRef]

18. Bartholy, J.; Pongrácz, R.; Gelybó, G.; Kern, A. What Climate Can We Expect in Central/Eastern Europe by 2071–2100? BT-Bioclimatology and Natural Hazards. In *Bioclimatology and Natural Hazards*; Střelcová, K., Mátyás, C., Kleidon, A., Lapin, M., Matejka, F., Blaženec, M., Škvarenina, J., Holécy, J., Eds.; Springer: Dordrecht, The Netherlands, 2009; pp. 3–14.

19. Škvarenina, J.; Tomlain, J.; Hrvoľ, J.; Škvareninová, J. Occurrence of Dry and Wet Periods in Altitudinal Vegetation Stages of West Carpathians in Slovakia: Time-Series Analysis 1951–2005 BT-Bioclimatology and Natural Hazards. In *Bioclimatology and Natural Hazards*; Střelcová, K., Mátyás, C., Kleidon, A., Lapin, M., Matejka, F., Blaženec, M., Škvarenina, J., Holécy, J., Eds.; Springer: Dordrecht, The Netherlands, 2009; pp. 97–106.

20. Melo, M.; Lapin, M.; Kapolková, H.; Pecho, J.; Kružicová, A. Climate Trends in the Slovak Part of the Carpathians BT-The Carpathians: Integrating Nature and Society Towards Sustainability. In *Bioclimatology and Natural Hazards*; Kozak, J., Ostapowicz, K., Bytnerowicz, A., Wyżga, B., Eds.; Springer: Berlin/Heidelberg, Germany, 2013; pp. 131–150.

21. Melo, M.; Lapin, M.; Damborská, I. Shifts in Climatic Regions in Mountain Parts of Slovakia. *Sustain. Dev. Bioclimate* **2009**, 42–43. [CrossRef]

22. Zlatník, A. *Lesnícka Fytológia Forestry Phytology*; SZN: Praha, Czech Republic, 1976.

23. Vilček, J.; Škvarenina, J.; Vido, J.; Nalevanková, P.; Kandrík, R.; Škvareninová, J. Minimal change of thermal continentality in Slovakia within the period 1961–2013. *Earth Syst. Dyn.* **2016**, *7*, 735–744. [CrossRef]

24. Zeleňáková, M.; Vido, J.; Portela, M.C.A.S.; Purcz, P.; Blišťan, P.; Hlavatá, H.; Hluštík, P. Precipitation Trends over Slovakia in the Period 1981–2013. *Water* **2017**, *9*, 922. [CrossRef]

25. Briedoň, V. Ein Beitrag zum Problem der Niederschlagesabhängigkeit von der Seehohe im Tschechoslowakischen Karpatengebiet. In *Príspevok k Meteorológii Karpát Contribution to Carpathian Meteorology*; Konček, M., Ed.; Slovenská Akadémia Vied Slovak Academy of Sciences: Bratislava, Slovakia, 1961; pp. 212–220.

26. Bochníček, O. *Klimatický atlas Slovenska Climate atlas of Slovakia*; Slovenský Hydrometeorologický ústav Slovak Hydrometeorological Institue: Bratislava, Slovakia, 2015.

27. Škvarenina, J.; Vido, J.; Minďaš, J.; Střelcová, K.; Škvareninová, J.; Fleischer, P.; Bošeľa, M. *Globálne zmeny klímy a lesné Ekosystémy Climate Change and Forest Ecosystems*; Technická univerzita vo Zvolene Technical University in Zvolen: Zvolen, Slovakia, 2018.

28. SHMI. *Report Containing Additional In-Formation with Respect to the Implementation of the GCOS Plan, Following the Established Reporting Guidelines FCCC/SBSTA/2007/L.14*; Slovak Hydrometeorological Institute: Bratislava, Slovakia, 2008; p. 210.

29. Vido, J.; Střelcová, K.; Nalevanková, P.; Leštianska, A.; Kandrík, R.; Pástorová, A.; Škvarenina, J.; Tadesse, T. Identifying the relationships of climate and physiological responses of a beech forest using the Standardised Precipitation Index: A case study for Slovakia. *J. Hydrol. Hydromech.* **2016**, *64*, 246–251. [CrossRef]

30. Vido, J.; Tadesse, T.; Šustek, Z.; Kandrík, R.; Hanzelová, M.; Škvarenina, J.; Škvareninová, J.; Hayes, M. Drought Occurrence in Central European Mountainous Region (Tatra National Park, Slovakia) within the Period 1961–2010. *Adv. Meteorol.* **2015**, *2015*, 1–8. [CrossRef]

31. Šustek, Z.; Vido, J.; Škvareninová, J.; Skvarenina, J.; Surda, P. Drought impact on ground beetle assemblages (Coleoptera, Carabidae) in Norway spruce forests with different management after windstorm damage—A case study from Tatra Mts. (Slovakia). *J. Hydrol. Hydromech.* **2017**, *65*, 333–342. [CrossRef]

32. McKee, T.B.; Doesken, N.J.; Kleist, J. The relationship of drought frequency and duration to time scales. In Proceedings of the 8th Conference on Applied Climatology, Anaheim, CA, USA, 17–22 January 1993; pp. 179–183.

33. Edwards, D.C. *Characteristics of 20th Century Drought in the United States at Multiple Time Scales*; Climatology Report No. 97-2. Atmospheric Science Paper No. 634; Colorado State University: Fort Collins, CO, USA, 1997; p. 155. 155p.

34. Hayes, M.J.; Svoboda, M.D.; Wilhite, D.A.; Vanyarkho, O.V. Monitoring the 1996 Drought Using the Standardized Precipitation Index. *Bull. Am. Meteorol. Soc.* **1999**, *80*, 430–438. Available online: http://digitalcommons.unl.edu/droughtfacpubhttp://digitalcommons.unl.edu/droughtfacpub/31 (accessed on 6 June 2020). [CrossRef]

35. Bak, B.; Labedzki, L. Assessing drought severity with the relative precipitation index [RPI] and the standardised precipitation index [SPI]. *J. Water Land Dev.* **2002**, *6*, 89–105.

36. Tirivarombo, S.; Osupile, D.; Eliasson, P. Drought monitoring and analysis: Standardised Precipitation Evapotranspiration Index (SPEI) and Standardised Precipitation Index (SPI). *Phys. Chem. Earth, Parts A/B/C* **2018**, *106*, 1–10. [CrossRef]

37. Heim, R.R. A Review of Twentieth-Century Drought Indices Used in the United States. *Bull. Am. Meteorol. Soc.* **2002**, *83*, 1149–1166. [CrossRef]

38. Vicente-Serrano, S.M.; Beguería, S.; López-Moreno, J.I. A Multiscalar Drought Index Sensitive to Global Warming: The Standardized Precipitation Evapotranspiration Index. *J. Clim.* **2010**, *23*, 1696–1718. [CrossRef]

39. Thornthwaite, C.W. An approach toward a rational classification of climate. *Geogr. Rev.* **1948**, *38*, 55–94. [CrossRef]

40. Yue, S.; Wang, C. The Mann-Kendall Test Modified by Effective Sample Size to Detect Trend in Serially Correlated Hydrological Series. *Water Resour. Manag.* **2004**, *18*, 201–218. [CrossRef]

41. Onyutha, C. Influence of Hydrological Model Selection on Simulation of Moderate and Extreme Flow Events: A Case Study of the Blue Nile Basin. *Adv. Meteorol.* **2016**, *2016*, 1–28. [CrossRef]

42. Onyutha, C. Statistical Uncertainty in Hydrometeorological Trend Analyses. *Adv. Meteorol.* **2016**, *2016*, 1–26. [CrossRef]

43. Gudmundsson, L.; I Seneviratne, S. Anthropogenic climate change affects meteorological drought risk in Europe. *Environ. Res. Lett.* **2016**, *11*, 044005. [CrossRef]

44. Lapin, M.; Gera, M.; Hrvol', J.; Melo, M.; Tomlain, J. Possible impacts of climate change on hydrologic cycle in Slovakia and results of observations in 1951–2007. *Biologia* **2009**, *64*, 454–459. [CrossRef]

45. Ionita, M.; Tallaksen, L.M.; Kingston, D.G.; Stagge, J.H.; Laaha, G.; Van Lanen, H.A.J.; Scholz, P.; Chelcea, S.M.; Haslinger, K. The European 2015 drought from a climatological perspective. *Hydrol. Earth Syst. Sci.* **2017**, *21*, 1397–1419. [CrossRef]

46. Hríbik, M.; Majlingová, A.; Škvarenina, J.; Kyselová, D. Winter Snow Supply in Small Mountain Watershed as a Potential Hazard of Spring Flood Formation BT-Bioclimatology and Natural Hazards. In *Bioclimatology and Natural Hazards*; Střelcová, K., Mátyás, C., Kleidon, A., Lapin, M., Matejka, F., Blaženec, M., Škvarenina, J., Holécy, J., Eds.; Springer: Dordrecht, The Netherlands, 2009; pp. 119–128.

47. Sleziak, P.; Szolgay, J.; Hlavčová, K.; Parajka, J. The Impact of the Variability of Precipitation and Temperatures on the Efficiency of a Conceptual Rainfall-Runoff Model. *Slovak J. Civ. Eng.* **2016**, *24*, 1–7. [CrossRef]

48. Marzen, M.; Iserloh, T.; De Lima, J.L.; Fister, W.; Ries, J.B. Impact of severe rain storms on soil erosion: Experimental evaluation of wind-driven rain and its implications for natural hazard management. *Sci. Total. Environ.* **2017**, *590–591*, 502–513. [CrossRef]

49. Hlavčová, K.; Kohnová, S.; Borga, M.; Horvát, O.; Šťastný, P.; Pekárová, P.; Majerčáková, O.; Danáčová, Z. Post-event analysis and flash flood hydrology in Slovakia. *J. Hydrol. Hydromech.* **2016**, *64*, 304–315. [CrossRef]

50. Millan, M.M. Extreme hydrometeorological events and climate change predictions in Europe. *J. Hydrol.* **2014**, *518*, 206–224. [CrossRef]

51. Šiška, B.; Takáč, J. Drought analyses of agricultural regions as influenced by climatic conditions in the Slovak Republic. *Idojárás* **2009**, *13*, 135–143.

52. Střelcová, K.; Kučera, J.; Fleischer, P.; Giorgi, S.; Gömöryová, E.; Škvarenina, J.; Ditmarová, L. Canopy transpiration of mountain mixed forest as a function of environmental conditions in boundary layer. *Biologia* **2009**, *64*, 507–511. [CrossRef]

53. Solberg, S. Summer drought: A driver for crown condition and mortality of Norway spruce in Norway. *For. Pathol.* **2004**, *34*, 93–104. [CrossRef]

54. Ďurský, J.; Škvarenina, J.; Minďáš, J.; Miková, A. Regional analysis of climate change impact on Norway spruce (*Picea abies* L. Karst.) growth in Slovak mountain forests. *J. For. Sci.* **2006**, *52*, 306–315. Available online: http://www.scopus.com/inward/record.url?eid=2-s2.0-33746599991&partnerID=40 (accessed on 6 June 2020).

Publisher's Note: MDPI stays neutral with regard to jurisdictional claims in published maps and institutional affiliations.

Article

Estimating the Effect of Deforestation on Runoff in Small Mountainous Basins in Slovakia

Michaela Danáčová, Gabriel Földes *, Marija Mihaela Labat, Silvia Kohnová and Kamila Hlavčová

Department of Land and Water Resources Management, Faculty of Civil Engineering,
Slovak University of Technology, Radlinského 11, 810 05 Bratislava, Slovakia;
michaela.danacova@stuba.sk (M.D.); marija.labat@stuba.sk (M.M.L.); silvia.kohnova@stuba.sk (S.K.);
kamila.hlavcova@stuba.sk (K.H.)
* Correspondence: gabriel.foldes@stuba.sk

Received: 15 September 2020; Accepted: 2 November 2020; Published: 6 November 2020

Abstract: The paper aims to assess the impact of deforestation due to windstorms on runoff in small mountain river basins. In the Boca and Ipoltica River basins, changes in forested areas were assessed from available historical and current digital map data. Significant forest losses occurred between 2004 and 2012. During the whole period of 1990–2018, forested areas in the Boca river decreased from 83% to 47% and in the Ipoltica River basin from 80% to 70%. Changes in runoff conditions were assessed based on an assessment of changes in the measured time series of the hydrometeorological data for the years 1981–2016. An empirical hydrological model was used to determine the design peak discharges before and after significant windstorms were estimated for different rain intensities and return periods. The regional climate scenario for the period 2070–2100 was used to assess the current impact of climate change and river basin deforestation on predicted changes in design floods in the coming decades. The effect of deforestation became evident in the extreme discharges, especially in future decades. In the Boca River basin, the estimated design floods increased by 59%, and in the Ipoltica River basin by 172% in the case of the 100-year return period.

Keywords: windstorms; forest area; land use change; climate change

---

1. Introduction

Environmental changes and their impact on hydrological regimes and the occurrence of extremes such as floods or droughts have been of topical interest in recent years all over the world. A hydrological regime is a set of natural and anthropogenic conditions that affect the surface and subsurface runoff from a river basin. The assessment of a runoff regime is undertaken for purposes of prevention or serves as a basis for the proposed measures within the river basin to alleviate extreme runoff situations.

In particular, land use changes and climate change have contributed to problems that have a direct or indirect impact on runoff regimes. Changes in land use have a strong effect on floods, as humans have heavily modified natural landscapes; large areas have been deforested or drained, thus either increasing or decreasing antecedent soil moisture and triggering erosion [1]. Among the various land cover types, the role of forests in catchment hydrology is highly consequential. According to many experimental studies, forest cover decreases discharges in river basins as a result of increased interception, transpiration and permeability of soils, and reduced soil moisture [1]. Such results can be found, e.g., in Brown et al. [2], where paired catchment studies were used for determining the changes in water yields on various timescales resulting from permanent changes in vegetation. Indirect effects of forest changes, which are caused mainly by the impacts of forest practices, include an increase in surface runoff and soil erosion on forest roads, and the development of gullies after deforestation, especially on steep terrains (e.g., Vose et al. [3]).

Deforestation is one of the oldest anthropogenic activities; one of the main reasons for it is due to changes in land use. In the case of forest catastrophes, whether as a result of a windstorm or an infestation of bark beetles, this is a serious problem. The main negatives of deforestation (either anthropogenic or from natural disasters) include the loss of biodiversity, soil erosion, landslides, and increased CO_2 in the atmosphere. As a result, deforestation is a serious global threat and one of the most significant environmental problems in the world. The impact of deforestation on a hydrological regime is highly variable and often difficult to explain, but it is often determined that deforestation increases and reforestation decreases annual flows [4]. The summary of paired watershed studies by Andreassian [5] has shown that deforestation can increase both flood volumes and flood peaks, and this effect is much more variable than its effect on a total flow. On the other hand, forested catchments have greater infiltration rates, which may decrease catchment runoff [6]. A study by Suryatmojo [7] shows that a loss of forest increases soil erosion and the flow of streams and also reduces water quality and soil fertility.

There are various options for assessing the development of and changes in land use. One of them is the use of digital topographic maps, and, in particular, satellite images; they contribute to the knowledge of a landscape and the ability to analyse changes in its use and the occurrence of landscape features [8]. Various approaches applied to analyze the effects of land use changes on runoff and flooding include the analysis of empirical data and mathematical modelling, which are dominant in current hydrological research, particularly in combination with GIS tools [9]. A review of several widely-used hydrological models and their applicability to simulate the impact of land use and climate change is provided in Dwarakish, Ganasri [10]. In this paper, the empirical event-based hydrological model SCS-CN model, which is relatively easy to use and yields satisfactory results, was applied in combination with a GIS environment.

According to the Intergovernmental Panel on Climate Change (IPCC) [11], the climate will presumably be more variable or extreme in the future with a potential increase in the frequency of heavy rainfall. Recent research shows that the increase is likely to occur in intensities of short-term rain in durations of less than one day, which may lead to an increase in the extent and frequency of flash floods. Climate models are widely used to assess the past and make forecasts for the future. Outputs from regional climate models are mainly used for the analysis, which are then compared with actual observations. An analysis of the evidence leading to an increase in the incidence of extreme short-term rain due to anthropogenic climate change, as well as a description of the current physical understanding of the association between extreme short-term rainfalls and the atmospheric temperature, is needed to help society adapt to anticipated future changes in short-term rains [12]. The anticipated increase in the incidence of extreme precipitation totals has been confirmed by various studies from around the world, e.g., Tebaldi et al. [13], Koutsoyiannis et al. [14], Kyselý et al. [15], Skaugen and Førland [16], Lapin et al. [17], Wang et al. [18], Gaál et al. [19], Pascale et al. [20], Mamoon et al. [21], Hanel et al. [22], Gera et al. [23], Nepal [24], Taibi et al. [25].

In Slovakia, there have been a number of significant natural disasters in the last twenty years. The causative agent of those disasters was primarily an abiotic factor, such as wind, snow or ice, or a combination thereof. The effects of forests and deforestation on the water balance in river basins in Slovakia has been studied in many recent works, e.g., Kostka and Holko [26]; Hlavčová et al. [27]; Hlásny et al. [4]; Kohnová et al. [28].

This paper is devoted to changes in land use due to deforestation in two small mountain river basins, i.e., the Boca and Ipoltica River basins in Slovakia, which have been affected by severe windstorms in recent years. As a result of these disasters, large areas of these river basins have become deforested. For this study, analyses of the land use maps of the deforested areas before and after the windstorms were compared during the period 1981–2018. Next, the long-term development of the time series of the hydro-meteorological characteristics measured such as discharges, precipitation, and air temperatures in the periods of 1981–2016 and 2005–2016 were further compared. The SCS-CN methodology was applied to estimate any changes in design floods before and after the disasters.

The development of and possible changes in the design floods in the deforested river basins in future decades was estimated using scenario outputs of the Community Land Model (CLM), a regional climate model.

The main objective of this study was to propose a methodology for estimating design floods in small mountainous deforested ungauged basins in the future decades that is based on the scaling of scenario designs of short-term rainfalls and a simple hydrological event-based model.

The following steps fulfilled these objectives: (1) analysis of the land use maps and identification of any changes in land use before and after the occurrence of the windstorms in the case studies. (2) Evaluation of the long-term regime of the time series of the hydrometeorological data, and where possible, identification of changes in runoff caused by climatic or anthropogenic influences. (3) Simulation of changes in design floods after the windstorms. (4) Estimation of potential changes in the design floods in the deforested river basins in future decades as a result of climate change.

2. Area of the Case Study

Many mountainous areas in Slovakia have been affected by severe windstorms in recent decades that caused significant deforestation in a large number of river basins. It is assumed that these changes also affected the runoff conditions in the case study area of the Boca and Ipoltica River basins (Figure 1). The Boca and Ipoltica River basins are located in the Low Tatras National Park of the district of Liptovský Mikuláš, which lies in northern Slovakia.

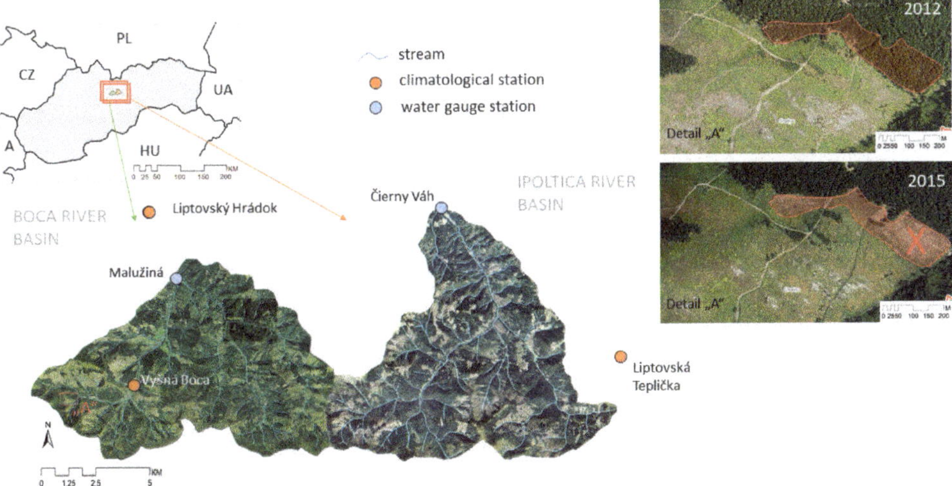

Figure 1. Location map of the basins analysed (© GKÚ, NLC; Slovakia, 2017–2019) and illustration of the state of the deforestation on the Boca River basin in 2015.

From a geomorphological point of view, the Boca and Ipoltica River basins belong to the Ďumbier Tatras subunit. The soil in these river basins is mostly silt and sandy loam, as well as loam and loamy sand in smaller quantities. Specific to this area are spruce forests.

2.1. Boca River Catchment

The Boca River basin, with its outlet at the Malužiná gauging station, has a basin area of 81.93 km^2; it is a left tributary of the Váh River. It springs in the Low Tatras at an altitude of approximately 1400 m above sea level. The average altitude of the basin is 1114 m a. s. l. and the average slope is 22.1 degrees (Figure 2).

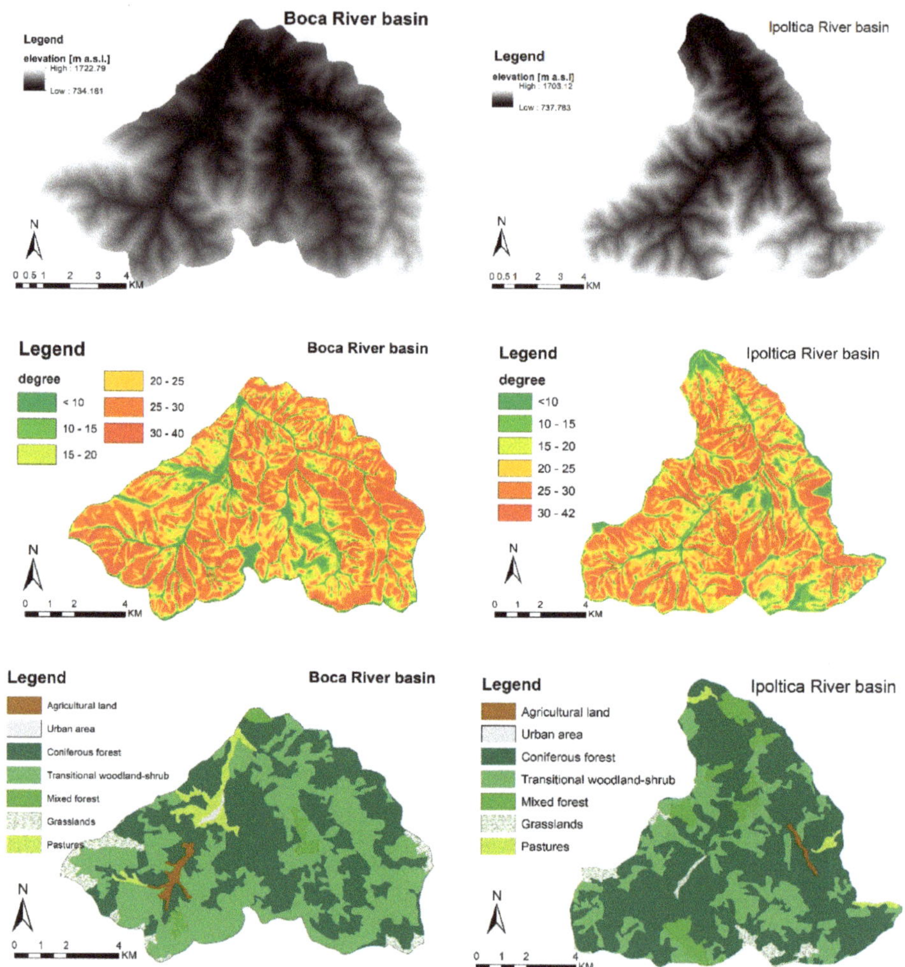

Figure 2. Elevation, slope map, and land cover classification for the Boca and Ipoltica River basins.

In 1990, 82.76% of the area was covered by forests, but by 2018, this area had decreased to 46.8%. The deforested areas have mostly been replaced by transitional woodland shrubs.

2.2. Ipoltica River Catchment

The Ipoltica River basin, with its outlet at the Čierny Váh gauging station, is a left tributary of the Čierny Váh River. The basin area is 86.25 km². It springs in the Low Tatras in the Kráľovohoľské Tatras at an altitude of about 1405 m a. s. l. The average altitude of the basin is 1118 m a. s. l., and the average slope is 22.3 degrees (Table 1). In 1990, forests used to cover 79.51% of this area. By 2018, the area with forests had been reduced to 70.34%; instead of forests, transitional woodland-shrubs started to spread (Table 2). There is a dense forest stand in the basin, which has no urban area.

Table 1. Basic characteristics of the basins analysed (P–precipitation, Q–discharge, T–air temperature–average values for the period 1981–2016).

River basin	Area	Elevation (m a.s.l.)			Slope (°)	P	Q	T
	(km²)	Min.	Max.	Mean	Mean	(mm)	(m³/s)	(°C)
Boca	81.93	734.2	1719.9	1140	22.1	827	1.38	7.1
Ipoltica	86.28	737.8	1703.1	1000	22.3	815	1.43	7.0

Table 2. The percentage of land use categories in 2018 for the Boca and Ipoltica River basins (CF–coniferous forest, MF–mixed forest, TW–transitional woodland-shrub, G–grasslands, AL–agriculture land, P–pastures, UA–urbanized area).

River Basin	Percentage of Land Use Category (%)						
	CF	MF	TW	G	AL	P	UA
Boca	44.85	1.95	42.52	3.68	1.32	3.22	0.30
Ipoltica	64.34	6.0	24.44	2.90	0.49	1.17	0.0

3. Materials and Methods

A number of significant windstorms have been recorded in the area of the Low and High Tatras since 1996 (Table 3). Additionally, uprooted trees have been attacked by bark beetles, which have a rapid reproductive capacity. The condition and health of the forests in the mountain areas with a predominance of spruce stands have deteriorated, especially after the most significant and widespread wind disaster of 2004 (known as "Elizabeth") [29].

Table 3. Overview of significant windstorms (disasters) of northern and central Slovakia.

Storm (Data of Its Occurrence)	Type of Natural Storm	Affected Region
Ivan (8 July 1996)	wind	Horehronie
Paulína (22 June 1999)	wind	Horná Nitra
Tamara (24–26 January 2001)	ice	Kriváň, Hnúšťa
Sabína (27–28 October 2002)	wind	High Tatras, Orava, Spiš
Klaudia (16–17 November 2002)	wind	High Tatras, Orava, Spiš
Alžbeta (19 November 2004)	wind	High and Low Tatras
Trojkráľová (January 2006)	snow	Orava, Low Tatras
Kyrill (18–19 January 2007)	wind	Low Tatras
Filip (23–24 August 2007)	wind	Gemer, Low Tatras
Žofia (May 2014)	wind	Low Tatras

Significant impacts resulting in changes in forest cover in the Boca River basin occurred in 2004 (Alžbeta), 2007 (Kyrill and Filip), and 2014 (Žofia). A bark beetle outbreak followed these windstorms. Alžbeta, the most severe windstorm, affected a large part of the High Tatras National Park and a substantial part of the Low Tatras National Park. The wind speed was 140 km/h with gusts up to 240 km/h [30]. In addition to a landslide, the storm also resulted in an extreme infestation of the undergrowth by insects; they attacked the areas of the forest affected by the calamity and continued spreading to a healthy part of the forest.

The Žofia windstorm damaged a large part of the Low Tatras. The gusts reached a speed of up to 100 km/h, which, together with intensive rainfall activity (141 mm per 24 h), caused great devastation to the area, including young forest stands. The soil saturated by the rainfall combined with the extreme wind force resulted in damaged trees and degraded vegetation over a large area [30].

3.1. Land Use Analysis

The CORINE Land Cover (CLC) is one of the most well-known European Environmental Agency databases. The CLC data are commonly used at various hierarchical levels of detail [31] and provide information on the biophysical characteristics of the Earth's surface. Observation satellites are used as the primary source data to derive land cover and land use information. Despite limitations in its spatial resolution, the database has become a primary spatial data source.

In this study, CLC data, specifically vector digital maps from 1990, 2006, 2012, and 2018 were used to describe changes in land cover; they were processed in an ArcGIS environment (Figures 3 and 4).

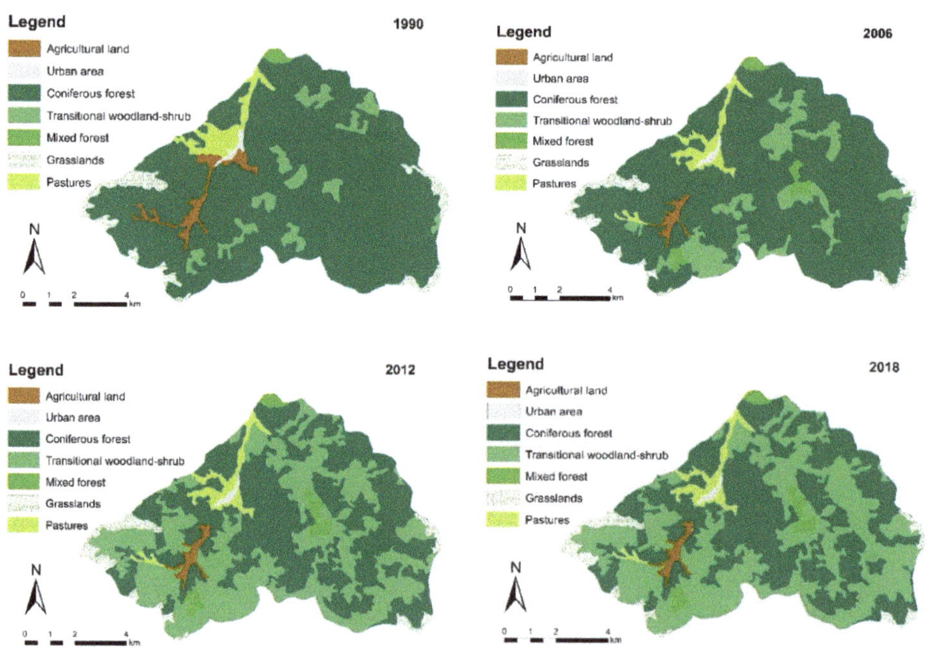

Figure 3. Boca River basin—CORINE Land Cover 1990, 2006, 2012, and 2018.

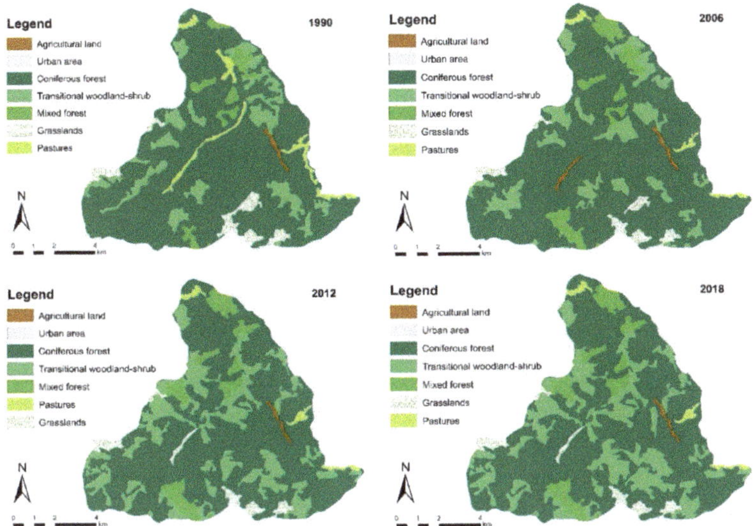

Figure 4. Ipoltica River basin—CORINE Land Cover 1990, 2006, 2012, and 2018.

3.2. Analysis of the Hydrometeorological Data

The hydrometeorological data used in our analysis includes daily average discharges, the air temperatures, and precipitation from the Slovak Hydrometeorological Institute (SHMI), for the period 1981–2016. Data from two hydrological stations ((Malužiná (ID 5336), Čierny Váh (ID 5530), and four climatological stations (Liptovský Hrádok (ID 11874), Vyšná Boca (ID 20160), Liptovská Teplička (20020), and Králová Lehota (ID 20140)) were evaluated. In this study, an estimate of changes in the runoff after demonstrated changes in the basin was made. The variance of the monthly data for the whole period of the time series 1981–2016 and the period 2005–2016 after the changes based on the differences in land use was evaluated.

Scaling of Short-Term Rainfall and Future Rainfall Scenarios

Subsequently, hourly, and daily precipitation data were used from the two climatological stations (Vyšná Boca (ID 20160) and Liptovská Teplička (ID 20020)) for the estimation of the IDF curves and short-term design rainfall. A simple scaling method is used to process rainfall data for a period of time shorter than one day. Simple scaling determines the design values for a duration shorter than one day and for a selected time period by using daily rainfall records that are commonly available. Applying simple scaling to the relationship between the IDF properties of short-term rainfall is possible. The determination of the rainfall scaling properties is based on the general shape of the following IDF formula in the form [32]:

$$i = \frac{a(T)}{b(d)} \tag{1}$$

where i is the rainfall intensity; T is the return period; d is the duration of the rainfall. The function a(T) can be determined from the probability distribution function of the maximum rainfall intensity, and b(d) is the duration function of the rain given by the formula:

$$b(d) = (d + \theta)^\eta \tag{2}$$

where θ, η are the parameters to be estimated ($\theta > 0$, $0 < \eta < 1$).

Simple scaling that used the scaling of the statistical moments was applied in this paper. The I_d and $I_{\lambda d}$ are annual maximum rainfall intensity series for the rainfall durations d and λd defined by [33]. The random variables I_d and $I_{\lambda d}$ have the following scaling property:

$$I_{\lambda d}^n \stackrel{dist}{=} \lambda^\beta I_d \tag{3}$$

where equality $\stackrel{dist}{=}$ is understood in the sense of the equality of the probability distributions, and β represents the scaling exponent. This property is usually referred to as "simple scaling in the strict sense" [34]. If $I_{\lambda d}$ has finite moments, $E\left[I_{\lambda d}^n\right]$ of order n, then the strict sense of the simple scaling in Equation (3) implies that $I_{\lambda d}^n$ and $(\lambda^\beta I_d)^n$ have the same probability distribution. Therefore, they have the same moments and are given by the following formula [35]:

$$E\left[I_{\lambda d}^n\right] = \lambda^{\beta n} E\left[I_d^n\right] \tag{4}$$

where βn represents the scaling exponent of the n-th order. To obtain the value of βn, Equation (4) can be transformed as

$$\log E\left[I_{\lambda d}^n\right] = \log E\left[I_d^n\right] + \beta_n \log \lambda \tag{5}$$

The scaling exponents were estimated with a linear regression from the slope between the logarithmic moment values and the scaling parameters for the different order of the moments. The scaling exponents, β_n can be estimated from the slope of the linear regression relationships between the log-transformed values of moment $\log E\left[I_{\lambda d}^n\right]$ and scale parameters $\log \lambda$ for the various orders of moment (n). If the scaling exponent and order of moment have a linear relationship, then β_n = $n\beta_1$, in which β_1 is the scaling exponent of order 1. This property is referred to as "wide sense simple scaling". If the above linear relationship does not exist, a multiscaling approach has to be considered [36].

3.3. Future Climate Change Scenarios

The data used in the analysis for the estimation of the future changes in short-term rainfall intensities and their subsequent impact on runoff extremities were created by a CLM simulation with a SRES A1B scenario, which is a semi-pessimistic scenario with a predicted increase in the global temperature of 2.9 °C by 2100. The data were provided by Dr. Martin Gera from Comenius University in Bratislava, Department of Astronomy, Physics of the Earth, and Meteorology.

CLM Scenario

The scenario was created as a collaborative project between scientists from several working groups from the USA, namely, the Terrestrial Sciences Section (TSS) and the Climate and Global Dynamics Division (CGD) at the National Center for Atmospheric Research (NCAR) and the Community Earth System Model (CESM), the Land Model, and the Biogeochemistry Working Groups. Ecological climatology concepts have been implemented in the model. Ecological climatology is a multidisciplinary structure that is used to understand the impacts of changes in vegetation on the climate. The scenario examines physical, chemical, and biological processes by which terrestrial ecosystems influence and are influenced by the climate on various spatial and temporal scales. The main motive is that terrestrial ecosystems are essential factors of the climate through their energy, water, chemical elements, and trace gases. The main parts of the model consist of surface heterogeneity, bio-geophysics, the hydrological cycle, biogeochemistry, ecosystem dynamics, and the human dimension. The CLM addresses several aspects that allow for the study of two-way interactions between human activities in the countryside and the climate, changes in land cover/land use, agricultural practices, and urbanization [37–39].

The CLM scenario data consists of hourly rainfall intensities for two time periods, i.e., a historical (1961–2020) and a future (2071–2100) period.

The seasonality and trend analyses were performed for two climatological stations, namely, Liptovská Teplička and Vyšná Boca. The Liptovská Teplička climatological station is located 903 m a. s. l.; the Vyšná Boca climatological station is located 948 m a. s. l. in the northern part of Slovakia. The area belongs to a slightly warm climatic region with a mountain climate and low-temperature inversions.

The results of the predicted rainfall intensities were compared to the actual measured rainfall data in hourly time steps, which were provided by the Slovak Hydrometeorological Institute for the 1995–2009 period for Liptovská Teplička. For the Vyšná Boca climatological station, only daily rainfall data for the 1981–2017 period were available.

3.3.1. SCN CN Methodology

The method of The Soil Conservation Service–Curve Number (SCS-CN) was developed by the United States Department of Agriculture–Natural Resources Conservation Service (USDA–NRCS), which was formerly called the Soil Conservation Service (SCS) [40,41]. It is used for estimating the volume of direct surface runoff characteristics for small basins in ungauged rural catchments where there are no measurements or observations of direct flows [42–45]. This method is used to predict direct surface runoff volume for a given rainfall event and to estimate the volume and peak rate of surface runoff [46]. A Curve Number (CN), which is the main parameter in this model, is based on an empirical study of runoff in small watersheds and hill slopes [47].

The primary reason for the method's widespread applicability and acceptability lies not only in the fact that it is empirical and simple to apply, but also in that it accounts for most runoff-producing watershed characteristics, e.g., soil types, land use/treatments, surface conditions, and antecedent moisture conditions [48].

As shown in Mishra et al. [40], the SCS-CN method is based on the following equations:

$$Q = \frac{(P-I_a)^2}{P-I_a+S} \quad \text{if } P > I_a \tag{6}$$

$$Q = 0 \quad \text{if } P \leq I_a \tag{7}$$

$$I_a = \lambda \times S \tag{8}$$

$$S = \frac{25400}{CN} - 254 \tag{9}$$

where: P–total rainfall (mm), Ia–initial abstraction (mm), Q–direct runoff (mm), λ–initial abstraction coefficient (-), S–maximum potential retention or infiltration (mm), CN–Curve Number (-).

When applying the CN method to calculate design floods in the Boca and Ipoltica River basins, the design rainfall was used as an input rainfall, and the initial abstraction coefficient was set to be equal to zero.

4. Results and Discussion

4.1. Land Use Analysis

Corine CLC maps were used to assess land-use changes from the period 1990–2018 (Figures 3 and 4). The area with forests was considerably reduced after 2006, which we can see in the visual analysis. This is the period after the greatest calamity, i.e., Alžbeta, in 2004 (note: the satellite maps are processed with a time interval of +/− 1 year).

Table 4 shows all the data obtained from the CLC digital maps. ArcGIS was applied as an evaluation tool.

Table 4. The percentage of land use categories from CLC data (1990, 2006, 2012 and 2018).

Land Use	Basin	Percentage of Land Use Category (%)			
		CLC 1990	CLC 2006	CLC 2012	CLC 2018
Urban area	Boca	0.41	0.3	0.3	0.3
	Ipoltica	0	0	0	0
Pastures	Boca	3.22	3.34	3.22	3.22
	Ipoltica	3.43	1.1	1.17	1.17
Agricultural land	Boca	2.78	1.21	1.32	1.32
	Ipoltica	0.53	0.87	0.49	0.49
Coniferous forest	Boca	82.24	77.09	47.90	44.85
	Ipoltica	76.14	76.09	66.93	64.34
Mixed forest	Boca	0.52	1.8	1.96	1.95
	Ipoltica	3.37	5.66	6.0	6
Grasslands	Boca	4.59	3.76	3.71	3.68
	Ipoltica	3.81	3.09	2.9	2.9
Transitional woodland-shrub	Boca	6.67	12.37	41.48	42.52
	Ipoltica	12.99	13.5	22.85	24.44

Significant changes in land use have been recorded in the Boca River basin. In 1990, 83.42% of the area was covered by forests; by 2018, this area had decreased to 46.8%. The deforested areas have mostly been replaced by transitional woodland shrubs. The Ipoltica River basin showed slight changes in comparison with the Boca River basin during the period 1990–2018. In 1990, forests used to cover 79.51% of this area. By 2018, the area with forests was reduced to 70.34%; instead, forests and transitional woodland-shrubs started to spread.

4.2. Analysis of Hydrometeorological Data

The long-term annual and monthly mean discharges and the trends in the selected discharge stations were compared. Trends were determined by a linear regression for the time period 1981–2016. The statistical significances of the trends were estimated by the Mann-Kendall non-parametrisation test. The comparison of the long-term data was provided between the period after the greatest calamity, i.e., Alžbeta (2005–2016) and before the calamity period (1981–2004). Subsequently, the comparison of the long-term data between the whole (1981–2016) and the post-deforestation (2005–2016) period was done.

4.2.1. Boca River Basin

The mean annual discharges showed an increasing trend for the Malužiná station outlet (Figure 5a). The linear trends in each month were detected too. A decreasing trend can be observed in April, June, and July. Increasing trends occurred in January, February, March, May, August, September, October, November, and December. The statistical analysis showed that the statistical significance, which was based on the Mann-Kendall test, achieved a 90% level of significance in all the months, except for February, March, October, November, and December. Comparisons of the mean monthly discharges between the pre- and post-deforestation periods (after the Alžbeta windstorm, 2004) show increases in discharges for all the months (Figure 5c). The highest differences are in the months of March and April. Comparisons of the mean monthly discharges between the whole and post-deforestation periods also show the differences (Figure 5d). The box plot (min., max., and median values) of the average monthly discharge is shown in Figure 5b.

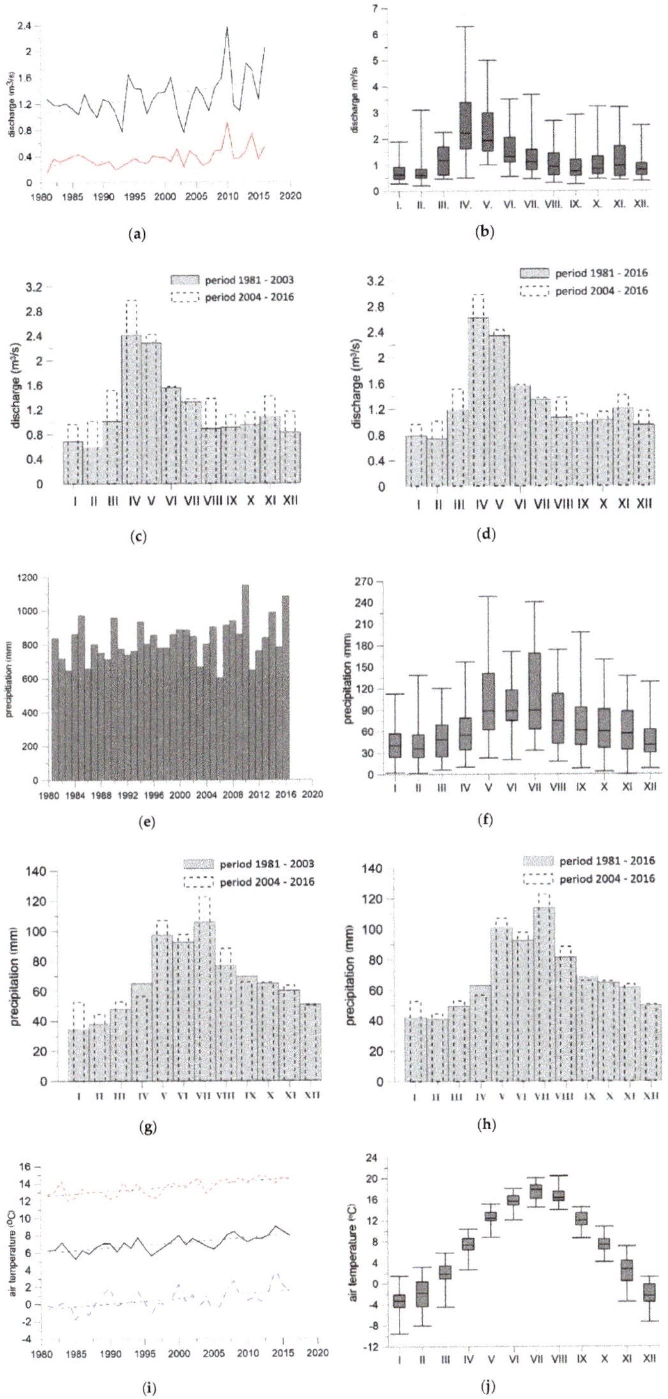

Figure 5. *Cont.*

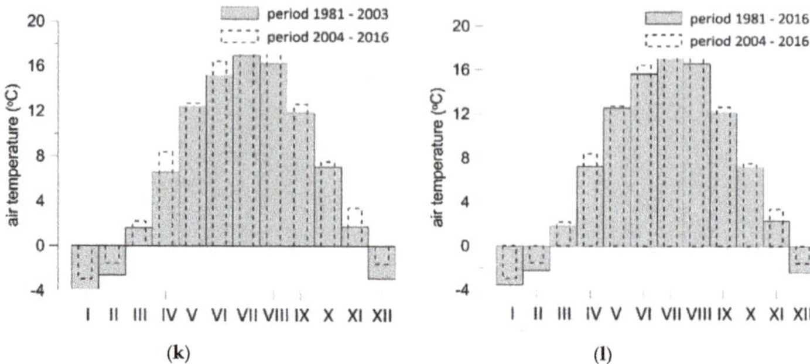

Figure 5. (**a–i**) Analysis of the hydrometeorological data–Boca River basin. (**a**) time series of annual discharges–average (black line), minimum (red line); (**b**) box plot of monthly discharges (1981–2016); (**c**) average monthly discharge for period 1981–2003 and period 2004–2016; (**d**) average monthly discharge for period 1981–2016 and period 2004–2016; (**e**) time series of annual precipitation totals; (**f**) box plot of monthly precipitation (1981–2016); (**g**) average monthly precipitation for period 1981–2003 and period 2004–2016; (**h**) average monthly precipitation for period 1981–2016 and period 2004–2016; (**i**) air temperature—annual (black line), summer season (red line) and winter season (blue line); (**j**) box plot of monthly air temperature (1981–2016); (**k**) average monthly air temperatures for period 1981–2003 and period 2004–2016; (**l**) average monthly air temperatures for period 1981–2016 and period 2004–2016.

A runoff regime is also associated with climatological conditions; therefore, the precipitation totals for the monthly and annual time steps were analyzed. The linear trend of the annual precipitation totals is increasing (Figure 5e). The box plot of the monthly precipitation totals is shown in Figure 5f, where the maximum precipitation totals in the summer months can be seen. The average monthly precipitation was also observed in different periods (before and after the calamity period, a comparison of the whole and post-calamity period), where, in addition to the months of April, September, and December, increased total precipitation was detected (Figure 5g,h).

The analysis of the air temperature revealed an increasing linear trend of the average annual temperature, as well as the temperatures in the summer and winter half-years (Figure 5i). The highest temperatures are in the summer months of June, July, and August. An increase in temperature was observed in all the months except for July when comparing the periods (Figure 5k,l).

4.2.2. Ipoltica River Basin

The annual discharges showed an increasing trend at the Čierny Váh station outlet (Figure 6a). Next, the linear trends of the discharges in each month were detected. Decreasing trends occurred in April, May, June, October, November, and December. Increasing trends were seen in January, February, March, July, August, and September. According to the statistical analysis, the Mann-Kendall test (with a 90% level of significance) achieved significance in all the months except May. Comparisons of the mean monthly discharges between the pre- and post-deforestation periods showed increases in discharges for all the months except May (Figure 6c). Comparisons of the mean monthly discharges between the whole and post-deforestation periods show a difference too (Figure 5d). The box plot of the average monthly discharge is shown in Figure 6b.

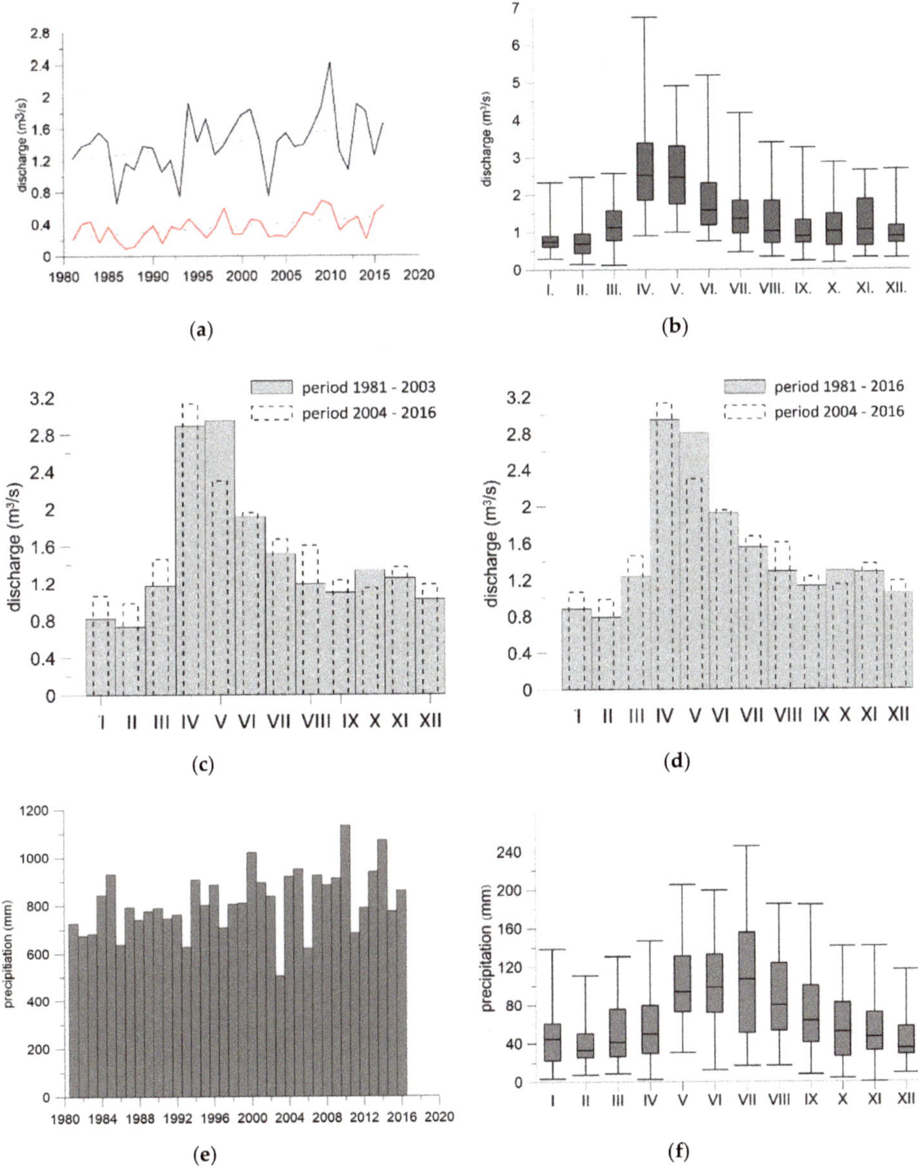

Figure 6. *Cont.*

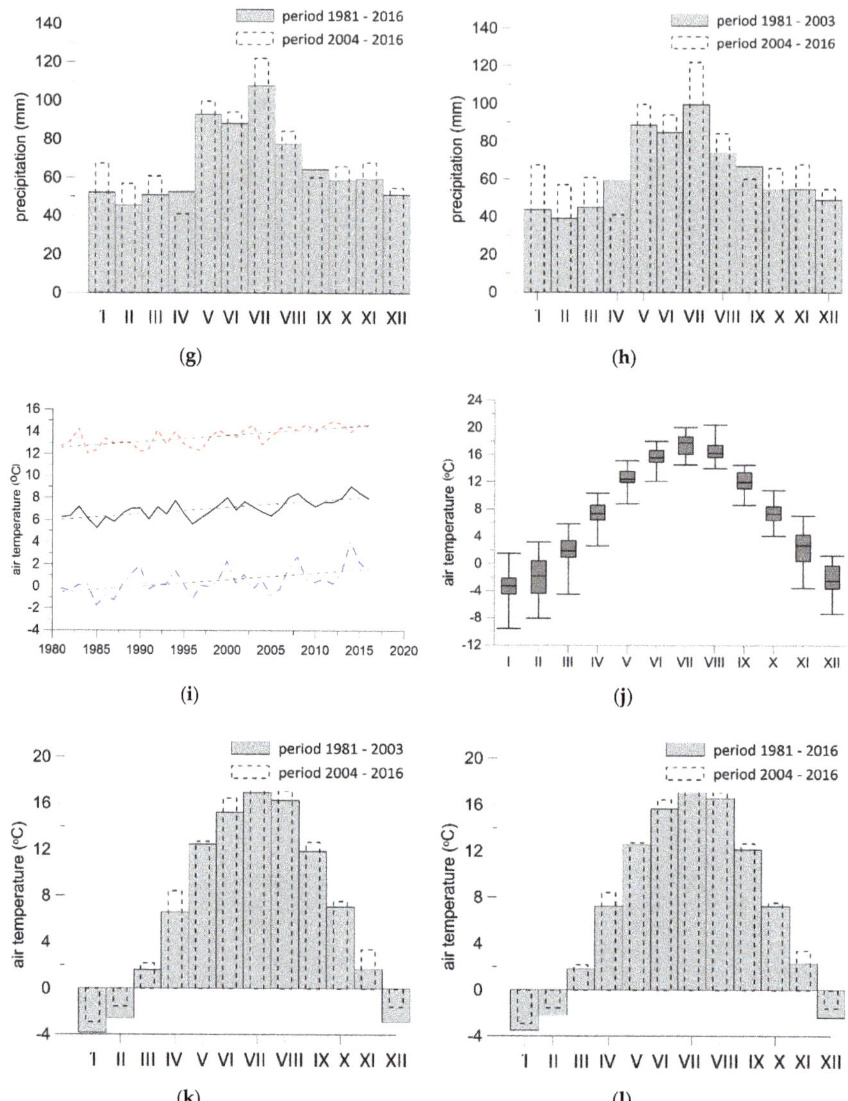

Figure 6. (a–i) Analysis of the hydrometeorological data–Ipoltica River basin. (a) time series of annual discharge—average (black line), minimum (red line); (b) box plot of monthly discharge (1981–2016); (c) average monthly discharge for period 1981–2003 and period 2004–2016; (d) average monthly discharge for period 1981–2016 and period 2004–2016; (e) time series of annual precipitation totals; (f) box plot of monthly precipitation (1981–2016); (g) average monthly precipitation for period 1981–2003 and period 2004–2016; (h) average monthly precipitation for period 1981–2016 and period 2004–2016; (i) air temperature—annual (black line), summer season (red line) and winter season (blue line); (j) box plot of monthly air temperature (1981–2016); (k) average monthly air temperatures for period 1981–2003 and period 2004–2016; (l) average monthly air temperatures for period 1981–2016 and period 2004–2016.

The linear trend of the annual precipitation totals is increasing (Figure 6e). The box plot of the monthly precipitation totals is shown in Figure 6f with the maximum precipitation totals. The average monthly precipitation was observed in two periods (between the pre and post-calamity period, a comparison of the whole and post-calamity period), where, in addition to the months of April and September, an increase in total precipitation was observed (Figure 6g,h).

The analysis of the air temperature revealed an increasing linear trend of the average annual temperature, as well as the temperatures in the summer and winter half-years (Figure 6i). The highest air temperatures were in the summer months of June, July, and August. An increase in air temperature was observed in all the months except for the month of July when comparing the periods (Figure 6k,l).

4.3. Analysis of Short-Term Rainfall Trends and Seasonality Changes

The short-term rainfall data analysis consists of the analysis of the trends and seasonality changes in the warm period for all the time periods selected and for the selected rainfall durations of 60, 120, 180, 240 and 1440 min.

At both of the analyzed climatological stations, i.e., Liptovská Teplička and Boca, the short-term extreme rainfall events with durations from one hour up to 1 day occurred in July, except for the historical period modelled in the Liptovská Teplička climatological station. For the future scenario, a shift in the rainfall extremes was seen in a later period of the month of July. For the seasonality characteristics of short-term rainfall intensities, Burn's vector methodology [49] was applied; it describes the variability of the date when an extreme rainfall event occurs, so that the direction of the vector corresponds to the expected day of the occurrence during the year, while its length describes the variability around the expected date of the occurrence of the extreme rainfall event. The results are shown in the unit circles (Burn's diagrams) and can be seen in Figures 7 and 8. The properties of the trend are determined by the Mann-Kendall [50,51] trend test. This method is used for determining and assessing the properties and significance of the trends in a selected quantity over time. The trend analysis showed increasing trends represented by + in the short-term rainfall intensities for the future period as well as for the actual observed data for a duration longer than 60 min, but all the trends were not significant at both of the climatological stations analyzed. The decreasing trends were observed in the actual observations in the 60 and 1440 min durations (Tables 5 and 6).

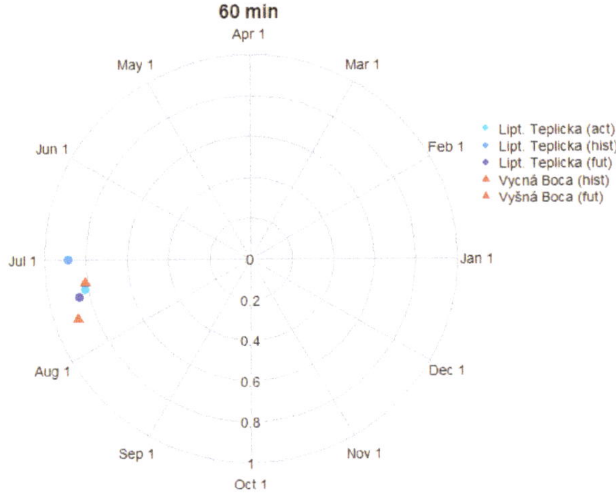

Figure 7. Burn´s diagram for a rainfall duration of 60 min for the Liptovská Teplička and Vyšná Boca climatological stations.

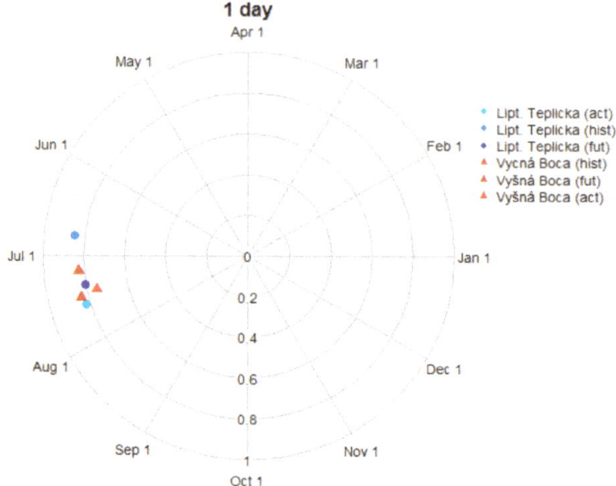

Figure 8. Burn´s diagram for a rainfall duration of one day at the Liptovská Teplička and Vyšná Boca climatological stations.

Table 5. Trend analysis of short-term rainfall intensities for the Liptovská Teplička climatological station—Boca River basin.

Rainfall Duration	Real Observations Actual Period 1995–2009	CLM Scenario Historical Period 1961–2000	CLM Scenario Future Period 2071–2100
60 min	-	+	+
120 min	+	+	+
180 min	+	+	+
240 min	+	+	+
1440 min	-	+	+

Table 6. Trend analysis of short-term rainfall intensities for the Vyšná Boca climatological station—Boca River basin.

Rainfall Duration	Historical Period 1961–2000	Future Period 2071–2100
60 min	+	+
120 min	+	+
180 min	+	+
240 min	+	+
1440 min	+	+

Simple Scaling Results

The estimation of the IDF curves of the rainfall intensities was performed using scaling exponents. The scaling exponents were determined by a simple scaling methodology with the base on a scaling of the statistical moments. The results are shown in Table 7, where the scaling exponents from the Liptovská Teplička and Vyšná Boca climatological stations are presented. For the Vyšná Boca climatological station, we could not derive the scaling exponent for the actual measured data due to the lack of actual measured short-term rainfall data. In this case, the scenario-based scaling exponent from the historical period was used for the subsequent analysis for downscaling the actual daily rainfalls from the Vyšná Boca climatological station. The scaling exponents have a declining character in the future, which is caused by less extreme events in the scenario data. Apart from these results, the IDF curve values are not lower for the future horizons, due to an increase in the daily precipitation totals that were used for downscaling. The downscaled values of the rainfall intensities were finally higher. As an example, the IDF curves for both stations and periodicity $p = 0.01$ are presented in Figures 9 and 10.

Table 7. Scaling exponents for the stations analyzed for the historical and future periods.

Station	Real Observations Actual Period 1995–2009	CLM Scenario Historical Period 1961–2000	CLM Scenario Future Period 2071–2100
Liptovská Teplička	0.762	0.669	0.6228
Vyšná Boca	-	0.6471	0.6169

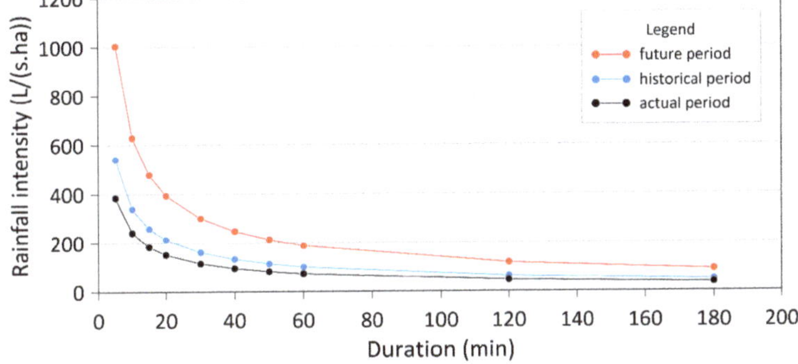

Figure 9. IDF curves for the Liptovská Teplička climatological station for the periodicity $p = 0.01$.

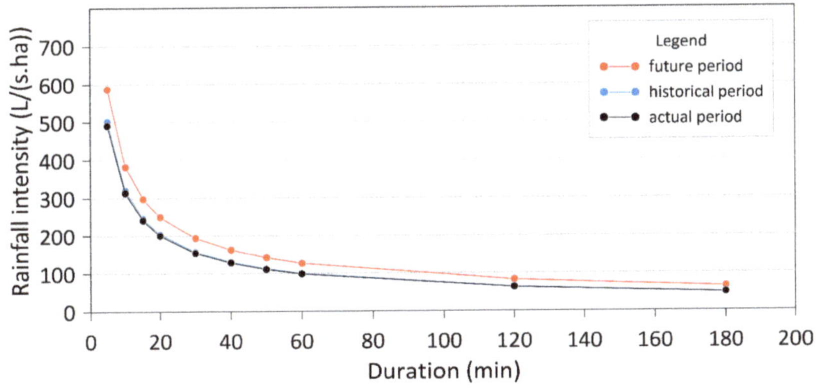

Figure 10. IDF curves for the Vyšná Boca climatological station for the periodicity $p = 0.01$.

4.4. Effect of Land Use and Climate Change on Design Flood

In the next step, the research focused on an analysis of the changes in extreme discharge caused by changes in land use and climate. The study was performed for the Boca and Ipoltica River basins. These areas have been affected by a number of severe windstorms in recent decades, which have had a significant impact on changes in the forest cover. The most significant windstorms occurred in 2004 (Alžbeta) and 2007 (Kyrill and Filip); later, bark beetle outbreaks occurred there.

When the SCS-CN method was applied for the design flood calculations, the initial abstraction coefficient was set equal to zero, and the short-term design rainfall was used as a total input rainfall.

The values of the main parameter CN (Tables 8 and 9), which depend on land surface characteristics and hydro-soil conditions, were selected from the CN table values [52]. Based on the soil type analysis, both river basins were classified in the B hydrological soil group.

Table 8. The selected values of the CN parameter for the Boca Basin (A–basin area; CN–Curve Number value; CN_w–weighted CN value).

Land Use Type	1990		2018	
	A (km^2)	CN (-)	A (km^2)	CN (-)
Coniferous forest	67.47	55	36.80	60
Mixed forest	0.43	55	1.60	55
Transitional woodland-shrub	5.01	58	36.52	65
Grasslands	3.77	58	3.04	58
Pastures	2.64	58	2.64	61
Agricultural land	2.28	78	1.08	78
Urban area	0.34	85	0.25	85
CN_w		56.18		62.40

Table 9. The selected values of the CN parameter for the Ipoltica River Basin (A–area; CN–Curve Number value; CN_w–weighted CN value).

Land Use	1990		2018	
	A (km^2)	CN (-)	A (km^2)	CN (-)
Coniferous forest	65.48	55	55.31	60
Mixed forest	2.89	55	5.16	55
Transitional woodland-shrub	11.17	58	21.87	65
Grasslands	3.29	58	2.49	58
Pastures	2.95	58	1.01	61
Agricultural land	0.46	78	0.42	78
CN_w		55.73		61.01

Effect of Changes in Land Use on Design Floods

This part of the research focuses on the changes in runoff caused by changes in land use for the period from 1990 (before the significant windstorms) to 2018 ("present"/latest available CLC land use). The design flood (runoff) Q_N was estimated for 1990 and 2018 and for return periods (N) of 10, 20, 50, and 100 years.

The design values of the short-term rainfalls from actual observations for the Vyšná Boca (Figure 11) and Liptovská Teplička (Figure 12) climatological stations were used for the analysis.

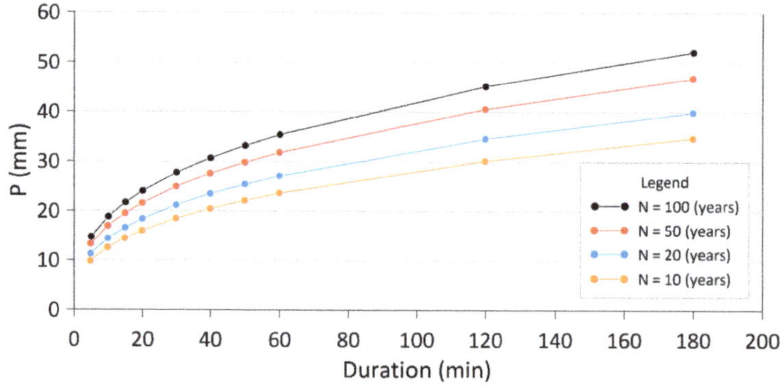

Figure 11. The actual design values of the short-term rainfall for the Vyšná Boca climatological station.

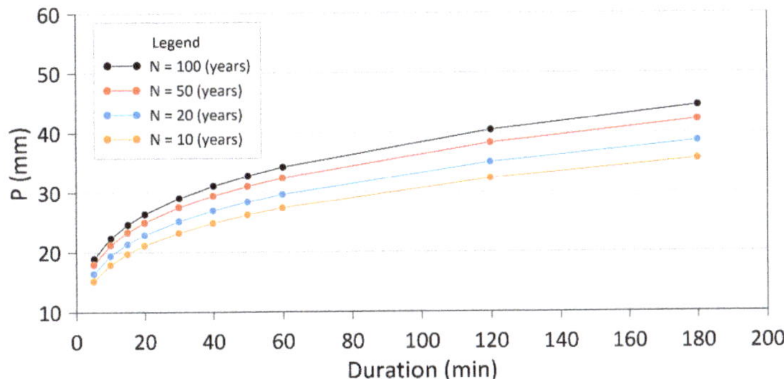

Figure 12. The actual design values of the short-term rainfall for the Liptovska Teplička climatological station.

The duration of the design short-term rainfall was selected according to the time of the concentration, which was calculated from the mean runoff velocity along the runoff path (determined according to Alena [53]). The time of concentration and runoff velocity is affected by the slope of the terrain and the land use and land cover along the runoff's path. In the case of the Boca River basin, the land use along the path has changed significantly over time; hence, the time of concentration/duration for 1990 (80 min) is different than that for 2018 (69 min). In the case of the Ipoltica River, the duration of the design short-term rainfall was determined to be equal to 84 min for 1990 and 2018.

Rainfall intensities for the 1941–1944 period from the Vyšná Boca climatological station were used for the estimation of the design flood in the Boca River basin. The data from the Liptovská Teplička climatological station consists of rainfall intensities for the 1995–2009 period and were used for the estimation of the design floods in the Ipoltica River basin.

The calculations and results of the design floods in the Boca River basin estimation using the design values of the short-term rainfall for the Vyšná Boca climatological station for the actual period are shown in Table 10.

Table 10. Estimation of the design floods for the Boca River basin using the actual design values of short-term rainfall (N–return period; P–design values of short-term rainfall; CN_w–weighted CN value; S–maximum potential retention; Q_N–the design flood).

N	1990				2018			
	P (mm)	CN_W (-)	S (mm)	Q_N (m³ s⁻¹)	P (mm)	CN_W (-)	S (mm)	Q_N (m³ s⁻¹)
10	26			34.60	25			45.42
20	30	55.18	198	44.88	29	62.40	153	58.71
50	35			60.37	33			78.63
100	39			73.66	37			95.64

When the results from Table 8 are compared, the values of the design floods increased by 31% in the case of the 10-year return period, by 31% in the case of the 20-year return period, and by 30% in the case of the 50 and 100-year return periods.

The calculations and results of the estimation of the design floods in the Ipoltica River basin using the design values of the short-term rainfall for the Vyšná Boca climatological station for the actual period are shown in Table 11.

Table 11. Estimation of the design floods for the Ipoltica River basin using the actual design values of the short-term rainfall (N–return period; P–design values of short-term rainfall; CN_w–weighted CN value; S–maximum potential retention; Q_N–the design flood).

N (year)	P (mm)	1990			2018		
		CN_W (-)	S (mm)	Q_N (m^3 s^{-1})	CN_W (-)	S (mm)	Q_N (m^3 s^{-1})
10	30			43.37			52.28
20	32			50.41			60.64
50	35	55.73	201.8	59.32	61.01	162	71.17
100	37			65.66			78.65

When the results from Table 11 are compared, the values of the design floods increased by 20.5% in the case of the 10-year return period, and by 20% in the case of the 20, 50, and 100-year return periods.

From the results in Tables 10 and 11, it can be concluded that the effect of the changes in land use considerably influenced the values of the design floods. In both river basins, the increase in the design floods was mostly caused by the changes in land use, more specifically by the deforestation. In the Boca River basin, the forests have decreased by 36% since 1990, while in the Ipoltica River basin, the forests decreased by 9%. The forests were mostly replaced by transitional woodland shrubs, which neither slow down the runoff nor absorb rainfall as well as a forest.

4.5. Effect of Climate Change on Design Floods

Subsequently, we focused on the changes in extreme discharges caused by climate change. The calculations of the design floods were performed using the latest available land use data (from 2018). The climate change is represented by data from the Regional Climate Model (RCM) scenario for the Vyšná Boca and Liptovská Teplička climatological stations. The rainfall data from the CLM simulation were provided by Dr. Martin Gera from Comenius University in Bratislava [17]. The RCM scenario used consists of the rainfall intensities for the future period (2070–2100). The RCM scenario selected for the simulation of the climate was the SRES A1B scenario, which is a semi-pessimistic scenario with an increase in the global warming temperature of about 2.9° by the year 2100. This scenario relates well to the current processes in the atmosphere.

The estimation of changes in the design floods (Q_N) is provided for the period 2070–2100 for the return periods (N) of 10, 20, 50, and 100 years (Figures 13 and 14).

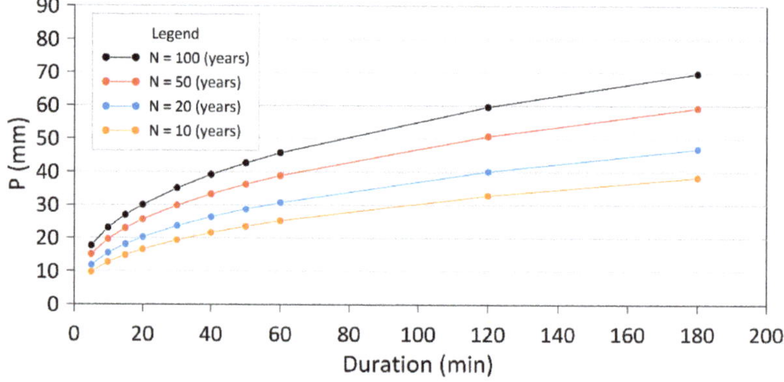

Figure 13. The future downscaled design rainfall intensities for the Vyšná Boca climatological station.

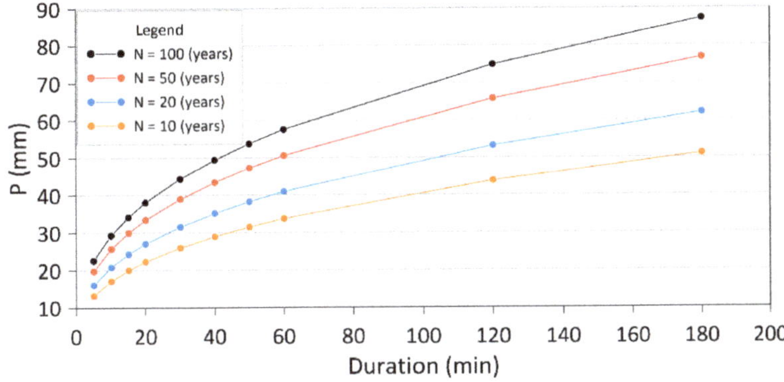

Figure 14. The future downscaled design rainfall intensities for the Liptovska Teplička climatological station.

The calculations and results of the estimation of the design floods using the downscaled data for the future scenarios for the Vyšná Boca and Liptovská Teplička climatological stations are shown in Table 12.

Table 12. Estimation of the design floods using the design values of the rainfall intensities from the future scenarios (N–return period; P–design values of short-term rainfall; CN_w–weighted CN value; S–maximum potential retention; Q_N–the design flood).

N (year)	Boca River Basin				Ipoltica River Basin			
	P (mm)	CN_W (-)	S (mm)	Q_N (m^3 s^{-1})	P (mm)	CN_W (-)	S (mm)	Q_N (m^3 s^{-1})
10	27			52.07	38			83.49
20	33			75.23	47			118.26
50	41	62.40	153	114.59	58	61.01	162	171.47
100	48			152.23	66			213.94

A comparison of the results from the calculations using the actual design values of the rainfall intensities and the design values of the rainfall intensities from the future scenario are shown in Table 13.

Table 13. Comparison of the design floods using the actual design values of the rainfall intensities and the design values of the rainfall intensities from the future scenarios.

N (year)	Boca River Basin		Ipoltica River Basin	
	Q_N (m^3 s^{-1})		Q_N (m^3 s^{-1})	
	Actual Period	Future Period	Actual Period	Future Period
10	45.42	52.07	52.28	83.49
20	58.71	75.23	60.64	118.26
50	78.63	114.59	71.17	171.47
100	95.64	152.23	78.65	213.94

When the results from the calculations of the design floods using the actual design values of the rainfall intensities and the design values of the rainfall intensities from the future scenarios are compared, it can be concluded that the results predict even more increased values of the design floods. In the Boca River basin, the estimated design floods increased about 15% in the case of the 10-year return period, by 28% in the case of the 20-year return period, by 45.7% in the case of the 5- year return period, and by 59% in the case of the 100-year return period. In the Ipoltica River basin, the estimated design floods increased by 60% in the case of the 10-year return period, by 95% in the case of the 20-year return period, by 141% in the case of the 50-year return period, and by 172% in the case of the 100-year return period. From the results we can conclude that the impact of the predicted climate change increases with the return period.

5. Discussion and Conclusions

Environmental changes, particularly changes in land use, overall land use, and climate change, and their impact on water regimes as well as the occurrence of floods have been issues of tropical concern in recent years. Forests play a crucial role in the partitioning of water into surface flow, subsurface flow, and evapotranspiration. Deforestation can strongly impair the hydrological functioning of forested systems.

In recent decades, many regions in Slovakia have been affected by severe windstorms that caused significant deforestation, especially in mountainous river basins. It is assumed that these changes also affected the runoff conditions in the case study area of the Boca and Ipoltica River basins located in the Low Tatras National Park of the district of Liptovský Mikuláš, which lies in northern Slovakia.

To examine the changes in land use due to deforestation in the Boca and Ipoltica River basins, we first analysed the CLC maps. Based on the analysis of the CLC maps from the period 1990–2018, changes in land use before and after the occurrence of the windstorms in these river basins were identified. From the comparison, we can conclude that after Alžbeta, the most significant windstorm in 2004, the areas of the forests (coniferous and mixed) decreased from 83% to 79% of all the basin areas in the Boca River basin. After the Kyril windstorm in 2006, the areas of forests decreased from 79% to 50% in the Boca River basin and from 82% to 73% in the Ipoltica River basin. During the whole period of 1990–2018, forested areas in the Boca river decreased from 83% to 47% (by almost 40%) and in the Ipoltica river basin from 8% to 70% (by almost 10%).

In the next step, an analysis of the hydrometeorological data was performed. An increasing trend in the average annual discharge was found in the Boca and Ipoltica River basins. The largest scatter was demonstrated in the months of April and May, which was to be expected due to the spring melting of the snow cover. Compared to the periods before and after the first significant calamity (as well as after the calamity and the whole period), a decrease in flows was detected in June in the Boca River basin, and a decrease in May and October occurred in the Ipoltica River basin. The analysis of the precipitation data revealed an upward trend in annual totals in both river basins. The largest variance within the year was recorded in the summer months. A comparison of the two periods before and after the calamity, as well as after the calamity and the whole period, showed lower total precipitation in April and October (in both river basins). The expected rising trend of temperatures was also found in this case (the increase in temperature is reflected in the greater evaporation, which indirectly affects the condition of the surface). An interesting finding was that the increase in the average monthly temperature was reflected in all the months except July, and the most significant difference compared to the two periods was in April. This probably also results in an increase in runoff from the river basin in the spring, especially in April. The cause of this behavior may be the earlier snow-melting of the river basin. Earlier snowmelt was also demonstrated in the study by Hríbik et al. [54] on unforested areas in central Slovakia. The direct relationship between the increase or decrease in total precipitation per runoff regime did not occur in any of the river basins.

Despite the fact that deforestation due to wind calamities in both river basins had no clear response in terms of changes in the measured discharge data in a monthly time step, the effect of deforestation became evident in the extreme discharges. The design values of the rainfalls from the actual observations of the analysed climatological stations and the Corine Land Cover land use map for 1990 and 2018 were used as inputs for the calculation of the changes in the design floods. The design floods (QN) were estimated for return periods (*N*) of 10, 20, 50 and 100 years, using the Soil Conservation Service—Curve Number method, ArcGIS software, and raster tools. From the results we can conclude that the effects of the changes in land use have considerably influenced the values of the design floods. In both river basins, the increase in the design flood was mostly caused because of changes in land use, more specifically by deforestation. In the Boca River basin, the forests have decreased by 36% since 1990, while in the Ipoltica River basin, the forests decreased by 9%. When the results from the calculations of the design floods using the actual design values of the rainfall intensities and the design values of the rainfall intensities from the future scenarios are compared, it can be concluded that the results predict even more increased values of design floods. In the Boca River basin, the estimated design floods increased by 59%, and in the Ipoltica River basin by 172% in the case of the 100-year return period. From the results we can conclude that the impact of climate change increases with the return periods. All the results presented can be used for water management planning and flood protection measures proposed in the basins studied.

Author Contributions: Conceptualization: M.D., G.F., M.M.L., S.K., K.H.; methodology, M.D., G.F., M.M.L., S.K., K.H.; formal analysis, M.D., G.F., M.M.L.; investigation, M.D., G.F., M.M.L.; writing-original draft preparation, M.D., G.F., M.M.L.; writing-review and editing S.K., K.H.; visualization, G.F., M.D. All authors have read and agreed to the published version of the manuscript.

Funding: This work was supported by the Slovak Research and Development Agency under Contracts No. APVV-19-0340, APVV-18-0347, the VEGA Grant Agency No. 1/0632/19. The authors thank the Agencies for the research support.

Conflicts of Interest: The authors declare no conflict of interest.

References

1. Rogger, M.; Agnoletti, M.; Alaoui, A.; Bathurst, J.C.; Bodner, G.; Borga, M.; Chaplot, V.; Gallart, F.; Glatzel, G.; Hall, J.; et al. Land use change impacts on floods at the catchment scale: Challenges and opportunities for future research. *Water Resour. Res.* **2017**, *53*, 5209–5219. [CrossRef]

2. Brown, E.A.; Zhang, L.; McMahon, A.T.; Western, W.A.; Vertessy, A.R. A review of paired catchment studies for determining changes in water yield resulting from alterations in vegetation. *J. Hydrol.* **2005**, *310*, 28–61. [CrossRef]

3. Vose, J.M.; Sun, G.; Ford, C.R.; Bredemeier, M.; Otsuki, K.; Wei, X.; Zhang, Z.; Zhang, L. Forest ecohydrological research in the 21st century: What are the critical needs? *Ecohydrology* **2011**, *4*, 146–158. [CrossRef]

4. Hlásny, T.; Kočický, D.; Maretta, M.; Sitková, Z.; Barka, I.; Konôpka, M.; Hlavatá, H. Effect of deforestation on watershed water balance: Hydrological modelling-based approach. *Lesn. Cas. For. J.* **2015**, *61*, 89–100.

5. Andreassian, V. Waters and forests: From historical controversy to scientific debate. *J. Hydrol.* **2004**, *291*, 1–27. [CrossRef]

6. Zhang, T.; Zhang, X.; Xia, D.; Liu, Y. An Analysis of land use dynamics and its impacts in hydrological processes in the Jialing River Basin. *Water* **2014**, *6*, 3758–3782. [CrossRef]

7. Suryatmojo, H. Rainfall-runoff investigation of pine forest plantation in the upstream area of Gajah Mungkur reservoir. *Procedia Environ. Sci.* **2014**, *28*, 307–314. [CrossRef]

8. Danáčová, M.; Hlavčová, K.; Labat, M.M. Možnosť posúdenia zmien využitia krajiny na odtokový režim v povodí toku Boca. *Czech J. Civil Eng.* **2019**, *5*, 40–46.

9. Váňová, V.; Langhammer, J. Modelling the impact of land cover changes on flood mitigation in the upper Lužnice basin. *Hydrol. Hydromech.* **2011**, *59*, 262–274. [CrossRef]

10. Dwarakish, G.; Ganasri, B. Impact of land use change on hydrological systems: A review of current modeling approaches. *Cogent Geosci.* **2015**, *1*, 1115691. [CrossRef]

11. Intergovernmental Panel on Climate Change. *Climate Change 2013. The Physical Science Basis. Contribution of Working Group I to the Fifth Assessment Report of the Intergovernmental Panel on Climate Change*; Cambridge University Press: Cambridge, UK; New York, NY, USA, 2013; ISBN 9781107661820.

12. Westra, S.J.; Fowler, H.J.; Evans, J.P.; Alexander, L.V.; Berg, P.; Johnson, F.R.; Kendon, E.J.; Lenderink, G.; Roberts, N.M. Future changes to the intensity and frequency of short-duration extreme rainfall. *Rev. Geophys.* **2014**, *52*, 522–555. [CrossRef]

13. Tebaldi, C.; Hayhoe, K.; Arblaster, J.M.; Meehl, G.A. Going to the Extremes. *Clim. Chang.* **2006**, *79*, 185–211. [CrossRef]

14. Koutsoyiannis, D.; Efstratiadis, A.; Mamassis, N.; Christofides, A. On the credibility of climate predictions. *Hydrol. Sci. J.* **2008**, *53*, 671–684. [CrossRef]
15. Kyselý, J.; Gaál, L.; Beranova, R. Projected Changes in Flood-Generating Precipitation Extremes Over the Czech Republic in High-Resolution Regional Climate Models. *J. Hydrol. Hydromech.* **2011**, *59*, 217–227. [CrossRef]
16. Engen-Skaugen, T.; Førland, E.J. Future changes in extreme precipitation estimated in Norwegian catchments. *Met Rep.* **2011**, *1*, 1–28.
17. Lapin, M.; Bašták-Ďurán, I.; Gera, M.; Hrvoľ, J.; Kremler, M.; Melo, M. New climate change scenarios for Slovakia based on global and regional general circulation models. *Acta Met. Univ. Comen.* **2012**, *37*, 25–74.
18. Wang, D.; Hagen, S.C.; Alizad, K. Climate change impact and uncertainty analysis of extreme rainfall events in the Apalachicola River basin, Florida. *J. Hydrol.* **2013**, *480*, 125–135. [CrossRef]
19. Gaál, L.; Beranová, R.; Hlavčová, K.; Kyselý, J. Climate Change Scenarios of Precipitation Extremes in the Carpathian Region Based on an Ensemble of Regional Climate Models. *Adv. Meteorol.* **2014**, *2014*, 1–14. [CrossRef]
20. Pascale, S.; Lucarini, V.; Feng, X.; Porporato, A.; Hasson, S.U. Projected changes of rainfall seasonality and dry spells in a high greenhouse gas emissions scenario. *Clim. Dyn.* **2015**, *46*, 1331–1350. [CrossRef]
21. Al Mamoon, A.; Joergensen, N.E.; Rahman, A.; Qasem, H. Design rainfall in Qatar: Sensitivity to climate change scenarios. *Nat. Hazards* **2016**, *81*, 1797–1810. [CrossRef]
22. Hanel, M.; Pavlásková, A.; Kysely, J. Trends in characteristics of sub-daily heavy precipitation and rainfall erosivity in the Czech Republic. *Int. J. Clim.* **2015**, *36*, 1833–1845. [CrossRef]
23. Gera, M.; Damborská, I.; Lapin, M.; Melo, M. *Climate Changes in Slovakia: Analysis of Past and Present Observations and Scenarios of Future Developments*; Springer Science and Business Media LLC.: Berlin, Germany, 2017; pp. 21–47.
24. Nepal, S. Impacts of climate change on the hydrological regime of the Koshi river basin in the Himalayan region. *J. Hydro-Environ. Res.* **2016**, *10*, 76–89. [CrossRef]
25. Taibi, S.; Meddi, M.; Mahé, G.; Tramblay, Y.; Feddal, M.A. Monthly rainfall variability simulated by MED-Cordex regional climate models on Algiers coastal basin in past and future climate conditions. In Proceedings of the 6th International Conference on Sustainable Agriculture and Environment, Konya, Turkey, 3–5 October 2019; Mithat, D., Ed.; pp. 469–471, ISBN 9786051841946.
26. Kostka, Z.; Holko, L. Role of forest in hydrological cycle—Forest and runoff. *Meteorol. J.* **2006**, *9*, 143–148.
27. Hlavčová, K.; Szolgay, J.; Kohnová, S.; Horvát, O. The limitations of assessing impacts of land use changes on runoff with a distributed hydrological model: Case study of the Hron River. *Biologia* **2009**, *64*, 589–593. [CrossRef]
28. Kohnová, S.; Rončák, P.; Hlavčová, K.; Szolgay, J.; Rutkowska, A. Future impacts of land use and climate change on extreme runoff values in selected catchments of Slovakia. *Meteorol. Hydrol. Water Manag.* **2019**, *7*, 47–55. [CrossRef]
29. Kunca, A.; Galko, J.; Zúbrik, M. Significant calamities in the forests of Slovakia over the last 50 years (Významné kalamity v lesoch Slovenska za posledných 50 rokov). In Proceedings of the 2014 23rd International Conference, Nový Smokovec, Slovakia, 23–24 April 2014; pp. 25–31.
30. Gubka, A.; Kunca, A.; Longauerová, V.; Maľová, M.; Vakula, J.; Galko, J.; Nikolov Ch Rell, S.; Zúbrik, M.; Leontovyč, R. *Windstorm Žofia from 15. 5. 2014. Guidelines of the Forest Protection Service Banská Štiavnica (Vetrová kalamita žolfia z 15.5.2014. Usmernenie Lesníckej ochranárskej služby Banská Štiavnica)*; National Forestry Center: Zvolen, Slovakia, 2014; p. 8. (In Slovak)
31. Druga, M.; Falťan, V.; Herichová, M. Proposal for modification of the CORINE Land Cover methodology for the purpose of mapping historical changes in landscape cover in Slovakia on a scale of 1:10 000—An example study of the historical cadastral area of Batizovce (Návrh modifikácie metodiky CORINE Land Cover pre účely mapovania historických zmien krajinnej pokrývky na území Slovenska v mierke 1:10 000–príkladová štúdia historického k.ú. Batizovce). *Geo. Cassoviensis IX* **2015**, *1*. (In Slovak)
32. Koutsoyiannis, D.; Foufoula-Georgiou, E. A scaling model of a storm hyetograph. *Water Resour. Res.* **1993**, *29*, 2345–2361. [CrossRef]
33. Koutsoyiannis, D.; Kozonis, D.; Manetas, A. A mathematical framework for studying rainfall intensity-duration-frequency relationships. *J. Hydrol.* **1998**, *206*, 118–135. [CrossRef]

34. Gupta, V.K.; Waymire, E. Multiscaling properties of spatial rainfall and river flow distributions. *J. Geophys. Res. Space Phys.* **1990**, *95*, 1999–2009. [CrossRef]

35. Menabde, M.; Seed, A.; Pegram, G. A simple scaling model for extreme rainfall. *Water Resour. Res.* **1999**, *35*, 335–339. [CrossRef]

36. Yu, P.-S.; Yang, T.-C.; Lin, C.-S. Regional rainfall intensity formulas based on scaling property of rainfall. *J. Hydrol.* **2004**, *295*, 108–123. [CrossRef]

37. CLM Community Land Model, Overview. Available online: http://www.cgd.ucar.edu/tss/clm/ (accessed on 15 December 2019).

38. CESM Community Earth System Model, Community Land Model. Available online: http://www.cesm.ucar.edu/models/clm/ (accessed on 15 December 2019).

39. Böhm, U.; Kücken, M.; Ahrens, W.; Block, A.; Hauffe, D.; Keuler, K.; Rockel, B.; Will, A. CLM—The climate version of LM: Brief description and long-term applications. *COSMO Newslett.* **2006**, *6*, 225–235.

40. Mishra, S.; Tyagi, J.; Singh, V.P.; Singh, R. SCS-CN-based modeling of sediment yield. *J. Hydrol.* **2006**, *324*, 301–322. [CrossRef]

41. USDA-SCS. Estimation of Direct Runoff from Storm Rainfall. In *National Engineering Handbook, Part 630—Hydrology*; US Dept. of Agriculture—Soil Conservation Service: Washington, DC, USA, 1954.

42. USDA-NRSA. *Urban Hydrology for Small Watersheds. Technical Release 55*; US Dept. of Agriculture—Natural Resources Conservation Service, Engineering Division: Washington, DC, USA, 1986; p. 164.

43. Walega, A.; Salata, T. Influence of land cover data sources on estimation of direct runoff according to SCS-CN and modified SME methods. *Catena* **2019**, *172*, 232–242. [CrossRef]

44. Labat, M.M.; Korbeľová, L.; Kohnová, S.; Hlavčová, K. Design of measures for soil erosion control and assessment of their effect on the reduction of peak flows. *Pollack Period.* **2018**, *13*, 209–219. [CrossRef]

45. Labat, M.M.; Rattayová, V.; Hlavčová, K. The impact of changes in land use on reduction in peak floods. *Acta Hydrol. Slovaca* **2018**, *1*, 69–77.

46. Boughton, W. A review of the USDA SCS curve number method. *Soil Res.* **1989**, *27*, 511–523. [CrossRef]

47. Fan, F.; Deng, Y.; Hu, X.; Weng, Q. Estimating Composite Curve Number Using an Improved SCS-CN Method with Remotely Sensed Variables in Guangzhou, China. *Remote. Sens.* **2013**, *5*, 1425–1438. [CrossRef]

48. Mishra, S.K.; Singh, V.P. *Soil Conservation Service Curve Number (SCS-CN) Methodology*; Springer Science and Business Media LLC.: Berlin, Germany, 2003; pp. 84–146.

49. Burn, D.H. Catchment similarity for regional flood frequency analysis using seasonality measures. *J. Hydrol.* **1997**, *202*, 212–230. [CrossRef]

50. Mann, H.B. Nonparametric Tests Against Trend. *Econometrica* **1945**, *13*, 245. [CrossRef]

51. Kendall, M.G. *Rank Correlation Methods*; Griffin: London, UK, 1975.

52. USDA-SCS. *Engineering Hydrology Training Series. Module 104—Runoff Curve Number Computations*; Study Guide. No. 2; US Dept. of Agriculture—Soil Conservation Service: Washington, DC, USA, 1989.

53. Alena, F. Soil erosion protection on vineyards, Methodological aid, State Amelioration Report, Bratislava. *Cuad. de Investig. Geogr.* **1990**, *43*, 18–21.

54. Hríbik, M.; Majlingová, A.; Škvarenina, J.; Kyselová, D. Winter Snow Supply in Small Mountain Watershed as a Potential Hazard of Spring Flood Formation. *Bioclimatol. Nat. Hazards* **2009**, 119–128. [CrossRef]

Publisher's Note: MDPI stays neutral with regard to jurisdictional claims in published maps and institutional affiliations.

Article

Carbon Balance and Streamflow at a Small Catchment Scale 10 Years after the Severe Natural Disturbance in the Tatra Mts, Slovakia

Peter Fleischer, Jr. [1,*], Ladislav Holko [2], Slavomír Celer [3], Lucia Čekovská [4], Jozef Rozkošný [1], Peter Škoda [5], Lukáš Olejár [6] and Peter Fleischer [1,7]

1   Forestry Faculty, Technical University in Zvolen, 960 03 Zvolen, Slovakia; xrozkosny@is.tuzvo.sk (J.R.); peter.fleischer@lesytanap.sk (P.F.)
2   Institute of Hydrology, Slovak Academy of Sciences, 841 04 Bratislava, Slovakia; ladislav.holko@savba.sk
3   Administration of Tatra National Park, 059 21 Svit, Slovakia; slavomir.celer@sopsr.sk
4   State Forest of Slovakia Enterprise, 975 66 Banská Bystrica, Slovakia; lucia.cekovska@lesy.sk
5   Slovak Hydrometeorological Institute Bratislava, 833 15 Bratislava, Slovakia; peter.skoda@shmu.sk
6   Faculty of Forestry, and Wood Sciences, Czech University of Life Sciences, 165 21 Prague, Czech Republic; olejar@fld.czu.cz
7   State Forest of Tatra National Park, 059 60 Tatranská Lomnica, Slovakia
*   Correspondence: xfleischer@is.tuzvo.sk

Received: 15 September 2020; Accepted: 12 October 2020; Published: 19 October 2020

Abstract: Natural disturbances (windthrow, bark beetle, and fire) have reduced forest cover in the Tatra National Park (Slovakia) by 50% since the year 2004. We analyzed carbon fluxes and streamflow ten years after the forest destruction in three small catchments which differ in size, land cover, disturbance type and post-disturbance management. Point-wise CO_2 fluxes were estimated by chamber methods for vegetation-dominated land-use types and extrapolated over the catchments using the site-specific regressions with environmental variables. Streamflow characteristics in the pre- and post-disturbance periods (water years of 1965–2004 and 2005–2014, respectively) were compared to identify changes in hydrological cycle initiated by the disturbances. Mature Norway spruce forest which was carbon neutral, turned to carbon source (330 ± 98 gC m^{-2} y^{-1}) just one year after the wind disturbance. After ten years most of the windthrow sites acted as carbon sinks (from -341 ± 92.1 up to -463 ± 178 gC m^{-2} y^{-1}). In contrast, forest stands strongly infested by bark beetles regenerated much slowly and on average emitted 495 ± 176 gC m^{-2} year^{-1}. Ten years after the forest destruction, annual carbon balance in studied catchments was almost neutral in the least disturbed catchment. Carbon uptake notably exceeded its release in the most severely disturbed catchment (by windthrow and fire), where net ecosystem exchange (NEE) was -206 ± 115 gC m^{-2}. The amount of sequestered carbon in studied catchments was driven by the extent of fast-growing successional vegetation cover (represented by the leaf area index *LAI*) rather than by the disturbance or vegetation types. Different post-disturbance management has not influenced the carbon balance yet. Streamflow characteristics did not indicate significant changes in the hydrological cycle. However, greater cumulative decadal runoff, different median monthly flows and low flows and the greater number of flow reversals in the in the first years after the windthrow in two severely affected catchments could be partially related to the influence of the disturbances.

Keywords: windthrow; bark beetle outbreak; chamber CO_2 flux upscaling; *LAI*; streamflow characteristics

---

1. Introduction

In addition to wood supply, forest ecosystems provide a multitude of benefits in terms of climate and water regulations, carbon sequestration, freshwater quality, soil protection, nutrient cycling, human health, recreation, biodiversity, wildlife habitats and many other. The ability of forest ecosystems to provide these ecosystem services (ES) is very much dependent on their status, structure and functioning [1]. In the last decades, forests have been globally affected by more frequent disturbances [2] due to which many ES are supposed to be threatened [3]. This trend is believed to continue in the future because of the ongoing climate changes [4–6]. Changes in the climate system will change an overall redistribution of precipitation [7] and alter the partitioning of carbon uptake and loss [8].

The role of forests in carbon sequestration has been acknowledged in many studies [9], forest and landscape projects [10]. Forest disturbances change carbon fluxes resulting from the difference between carbon uptake by vegetation (gross primary photosynthesis, *GPP*) and loss by ecosystem respiration (RE). This difference, the net ecosystem exchange (NEE), indicates whether the ecosystem is sequestering or emitting carbon to the atmosphere. Initially, most disturbances shift ecosystems to being a carbon source, while recovery is usually associated with greater carbon storage [11–15]. The duration and dynamics of these fluxes depend on the kind of disturbance and its intensity [16,17]. Recently, climate change-induced changes in carbon sequestration have been intensively studied (e.g., [18–21]), and the results show the role of site-specific conditions and the influence of silvicultural methods [22,23]. Less attention has been given to post-disturbance forest management impact on carbon fluxes, although the role of reforestation in carbon sequestration is widely recognized [9].

The most commonly used method for ecosystem or landscape-scale estimation of carbon balance is the eddy-covariance (EC) technique [24,25]. Despite recent technical and methodological advances [26] this method has still some limitations, especially in complex terrains. An alternative approach, particularly suitable in heterogeneous landscapes, is the chamber method which is low-cost and easy to apply. Chamber measurements range from leaf to entire plant flux estimations. They were successfully used for validation of the EC technique [27]. However, small spatial coverage of the chamber method requires a large effort to collect sufficient data needed to develop simple empirical equations to scale up the results to the entire ecosystem [28]. At the same time, the chamber-based flux estimation is subject to uncertainties related to site (soil, microclimate) heterogeneity, that are further increased by heterogeneous disturbance patterns.

Plant assimilation and evapotranspiration are regulated by canopy stomatal conductance. Plant canopy in carbon-water coupling models is often abstracted as a big leaf. Such a simplification in some implications is in conflict with the clumped canopy structure. A two-big-leaf scheme and two-leaf scheme that stratify a canopy to sun-exposed and shaded leaves have been developed to make the big leaf concept more applicable [29].

Forest disturbances do not affect only carbon fluxes. They have a direct effect on the main components of catchment water balance, i.e., precipitation (interception) and evapotranspiration (e.g., [30]). Consequently, runoff generation, snow accumulation and melt, groundwater recharge or catchment runoff characteristics can be altered, e.g., [31–33]. Catchment runoff is an integrated result of hydrological processes taking place in a catchment. Runoff data series are therefore often analyzed to identify the effects of forest disturbances on catchment hydrology by comparing flow characteristics before and after the disturbance. A number of studies analyzing the influence of deforestation concluded that peakflow or low flow characteristics temporarily increased after deforestation, e.g., [34–36] and that deforestation had to affect a larger area (e.g., at least 20% of the catchment) to show the effects in runoff records [34,37]. These conclusions were often obtained by comparing flow characteristics in paired catchments, in which one catchment was subjected to forest harvesting. Alila et al. [38] contend that in paired catchments the effects of forest harvesting on floods should be evaluated considering simultaneous changes in both magnitude and frequency of the floods. In their study, floods are understood as a subset of peakflow frequency distribution containing flows with the magnitude exceeding channel capacity and return interval of 1–10 years or more.

In addition to peakflow, low flow or water yields traditionally used in the forest harvesting impact studies, many indicators were developed in environmental flow studies to characterize impacts of river regulations. The commonly used IHA tool (Indicators of Hydrologic Alteration) [39] calculates 67 characteristics based on data series of daily discharge. They characterize magnitude and duration of monthly, annual and extreme water conditions. Olden and Poff [40] examined 171 indices describing magnitude, frequency, duration and timing and rate of change in flow events (including those from IHA) using long-term flow records from 420 sites across the continental USA. They found out that the majority of indices were highly inter-correlated and proposed sets of selected indices that adequately characterize flow regimes in a non-redundant manner for different stream types. Gao et al. [41] concluded that metrics termed ecodeficit and ecosurplus [42] based on a flow duration curve can provide a good overall representation of the degree of alteration of streamflow time series.

Even though carbon and water balance are coupled processes, they are rarely compiled at a catchment scale [8]. Such an analysis is missing also in the Tatra Mountains literature where intensive research of multiple disturbing factors impact on the forests has resulted in numerous studies and publications. Our work attempts to fill the gap in existing studies by evaluating spatially integrated information on ecosystem dynamics at a landscape scale.

The main aim of this study is to evaluate carbon balance and to examine streamflow characteristics in three small catchments in the Slovak part of the Tatra Mountains ten years after the extraordinary wind disturbance that initiated vast forest destruction in the area. Secondary objectives of this study are: (i) to estimate CO_2 fluxes with chamber and biometric methods in different ecosystem (landscape) types and under different post-disturbance management, (ii) to extrapolate C balance from point-wise to the catchment scale, and iii) to evaluate changes in catchment runoff. We hypothesized that (a) intact forest acts as a carbon sink, (b) forest disturbances stimulate immediate large carbon losses from forests and subsequent gradual recovery of carbon balance, and (c) disturbances change certain flow characteristics.

2. Material and Methods

2.1. Study Sites

Tatra Mountains is the highest mountain range of the Carpathians and a regional water tower of northern Slovakia and southern Poland. The majority of the territory is protected within the Tatra National Park, Slovakia (TANAP) and Tatrzański Park Narodowy, Poland. The TANAP protects also natural and seminatural predominantly Norway spruce forest that covers approximately 40,000 ha. Roughly 30% of the forests is strictly protected in the no management zone. A downslope wind locally named "bora" sporadically hits forest stands on the lee side of the mountains [43] and causes large destructions. In November 2004, more than 12,000 ha of mature forests was destroyed by such an extremely strong windstorm. Smaller windstorms and bark beetle outbreaks initiated by the 2004 windthrow damaged additional 7000 ha of the forests in the following years. In total, 50% of mature forest stands were damaged during the 2004–2014 period in the Tatra National Park (www.lesytanap.sk, accessed on 1 July 2020).

Three small catchments with long streamflow records were selected for this study (Figure S1). The catchments are located on the southern (lee) slopes of the Tatra mountains and differ in their areas and proportion of ecosystem types (Figure 1). The 2004 windfall significantly reduced forest cover in two catchments (the Velický and Slavkovský Creek catchments, hereafter denoted as VP and SP), while forests in the third catchment remained almost unaffected by the windfall (the Mlynický Creek catchment, hereafter MP). The catchments are built by crystalline rocks that are partially covered by Quaternary sediments and by glaciofluvial sediments (moraines) at higher and lower elevations, respectively. Soils are generally stony, well drained, mostly shallow, very acid and poor in nutrients [44,45]. Lithic Leptosols and Hypersceletic Regosol are dominant above tree line. Haplic Podzols and Dystric Cambisols (WRB classification) are dominant forest soil types. Extremely rocky soils (rankers) cover ca 30% of the forest area [45].

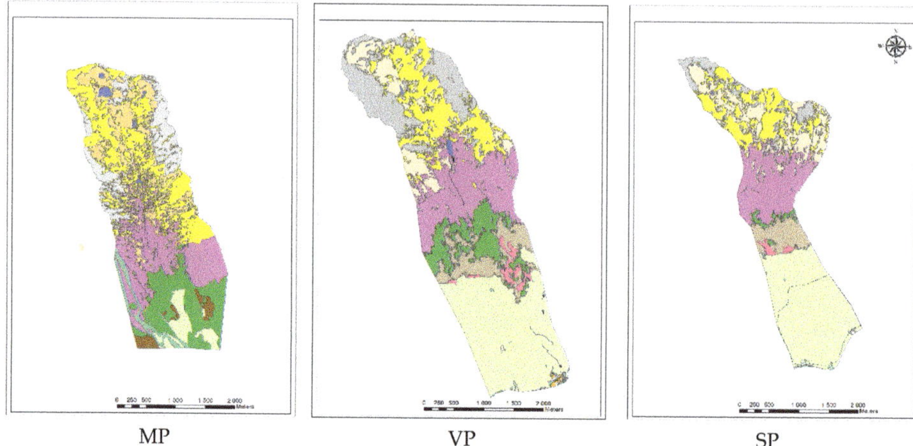

MP   VP   SP

Figure 1. Ecosystem types (legend in Table 1) in the carbon study part of the catchments; MP (area to stream gage 83 km^2, mean altitude 991 m a.s.l., mean slope 9.2°), VP (area 58 km^2, mean altitude 1094 m a.s.l., mean slope 9.3°) and SP (area 43 km^2, mean altitude 1017 m a.s.l., mean slope 8.6°).

Table 1. Total catchment area (km^2), portion of deforestation in 2004 and carbon study areas (ha) with proportion (%) of ecosystem types in the mountain parts of study catchments in 2015; MP—the Mlynický Creek catchment, VP—the Velický Creek catchment, SP—the Slavkovský Creek catchment; ROC—rock slopes, DEB—debris and moraines, WAT—water, ALM—alpine meadows; DWP—dwarf pine; IPS—area affected by the bark beetle outbreaks; REF—mature forest unaffected by the 2004 windthrow; EXT—extracted windthrown area; FIR—windthrown area affected by subsequent fire; NEX—non-extracted windthrown area; REX—extracted 1-year-old windthrow.

Catchment	Total Area [km^2]	Total Forest Reduction [%]	C Study Area [ha]	ROC	DEB	WAT	ALM	DWP	IPS	REF	EXT	FIR	NEX	REX
Color in Figure 1														
MP	83.0	5.0	758.7	11.6	16.6	0.8	25.7	21.2	2.1	14.6	4.6	0.0	0.0	0.01
VP	58.0	20.0	1031.7	13.3	10.0	0.3	14.1	16.8	7.3	8.1	27.5	0.0	1.7	0.01
SP	43.0	32.0	612.1	5.6	15.1	0.1	16.0	18.7	6.3	1.4	7.9	27.7	0.9	0.01

Vegetation at the highest elevations (above 1700 m a.s.l.) is sparse and dominated by *Agrostis pyreneica*, *Juncus trifidus*, *Avenella flexuosa*, *Avenula versicolor*, *Luzula sudetica*. Mountain pine (*Pinus mugo*) forms a continuous belt above the tree line (1550 m a.s.l.) and grows up to 1700 m a.s.l. with *Avenella flexuosa* and *Calamagrostis villosa* under the canopy. The area between 900 and 1500 m a.s.l. was almost completely covered by mature forests (*Lariceto-Piceetum* and *Sorbeto-Piceetum* communities) before the disturbances. Norway spruce (*Picea abies*) was the dominant tree species (covering 70–90%), while European larch (*Larix decidua*) covered 10–30% of the forest in the studied catchments [44].

The long-term annual average temperature ranges from 5.3 °C (850 m a.s.l.) to 1.7 °C (1750 m a.s.l.) Annual precipitation total ranges from 800 mm up to 1350 mm with maximum in summer months [45]. Study catchments are shown in Supplement, Figure S1. For carbon study, only the upper parts (micro catchments) of the entire catchments were chosen, as presented in Figure 1.

The carbon study areas feature typical land cover types that are hereafter termed as types of ecosystems. They are represented by tarns and streams, alpine meadows, dwarf mountain pine stands, undisturbed mature Norway spruce forests, managed 10-year-old windthrow, unmanaged 10-year-old windthrow, managed 1-year-old windthrow, standing/lying trees killed by bark beetle, burnt windthrow and non-vegetation types (rocks, roads, buildings, etc.) Ecosystem types were classified by ArcGIS from the aerial orthophoto maps created in 2015. The areas of ecosystem types in the catchments used in the carbon balance evaluation are given in Table 1.

Vegetation (ALM, DWP, mature intact REF forest, and disturbed IPS, EXT and NEX ecosystem types) together covered 72% of the catchment area in MP, 76% in VP and 79% in SP. Pre-disturbance forest cover was 160.7 ha in MP, 461.1 in VP and 270.2 ha in SP. Until 2014 forest cover decreased from 21% to 14% in MP, from 45% to 8% in VP and from 44% to 1% in SP. Thus, the most heavily affected forest stands were in the SP catchment, where total forest cover was reduced by 98%.

Most of the forests affected by disturbances were managed. Post-disturbance forestry operations focused on slash harvest and reforestation. Only small patches remained unmanaged, namely 1 ha in MP, 9 ha in VP and 7.2 ha in SP catchment. These numbers include also forests killed by consequent bark beetle attack in the no management zone of TANAP. In 2014, strong winds damaged forest edges of the 2004 windthrow. This event offered an opportunity to study carbon dynamics immediately after the disturbance. We use the abbreviation REX to refer to this type of ecosystem, which is not shown in Figure 1 due to the limited size of patches and spatially distributed occurrence of this ecosystem type.

The conditions in our study differ from studies devoted to impacts of forest harvesting on streamflow. In our case the primary forest disturbance was natural and instantaneous (it occurred within a few hours on 19 November 2004), and was followed by subsequent disturbances (bark beetle outbreak, local fires and other smaller windthrows). The urbanization of the disturbed area is low, and there is no flow regulation. As it is quantified above, the windthrow damaged a large forested area. Due to the timing of the disturbance and the beginning of winter, most of the wood windthrown wood remained on the ground until spring 2005. Thus, the highest potential flood hazard related to forest removal was approximately in the period between autumn 2005 and winter 2006. The flood hazard declined in the following years due to vegetation regrowth (both natural and managed). Holko et al. [46] compared water balance, minimum and maximum runoff, runoff thresholds, number of runoff events and their selected characteristics, runoff coefficients, and flashiness indices before (starting with the water year 1962) and after the windfall (until the water year 2007) in eight catchments of the disturbed area. The impact of deforestation was not clearly manifested in the analyzed hydrological data. Analysis of baseflow variability [47], runoff response to heavy rainfall [45] and flow duration curves [48] resulted in the same conclusions although [48] concluded that discharges with the highest probability of exceedance (Q90% and Q80%) in the decade following the windfall (2005–2014) occurred more frequently than in period 1965–2004. Most of the above analyses were based on a shorter time series of data after the first disturbance (windthrow). In this article we used additional indices that were not used in the previous studies. Longer data series also allowed additional analysis of the occurrence of peakflows with different return periods.

Evaluation of carbon balance was based on manual CO_2 fluxes measurement by chamber methods during the growing season 1 April–31 August 2014. Meteorological and phenological data were recorded all year round. Changes in catchment streamflow were evaluated using daily discharge data from water years 1965–2014 measured in the national network operated by the Slovak Hydrometeorological Institute.

2.2. Meteorological and Hydrological Data

Climatic data are commonly used as a proxy for estimation of carbon fluxes. Hourly measurements of air temperature and humidity at a standard height of 2 m above the terrain (Hygroclip, Switzerland), soil temperature at 2 cm and 10 cm depths (107 Campbell, UK), soil moisture at 10 cm (theta Ml2x, Delta UK), photo synthetically active radiation (*PAR* Quantum Skye, Ireland) and precipitation (Davis,

CA, USA) were recorded by the Campbell data loggers (Campbell, UK). Mobile meteorological stations were located at 1150 m a.s.l. in all studied catchments. Additionally, instant soil temperatures (at 2 cm and 10 cm depths), soil moisture (0–6 cm) and *PAR* were recorded concurrently with CO_2 flux measurements. Daily discharges measured at catchment outlets in period 1965–2014 were provided by the Slovak Hydrometeorological Institute.

2.3. Vegetation and Phenology

Qualitative (species composition) and quantitative (biomass, coverage) vegetation parameters are the most important predictors of plant assimilation. Species composition in relevant ecosystem types was derived as a species-specific contribution to total leaf area index (*LAI*). Species-specific and total *LAI* were used for a big leaf model construction. In this study the big leaf model is three-dimensional with distinct vertical distribution of *LAI* and available *PAR*.

Plots for vegetation sampling were established along 21 transects, each 540 m long, located in the DWP, IPS, REF, EXT, FIR, NEX and REX ecosystem types. The plots were set up every ten meters and each plot consisted of three squares of different size: 0.04 m^2 for grass and herbs, 0.25 m^2 for shrubs and 25 m^2 for trees. During the peak of the growing season (mid-June–end of July), grasses, herbs and shrubs were clipped, scanned, dried at 60 °C for 48 h and weighted in the laboratory. First, we estimated species-specific leaf area (SLA) from a regression between the foliage biomass and the foliage area. Plant leaf area of scanned leaves (40–50 samples for each species, each sample contained 10–50 leaves) was calculated by the ImageJ software [49]. Species-specific leaf area (*LAI*) was derived by multiplying foliage biomass and SLA. Sum of species-specific *LAI* gave total site *LAI*. Tree diameters near the ground surface and height of each tree were measured with digital clipper and telescopic sticks, respectively. Tree foliage biomass was derived from biomass regressions proposed by [50–53]. Vertical distribution of *LAI* on dominant plant species was measured with Licor 2200 Plant Canopy Analyser in 10 cm vertical categories from the top of the canopy to the ground. Licor 2200 was also used for the estimation of temporal changes along established vegetation transects (on 5 × 5 m squares) and verification of total *LAI*.

The dates of bud break, first and full leaves development were recorded and kindly provided by the phenology observation stations of the Slovak Hydro Meteorological Institute located in the MP catchment.

2.4. Carbon Balance

Carbon balance (net ecosystem exchange, NEE) refers to the amount of carbon captured by plants. We estimated NEE as a difference between assimilation (*GPP*) and ecosystem respiration (RE). Assimilation was estimated during growing season (1 April–31 August). Respiration was measured all year round, except for the snow cover period (mid December–early March). Negative sign in NEE represented carbon uptake, while positive sign for carbon efflux. Spatial extrapolations of leaf, plant and point-wise carbon flux data to the catchment scale were based on regressions with actual vegetation. Temporal extrapolation of instant carbon fluxes was based on their regression with meteorological data and plant phenological development [54].

2.4.1. Assimilation—GPP

GPP was estimated as a sum of net exchange (NEE) and respiration (RE) directly measured by the chambers. We used leaf or whole plant gas measurement for dominant plant species accordingly to the plant size and shape. *GPP* refers to the amount of CO_2 captured by leaves (μmol C m^{-2} s^{-1}) which is usually mainly controlled by available *PAR* [μmol photons m^{-2} s^{-1}]. This response is expressed by the light response curves which were constructed for dominant plant species. The light response curve defines light saturated assimilation (*GPP$_{max}$*) and quantum yield (α) which naturally change during season. For this purpose, we repeated field measurements of *GPP$_{max}$* and α of most plant species on a 2-week basis, and once a month for conifer trees. *GPP$_{max}$* was derived from plotted light curves and α was calculated using the nonlinear regression (Statistica 12). The Michelis – Menten type of regression [54] was used for calculation:

$$GPP = \frac{GPP_{max} \times \alpha \times PAR}{GPP_{max} + \alpha \times PAR} \times LAI \tag{1}$$

GPP was calculated on an hourly basis. As incoming *PAR* changes continuously, we used 60 min averages to calculate hourly and consequently daily and total (seasonal) *GPP*. As the shaded leaves inside the canopy receives only portion of incoming *PAR*, we divided the canopy into 10 cm vertical sections and measured incoming *PAR* from the canopy top to the ground (100 measurements per species) by Quantum sensor. Similarly, *LAI* vertical distribution across the canopy was measured by Licor 2200. Empirical models of vertical *LAI* distribution and *PAR* dissipation for dominant species were applied to correct *GPP* derived from leaf measurements.

Leaf measurements: Licor 6400XT device was used for the gas exchange measurement on wide leaf plants. The samples (20 leaves per species, equally at both sunny and shaded sides) were exposed to different intensities of *PAR* (0, 50, 100, 200, 400, 800, 1200, 2000 µmol photons m^{-2} s^{-1}) to construct the light response curves. The leaf CO_2 net exchange was measured biweekly on dominant species; the parameters are shown in Table S1.

Plant measurements: Customs built plexiglass chambers of 16 and 60 dm^3 were used to measure the whole plant gas exchange on narrow leaf plants (grasses). The chambers were equipped with GMP 343 CO_2 sensor (Vaisala, Finland), *PAR*, temperature and air humidity probes and a small fan. Incoming radiation was modified with a variable combination of shading nets. Each measurement lasted two minutes. Only the linear part of the CO_2 concentration change was used for the calculation of CO_2 flux following the ideal gas law [55].

Mature forest estimation: The *GPP* in mature spruce forest was calculated with annual wood stock increment (m^3 ha^{-1}). Average annual wood stock increment was derived from permanent research plots located in studied catchments [56]. Allometric and biomass expansion factors were used to transform wood stock increment to net primary productivity (*NPP*) according to [7] and [57]:

$$NPP = V \times D \times BEF \times k \times c \tag{2}$$

where *V*—annual wood stock increment, *D*—wood density, *BEF*—expansion factor for above-ground biomass, *k*—coefficient for below-ground biomass, and *c*—portion of C in wood.

The *GPP* is equal to *NPP* divided by carbon use efficiency coefficient (CUE), *GPP = NPP/CUE*. We used the CUE value for Norway spruce according to [58].

2.4.2. Ecosystem Respiration—RE

Ecosystem respiration (RE) is the sum of soil and aboveground plant respiration. Soil respiration (SR) was measured on fixed collars with a PVC custom-built dark closed chamber (diameter 30 cm, height 10 cm) equipped with Vaisala GMP 343 CO_2 probe and a small fan. The CO_2 concentration was measured every five seconds for five minutes and corrected for air pressure, temperature and humidity. Soil temperature and soil moisture were recorded along with the SR measurement. SR was measured every two weeks along fixed transects (for details see [54]). For this study we used the location and the number of sampling points (collars) as shown in Table 2.

Measured SR values were fitted with an exponential function based on soil temperature [59]:

$$SR = k \times e^{a \, X \, T} \tag{3}$$

SR was calculated for each individual sampling point (collar), where *T* is soil temperature in 10 cm, *k* and *a* are regression parameters of the function derived by nonlinear regression using Statistica 12. Only statistically significant parameters ($p < 0.05$) were employed for SR modelling. The annual SR model was based on extrapolation of instant SR values using continuously measured soil temperature in 10 cm depth. Each ecosystem type was represented by an average calculated from all measured points.

Table 2. Numbers of sampling points for estimation of soil respiration in different ecosystems.

Ecosystem Type	Abbreviation	Number of Points	Altitude (m a.s.l.)
Alpine meadow	ALM	3	1730
Dwarf pine	DWP	5	1550
Bark beetle attacked forest	IPS	10	1180
Mature undisturbed forest	REF	20	1200
Windthrow extracted	EXT	24	1100
Burnt windthrow	FIR	8	1100
Windthrow non-extracted	NEX	8	1100
1-year-old windthrow	REX	8	1100

Leaf respiration was measured with Licor6400 XT as species-specific leaf net CO_2 exchange under dark conditions (*PAR* set to 0). Stem (wood) respiration was measured on collars permanently fixed to the trunks (n = 5) on trees with average height and diameter. The measurements were realized on a 2-week basis. Arithmetic means of measured values were used to derive regressions with air temperature similarly as for soil respiration and to calculate component (leaves, stem) and annual ecosystem respiration.

2.4.3. Flux Data Validation

Confidence intervals for estimated CO_2 fluxes were calculated with the Monte Carlo method as proposed by Knohl [60]. First, we calculated the regressions in Equation (3) for all components of ER (soil, stem, leaf) and their standard deviations (SD). Second, we calculated 5000 seasonal sums of respiration based on measured temperatures and 5000 combinations of k and a parameters sampled by the replacement from bivariate normal distribution defined by the values and SD of k and a. A similar approach was applied to estimate *GPP* uncertainty. The GPP_{max} and α parameters in Equation (1) were estimated using nonlinear regression. The seasonal sum of *GPP* based on hourly *PAR* for different plant species was calculated with 5000 combinations of GPP_{max} and α parameters sampled using the same method as described above. The overall catchment scale confidence intervals for NEE were calculated as a square root of the weighted sums of partial ecosystem level CIs.

2.5. Streamflow Characteristics

Daily discharge data from the water years 1965–2014 were analyzed. The MP catchment represented a small mountain headwater catchment that was unaffected by the 2004 windthrow while the VP and SP represent catchments that were significantly affected by the windthrow and consecutive bark beetle outbreak and fire disturbances. Having data from ten water years after the windthrow, we first examined decadal cumulative runoff from 1965 to 2014 and compared it to cumulative precipitation from the only two precipitation stations existing in the study area that provide representative data from higher elevations (Skalnaté Pleso and Kasprowy wierch).

The IHA indicators were also examined visually (magnitudes, ranges, temporal trends and abrupt changes) with the aim to identify the influence of the disturbances on the low and high flow characteristics (annual minima: 1-day, 3-day, 7-day and 30-day annual minima, 1-day, 3-day and 7-day annual maxima), baseflow index, environmental flow components (frequency and duration of high and low flow pulses), rate and frequency of flow changes (rise, fall, number of reversals). High flows were defined as flows that exceeded 75% of all daily flows. Low flows were flows equal to or less than 50% of all daily flows. The number of reversals represents the number of flow changes from falling to rising and vice versa per year. Median values of all characteristics in the decades of 1965–2014 were compared as well.

Characteristics recommended by Olden and Poff [40] for perennial streams with snow and rain regime (Table 3) were then calculated and analyzed as well. Except for the number of flow reversals, those characteristics were not used in the study area in our earlier studies.

Table 3. Selected indicators of flow regime changes proposed for the evaluation of perennial streams with snow and rain regime by [40].

Code	Definition
MA3	coefficient of variation in daily flows
MA44	variability in annual flows divided by median annual flows; the variability is calculated as 90th minus the 10th percentile
ML13	coefficient of variation in minimum monthly flows
ML22	mean annual minimum flows divided by catchment area
MH17	mean of the 25th percentile from the flow duration curve divided be median daily flow across all years
MH20	mean annual maximum flow divided by catchment area
FL3	total number of low flow spells (threshold equal to 5% of men daily flow) divided by the record length in years
FL2	coefficient of variation in the low flow pulse count (below 25th percentile)
FH3	high flow pulse count (high flow pulse is 7 times the median daily flow)
FH5	mean of high flow events per year (high flow is 1 times the median flow)
DL6	coefficient of variation in annual minima (1-day annual minima)
DL13	mean annual 30 day minimum divided by median flow
DH12	mean annual 7-day maximum divided by median flow
TL1	mean Julian date of the 1-day annual minimum flow over all years
RA9	coefficient of variation in number of reversals
RA8	number of positive and negative changes in water conditions from one day to another (flow reversals)

Our final group of streamflow analyses focused on the occurrence of flows with specified probability of exceedance estimated from the flow duration curves. A flow duration curve is a cumulative frequency curve that characterizes probability of exceedance of specified discharges during a given period [61]. First, decadal median flow duration curves in the period 1965–2014 were calculated and compared between decades. Partial duration peakflow series comprising peak discharges of the four greatest runoff events in each hydrological year provided the flow duration curves for the study period 1965–2004 (i.e., before the windthrow disturbance). Then, the occurrence of peakflows with 10, 50 and 90% probability of exceedance was examined over the entire study period 1965–2014. It should be noted that our maximum peakflows may not inevitably represent floods as understood by [38] above because we did not have records about the events that exceeded the capacity of study channels.

3. Results

3.1. Weather, Vegetation and Phenology

In 2014, air temperature in the study region was exceptionally high compared to the long-term observations at nearby meteorological stations. The annual average temperature at 850 m a.s.l. was 7.5 °C (2.1 °C above the 1930–1960 average). Mean annual air temperature on the Lomnický štít peak (2633 m a.s.l) which is the highest meteorological station in the Tatra Mountains was 2.2 °C above the norm. The annual precipitation total of that year (963 mm) was 14% above the long-term norm at lower altitudes (850 m a.s.l.), while at higher altitudes it was equal to the long-term norm (1257 mm). The difference in annual mean soil temperature between the lowest (1150 m a.s.l.) and the highest (1750 m a.s.l.) locations was only 0.4 °C. Soil moisture (θ_{vol} %) exceeded 30% throughout the year except for the second half of August.

Species share in studied ecosystems was derived from their relative leaf coverage. The average *LAI* of alpine meadows dominated by *Avenella flexuosa* (50%) and *Luzula sudetica* (25%) was 1.6 (±0.4).

Average *LAI* of *Pinus mugo* was 3.4 (±0.6). Sparse grasses beneath dwarf pine canopy were neglected. Canopy *LAI* of the undisturbed mature forest (REF) was 3.6 (±0.9) and forest floor vegetation *LAI* was 1.0 (SD ± 0.3). Species shares in disturbed ecosystems are presented in Figure 2.

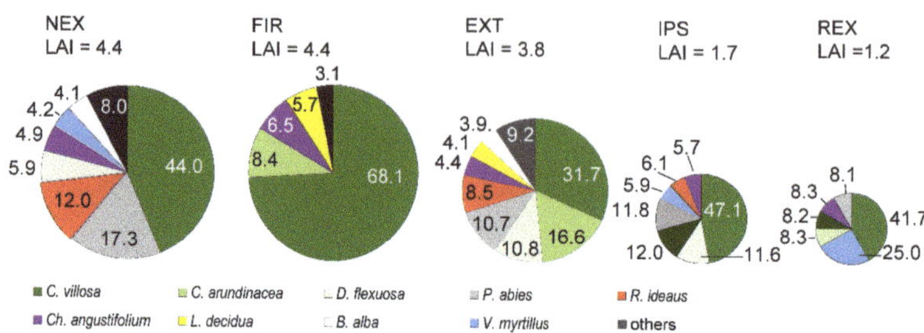

Figure 2. Total leaf area index (LAI) (m² m⁻²) and species share [%] for ecosystem types on disturbed sites, where NEX—non-extracted windthrow, FIR—burnt windthrow, EXT—extracted windthrow, IPS—forest killed by bark beetle, REX—one-year-old extracted windthrow.

The values represent the maximum of measured *LAI* in mid-June. Species with a share below 4% are not shown. Ground vegetation started to develop in late April. Leaf area increased rapidly until mid-June, after which it remained relatively stable. The earliest senescence was observed on *Chamaerion angustifolium* in mid-August. Near timberline vegetation started to develop after snow melt in the middle of May. First leaves on broadleaved tree species (*Betula verucosa, Salix caprea, Sorbus aucuparia, Populus tremulae*) growing at 1200 m a.s.l. occurred in early May (5–10) and fully developed leaves in early June (5–15)). At 1400 m a.s.l., the dates were postponed by 7–10 days. Needles on *Larix decidua* fully developed in early May.

The greatest vegetation cover was found on NEX and FIR (*LAI* 4.4) followed by EXT (3.8). The difference between them was insignificant (*p* = 0.054). As expected, lower *LAI* was found on IPS and REX sites (1.7 and 1.2, respectively, *p* = 0.14). Generally, grasses dominated on all disturbed sites. Contribution of trees to total *LAI* was relatively low. It was zero on REX, 11% on FIR, and 26% on EXT and NEX sites. The broadleaved species slightly prevailed among trees (60%). *LAI* notably varied during the season. The average seasonal course of *LAI* in the EXT ecosystem is shown in Figure 3.

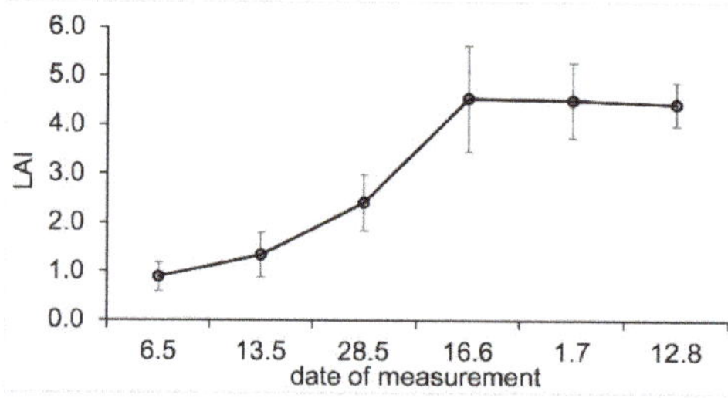

Figure 3. Seasonal changes of LAI on ten-year-old windthrow, between May and August 2014; the dots represent means from 25 plots, and the whiskers show SD.

As the site conditions and species composition in the studied catchments were rather similar, we expected the same temporal changes of *LAI* in all ecosystem types. For *GPP* modelling purposes, the *LAI* values between two subsequent measuring dates were linearly interpolated to obtain a time series with 60 min time step.

Vertical distribution of *LAI* and reduction of *PAR* along a canopy yielded very similar results for dominant plant species. Average courses of relative *LAI* distribution and *PAR* availability along canopy (0% = top of canopy, 100% = ground) are presented in Figure 4a,b.

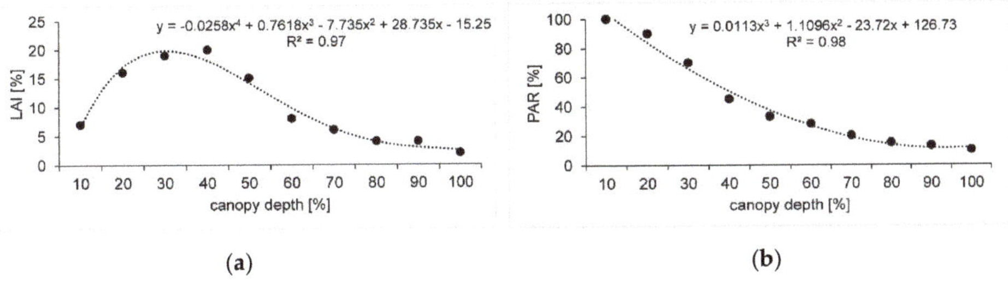

(**a**)　　　　　　　　　　　　　　　　(**b**)

Figure 4. Relative distribution of LAI (**a**) and PAR dissipation (**b**) in relation to relative canopy depth; black dots—average values for all measured species, line-fitted model.

3.2. Carbon Fluxes

3.2.1. Assimilation (GPP)

Light response curves (LRC) for selected species are shown in Figure S2. Early successional stages species (*Calamagrostis villosa*, *Chamaerion angustifolium*, *Calamagrostis arundinacea*, *Salix caprea*, *Rubus ideaus*) showed significantly higher ($p < 0.01$) values of light saturated *GPP* (*PAR* 1500 µmol photons m^{-2} s^{-1} *PAR*) than shade-tolerant species typical for spruce forests (*Picea abies*, *Vaccinium myrtillus*, *Avenella flexuosa*). The maximum rate of *GPP* for grasses, herbs and broadleaved trees culminated in late May, while conifers culminated in late June.

GPP_{max} and α derived from the LRC for seasonal modelling of photosynthesis according to Equation (1) are given in Supplement Table S1. The *GPP* derived from species-specific LRC was further summarized in 10 cm layers reduced by available *PAR* and *LAI* fractions as shown in Figure 4 and calculated for ecosystem types according to vegetation characteristics (*LAI* and species).

The highest seasonal *GPP* was fixed in FIR ecosystem type (1,420 ± 33.0 gC m^{-2}). Both managed (EXT) and unmanaged (NEX) types assimilated similar amount of C equal to 1198 ± 24.5 and 1250 ± 29.1 gC m^{-2} respectively. Dwarf pine fixed 886 ± 17.6 gC m^{-2}, and spruce forest damaged by bark beetle (IPS) assimilated 605 ± 19.4 gC m^{-2}. *GPP* on one-year-old windthrow (REX) and on alpine meadows was 425 ± 16.5 and 230 ± 8.2 g C m^{-2}, respectively, as presented in Figure 5.

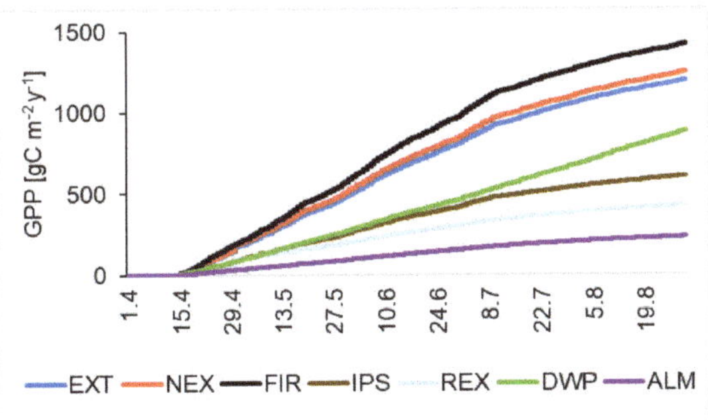

Figure 5. Seasonal (April 1–August 31) sum of *GPP* for studied ecosystem types, where EXT—extracted windthrow, NEX—non-extracted windthrow, FIR—burnt windthrow, IPS—forest killed by bark beetle, REX—extracted one-year-old windthrow, DWP—dwarf pine, ALM—alpine meadow.

According to the State Forest of TANAP records, the average annual wood stock increment (2012–2013) in *Lariceto-Piceetum*, prevailing forest community in studied catchments, was 8.0 m^3 ha^{-1} y^{-1}. Using *BEF* for aboveground (1.3) and belowground spruce biomass (1.2) and CUE (0.35), the estimated *GPP* was 892 gC m^{-2} y^{-1}. If ground vegetation assimilation (330.6 ± 97.8 gC m^{-2} y^{-1}) was added, total *GPP* in undisturbed REF forest was 1222.6 gC m^{-2} y^{-1}.

3.2.2. Ecosystem (RE) and Component Respiration

The highest ER was on REF (1148 ± 120.1 gC m^{-2} y^{-1}) and IPS (1100 ± 175.0 gC m^{-2} y^{-1}), followed by FIR (957 ± 175.0 gC m^{-2} y^{-1}) and NEX (909 ± 87.4 gC m^{-2}) sites. Soil carbon efflux was the largest component of ecosystem respiration. From all the ecosystem types, the largest annual soil carbon efflux equal to 953 gC m^{-2} was in the REF ecosystem. The second largest flux (905 gC m^{-2} y^{-1}) was emitted from the IPS type. Soil on the FIR site respired 617 gC m^{-2} y^{-1} and from managed windthrow (EXT) 596 gC m^{-2} y^{-1}. Soil respiration from *Pinus mugo* stands and alpine meadows was equal to 508 and 250 gC m^{-2} y^{-1}, respectively.

The average proportion of soil respiration to total ER in studied ecosystems was 74% and ranged from 67% (NEX) to 82% (IPS). Soil respiration showed the highest correlation (avg R^2 = 0.64) with temperature measured at 10 cm soil depth. The correlation between air temperature and leaf respiration was of similar magnitude (R^2 = 0.61), while the correlation of air temperature to wood respiration was the lowest (R^2 = 0.38).

3.2.3. Carbon Balance (NEE)

NEE as a difference between seasonal *GPP* and annual RE indicates whether the ecosystem or watershed sequesters or emits C to the atmosphere. The *GPP*, RE and NEE fluxes in the studied ecosystems are presented in Table 4.

Table 4. Average annual carbon fluxes (gC m^{-2} y $^{-1}$) and CI (95%) for different ecosystem types.

Flux/Ecosystem	ALM	DWP	IPS	REF	EXT	FIR	NEX	REX
GPP	−230 ± 8.2	−886 ± 17.6	−605 ± 19.4	−1223 *	−1198 ± 24.5	−1420 ± 33.0	−1250 ± 29.1	−425 ± 16.5
ER	220 ± 43.7	695 ± 90.3	1100 ± 175.0	1148 ± 120.1	846 ± 77.4	957 ± 175.0	909 ± 87.4	755 ± 96.4
NEE	−10 ± 44.5	−191 ± 91.9	495 ± 176.0	−75 ± 120.1	−352 ± 81.2	−463 ± 178.1	−341 ± 92.1	330.0 ± 97.8

* indicates that confidence intervals for the REF were not calculated as a different method was used for *GPP* estimation.

Most of the ecosystems were carbon sink. The largest sink was vegetation on FIR (−463 ± 178.1 gC m^{-2} year^{-1}), followed by the EXT and NEX ecosystems. The differences among wind disturbed sites were low mostly due to large variation of SR. Undisturbed mature REF type was carbon neutral (−75 ± 120.1 gC m^{-2} year^{-1}, an asterisk indicates that confidence intervals for the REF were not calculated as a different method was used for *GPP* estimation). IPS was the largest carbon source (495 ± 176 gC m^{-2} year^{-1}) followed by one-year-old windthrow REX (330 ± 97.8 gC m^{-2} year^{-1}). NEE in IPS and REX remarkably differed from all other ecosystems.

GPP, RE and NEE fluxes (tC ha^{-1} year^{-1}) in three catchments are presented in Figure 6.

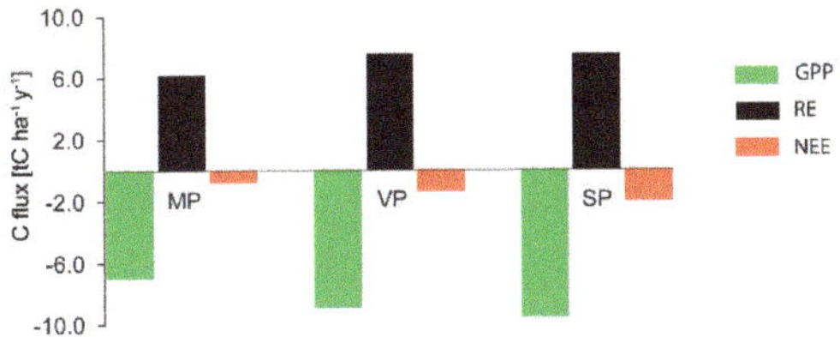

Figure 6. Annual carbon fluxes (tC ha^{-1} y^{-1}) on a catchment scale where MP—Mlynický Creek catchment, VP—Velický Creek catchment, SP—Slavkovský Creek catchment, green bars—*GPP*, black bars—RE and red bars—NEE. Negative values of NEE indicate carbon sink.

As the *GPP* and RE increased in order SP > VP > MP, the same order applied for NEE. NEE in MP catchment (-80 ± 73 gC m^{-2} y^{-1}) indicates that carbon balance was almost neutral. The VP (-136 ± 84 gC m^{-2} y^{-1}) and SP (-206 ± 115 gC m^{-2} y^{-1}) acted in 2014 as mild and strong carbon sinks, respectively.

3.3. Streamflow Characteristics

Decadal cumulative precipitation and runoff (Figure 7) revealed greater cumulative runoff in the VP and SP catchments in the decade after the windthrow (2005–2014). Similar increase is not visible in the MP catchment which was very little affected by the windthrow.

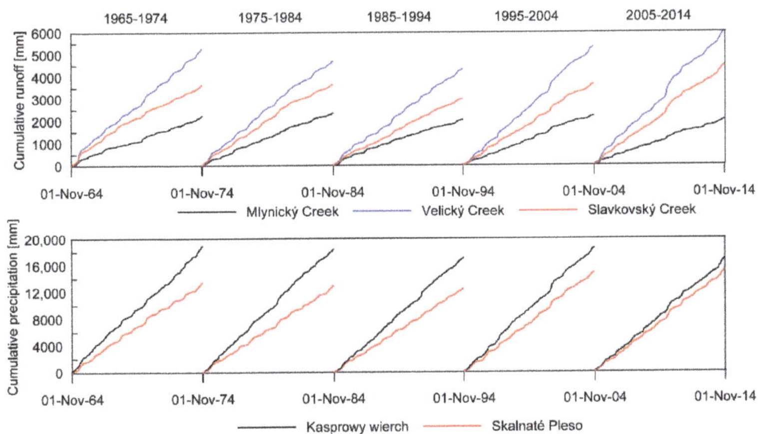

Figure 7. Cumulative runoff and precipitation in decades from 1965–1974 to 2004–2014; precipitation data adopted from Holko et al., 2020.

Median monthly flows and low flows in the period 2005–2014 increased in almost all months of a year in the SP catchment (Figure 8), while the opposite was observed in the MP catchment in the warm period of a year (May–October), except for August. The smallest changes were found in the VP catchment where they were more pronounced in monthly low flows. Figure 8 also documents seasonal variability of flows in the studied catchments that is probably not related to windthrow. For example, annual maxima in the MP catchment in the 1965–1974 and 2005–2014 decades were clearly related to the snowmelt period (April), while in other decades and other two catchments they occurred later in spring (May, June). A shift in the occurrence of seasonal flow maxima to the summer period (July or even August) is observed in the last two decades (from 1995) of the study period in the VP and SP catchments.

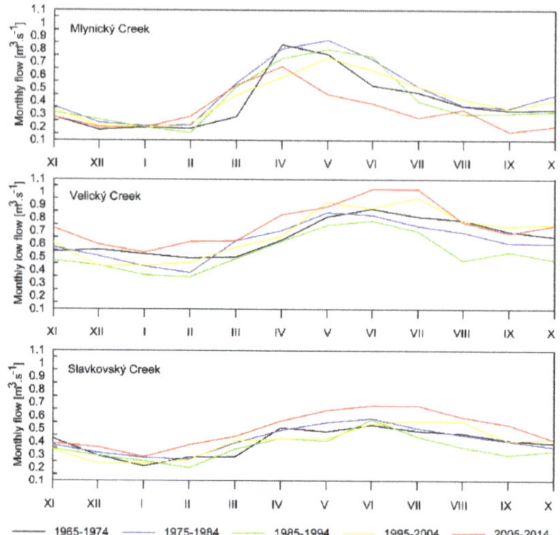

Figure 8. Medians of monthly flow characteristics observed in the decades from 1965–1974 to 2005–2014.

The indicators listed in Table 3 except for the number of flow reversals did not reveal the influence of the disturbances. In water years 2005 and 2006, the number of flow reversals increased in the VP and SP catchments (Figure 9).

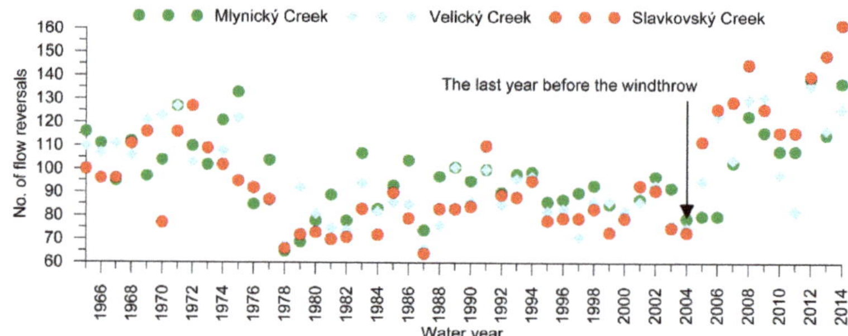

Figure 9. Number of flow reversals per year in the study catchments; the windthrow occurred at the beginning of the water year 2005.

Strong increase occurred in the MP catchment in 2007 (interestingly, the number of flow reversals decreased in 2007 in the VP catchment). Variability of this indicator after 2007 became similar in all studied catchments.

In the 2005–2014 decade, flow duration curves (FDCs) in the VP and SP catchments were shifted in the opposite direction compared to the FDC calculated for the MP catchment (Figure 10).

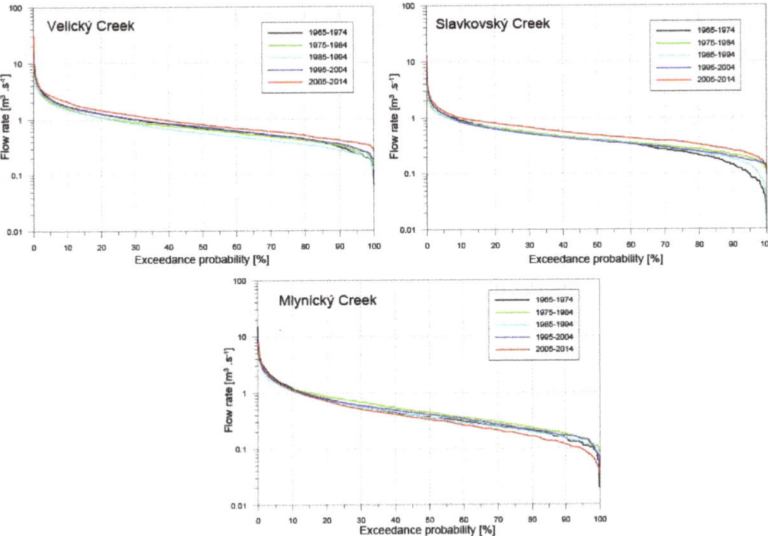

Figure 10. Median decadal flow duration curves.

Partial duration peakflow data did not indicate more frequent occurrence of higher discharges. Peakflows of the greatest annual runoff events in the VP and SP creek catchments became better correlated after the disturbance, i.e., in the 2005–2014 decade (Figure 11).

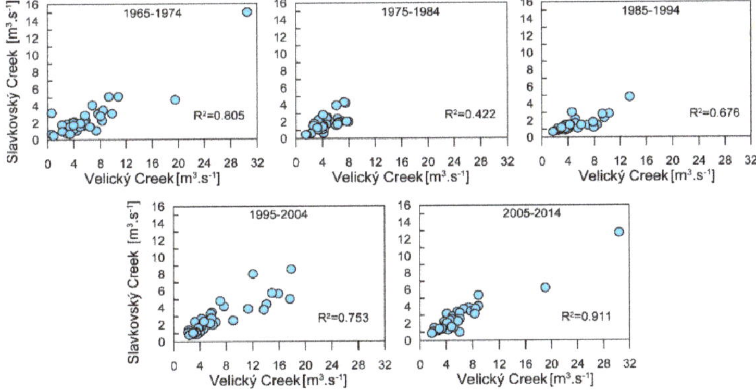

Figure 11. Relationships between peakflow discharges (four greatest runoff events in each year) in the VP (horizontal axis) and SP (vertical axis) catchments in the decades from 1965–1974 to 2005–2014.

4. Discussion

Forest potential to sequester carbon in biomass and soil and thus to mitigate climate change is of high interest. In reality, a forest, as well as any other vegetation, concurrently fixes and releases carbon. The strength of carbon sink depends on many factors, e.g., species composition, soil properties, age of vegetation, local climate, etc. [62,63]. Disturbances influence forest growth dynamics, mortality and decomposition processes and therefore carbon cycling [64]. Disturbed forests usually become a carbon source, at least temporarily [26,65,66] while disturbed biomass decomposes and productivity decreases. One of the key questions is how long after a disturbance it takes until forest ecosystems recover and become carbon neutral and eventually begin to act as carbon sinks.

Our chamber-based measurements confirmed our expectation that forest ecosystems acted as a carbon source after the disturbance. NEE in the one-year-old windthrow (REX) was 330 gC m^{-2} y^{-1}. The 2014 windthrow destroyed dense mature stands with sparse forest floor vegetation which led to strong decline of *GPP*. Due to canopy destruction also the autotrophic component of ER declined but the impact on NEE was much less pronounced than that on *GPP* reduction. These data coincide with the EC tower measurements in 2006–2007 [67]. The authors reported strong carbon efflux 2–3 years after the 2004 forest destruction from the same wind-disturbed sites that we analyzed in this paper. Direct measurements of carbon fluxes on windthrown sites are rare, but the impact is often compared to that of logging impact. Rebane et al. [64] in a review of post disturbance carbon balance stated that one year after harvesting the annual efflux usually ranged from 520 to 1000 gC m^{-2} y^{-1} and three years after the disturbance forests emitted 120 gC m^{-2} y^{-1}.

Dramatic post disturbance NEE changes were also revealed by our data. We found that ten years after the disturbance the ecosystems affected by the windthrow were either carbon neutral or already became a substantial carbon sink. Niu et al. [14] analyzed the time needed for carbon recovery following different disturbances and pointed out that recovery time was related to disturbance severity with more severe impact requiring longer recovery times. They stated that relatively fast turnover from carbon source to carbon sink was assigned with windthrow (23.5 ± 5 years). The longest recovery time (101 ± 28 years) was needed after wildfires. In our study the burnt site recovered the fastest. The carbon balance estimates at managed, unmanaged and burnt windthrown sites indicate that carbon neutrality was already exceeded in previous years as the NEE values reached −352, −341 and −463 gC m^{-2}, respectively. Changes in NEE in the years after the disturbance are expected, while their direction and magnitude depend upon the balance between the source effects of the load of decomposing wood [60,68] and the sink effects of the recovering vegetation. Recovery of carbon fluxes after such an extreme windthrow that occurred in the Tatra Mountains, which almost completely destroyed thousands of hectares, was surprisingly fast. Vigorous vegetation recovery on disturbed sites was already reported in our earlier papers [54]. According to Zhu et al. [69], if forest recovery is a dominant mechanism, then the current carbon sink is expected to saturate as forests age and vegetation reaches its late successional stage. This is the very likely future of carbon fate in our study catchments.

The expected difference in carbon balance between the forest that was managed and unmanaged after the disturbance was proven. Despite the fact that EXT exhibited greater species diversity as a consequence of replanting and NEX had more trees [70], the *LAI* values at the two sites were similar. Removal of undesirable plant species on the EXT site during weed control regularly reduced plant canopy and thus reduced potential carbon sequestration. On the other hand, greater plant diversity, especially of trees on the EXT site, supports forest resilience and future climate change mitigation potential. Dead wood intentionally left on the NEX site has not contributed to ecosystem respiration yet due to still weak decomposition rate of fallen logs. This is the likely reason why our data are in contrast with [71] or [72] who reported higher CO_2 efflux from unmanaged sites than from sites where dead wood was removed.

Contrary to wind disturbed sites, the IPS site acted as a large carbon source in 2015 (NEE = 495 gC m^{-2} y^{-1}). Besides reduced assimilation, increased soil respiration was the main factor. Total respiration was the highest at the IPS site from all disturbed sites (1100 gC m^{-2} y^{-1}). We attribute this result to elevated heterotrophic respiration. This is in agreement with [73] who reported increased soil respiration after bark beetle outbreak due to elevated microorganism activity stimulated by increased radiation trough open canopy and thus higher soil temperature. According to [64], carbon recovery in a forest disturbed by insects (bark beetles) should be less than 10 years, which is probably not our case. We see the reason for the large carbon efflux in our study in the fact that the studied IPS ecosystems are located outside the reach of downslope winds. Due to less frequent disturbances, the soil is covered with thicker humus horizons and thus contains more carbon. More light, elevated temperature and moisture stimulate more intensive decomposition and carbon efflux. Higher content of organic compounds in soil as well as less evident pit and mound microtopography indicate that the forest that is currently attacked by the bark beetles was rarely affected by the disturbances in the past [74]. On the other hand, [75] reported respiration decline after a bark beetle attack. They found decreasing carbon input from assimilation more important for forest carbon balance than increasing respiration of dead material. Regardless, large stock of labile carbon poses a risk of carbon release to the atmosphere under climatic warming [13,76]. Soil temperature was identified as a key environmental factor for respiration on wind-disturbed sites. The highest correlation was with soil respiration, the weakest with stem respiration. Don et al. [77] reported increased soil temperature by 4 °C and higher decomposition rates on the bare soil just after the disturbance in the Tatra Mountains compared to the intact forest. Fast decaying fine material (needles, bark) after the bark beetle attack probably also supported the increase of heterotrophic respiration, as stated by [11]. Heterotrophic respiration could also profit from increased soil moisture because of suppressed tree transpiration. Although ten years after the disturbance soil moisture did not notably differ between the disturbed and intact forest stands, it played only a marginal role in soil respiration rate. We suppose that sufficient soil moisture (around 30% Θ_{vol}) during the entire study period has not limited respiration.

Carbon sequestration on wind-disturbed sites massively exceeded the performance of the intact mature Norway spruce forest (-75 ± 120 gC m^{-2} y^{-1}). Higher carbon sequestration in post-disturbance forests is rather common. For example, [14] showed primary productivity higher by 35% than the pre-disturbance values. In our study, the post-disturbance carbon sequestration was three times higher than before forest destruction. Nevertheless, intact forest NEE values must be interpreted with caution. The *NPP* and consequently the *GPP* flux were estimated by the biometric method, while in other ecosystem types they were estimated by gasometrical methods. Calculated NEE values were much lower than those reported for comparable temperate forests. Intact forests in central Europe are generally expected to act as a carbon sink [78,79]. According to [10], European forests potentially sequester carbon from -115 to -410 gC m^{-2}. Etzold et al. [80] reported -153 gC m^{-2} y^{-1} from comparable Norway spruce stands in the Alps. Thus, the performance of mature forests in our study can be classified as carbon neutral, rather than a small sink. One of the possible explanations for low *NPP* production in our study might be less intense growth of mature trees. Mean annual increment of spruce forests in central Europe is 14 m^3 ha^{-1} [10]. At our study sites it reached only 8 m^3 ha^{-1} per year. It is obvious that NEE declines with age [81]. Direct measurements using eddy covariance method showed that middle age stands are stronger carbon sinks than very old stands [64,82]. However, forest age is not the case in our study as observed trees are 80–120 years old and the explanation of small *GPP* and NEE requires further analysis.

Upscaling of CO$_2$ fluxes from leaf to ecosystem/landscape scales is always challenging. However, it is necessary because quantification of source and sink relationships at a landscape level is inevitable to understand the relationships between NEE and the global carbon cycle [83]. Cleary et al. [27] compared leaf versus ecosystem fluxes in a grass-dominated ecosystem and found higher NEE with a leaf measurement. NEE declined in whole plant chambers reported [84] as a consequence of self-shading. Occasionally, the authors found twofold greater *GPP$_{max}$* in sparse than in dense populations. In our study we found similar differences between *GPP* measured by Licor XT6400 at a leaf scale versus whole plant in the closed chamber. Leaf-scale measurement yielded on average 50% higher *GPP* values unless empirical models for vertical *LAI* distribution and *PAR* dissipation were applied. After the correction, the difference declined below 12%. We applied this correction in *GPP* calculation when the *LAI* exceeded 2.5 m^2 m^{-2} and the shading effect became evident.

The analysis of streamflow characteristics did not unambiguously indicate the influence of forest disturbances. Only a few characteristics exhibited different behavior after the disturbances. Greater cumulative runoff in the most severely affected VP and SP catchments in 2005–2014 decade could be related to the disturbances. Górnik et al. [85] documented an increase in the number of days with precipitation of 40–60 mm in the warm period of the year (May–October) since 2001 and the unusually wet year 2010 was observed in the Tatra Mountains. Greater cumulative decadal runoff compared to the 1975–1984 and 1985–1994 decades occurred also before the windthrow, i.e., in the decade 1995–2004.

However, the values calculated for the VP and SP catchments at the end of 2005–2014 decade exceeded analogical values from all previous decades. This was not observed in the MP catchment with forest cover almost unaffected by the windthrow. Cumulative point precipitation at the end of the decade 2005–2014 is not greater than in the previous four decades. These data indicate the influence of forest disturbances. On the other hand, Figure 7 shows that the greatest increase in cumulative runoff occurred approximately between April and December 2010 and another significant increase was observed since April 2014, i.e., about five and nine years after the windthrow. The winters of 2010 and 2014 were both exceptionally snow-poor [86] and summer 2010 was very wet. It is therefore possible that the increased runoff after the snow-poor winters was related more to the snowmelt-precipitation regime in springs 2010 and 2014 than to the effects of forest disturbances.

The number of flow reversals is another indicator that increased in the VP and SP catchments in the years following the windthrow while it remained stable in the MP catchment in 2005 and 2006. Although the MP catchment was very little affected by the windthrow, a strong increase in the number of flow reversals was observed in 2007 (Figure 9). This could be related to massive forest dieback and following cutting. It is worth noting that greater numbers of flow reversals occurred also in the 1960s although some values after the windthrow were greater. The highest values that occurred in the SP catchment in 2013 and 2014 are probably not related to the windthrow. Because the variability of the indicator is similar in all studied catchment has been similar since 2008, we assume that the high values in 2012–2014 were caused by meteorological causes rather than by forest disturbances.

Although median decadal flow duration curves in the VP and SP catchments after the windthrow are slightly shifted to greater discharges at almost all probabilities of occurrence, our earlier analysis of the annual FDCs indicated that the shift was probably not caused solely by the disturbances, but also by meteorological conditions in the years 2005, 2010 and 2014 [48]. Similarly, higher correlations among peakflow discharges in the VP and SP catchments in the 2005–2014 period (Figure 11) were more influenced by meteorological conditions than by the effects of forest disturbances. Greater ranges of measured peak discharges given by higher precipitation (e.g., in June 2010) resulted in higher coefficients of determination between the discharges in the neighboring catchments of the Velický and Slavkovský Creek.

5. Conclusions

Our study revealed that ten years after extremely severe disturbances, the carbon balance at a catchment scale was neutral or even positive. Highest carbon sequestration occurred at the most heavily disturbed sites. Positive carbon balance reflects very intensive growth of successional vegetation in recent warm and sufficiently moist climate. Only small patches of the forest killed by bark beetle acted as a carbon source. Both managed and unmanaged sites in our study catchments acted as carbon sinks. Nevertheless, greater species diversity on the managed sites could potentially guarantee higher forest resilience and climate change mitigation effect under future, hardly predictable, conditions. The findings that the intact mature forest was carbon neutral requires further and more detailed study to identify the reasons.

Possible reasons for unclear effects of the disturbances on catchment hydrology were suggested by [45]. Most of the forest damage, while quite extensive, occurred in the middle sections of the catchments. The headwater areas, where most of the runoff is formed and where little forest existed before anyway, were not damaged by the windthrow. Windthrow deforested areas were built by moraines with high infiltration capacity. This disturbance affected relatively smaller percentages of catchment areas since it hit the catchments in the east-western direction, while the catchments are generally north-south oriented. Thus, not all the forests were damaged by the windthrow (although the subsequent disturbances in the SP catchment destroyed 98% of the original forests). Additional analyses conducted in this article confirm our previous conclusion about the lack of suitable indicators of changes in catchment hydrology in headwater mountain catchments with areas of several tens of square kilometers in which the forest disturbances affected up to 20–32% of the catchment area.

Supplementary Materials: The following are available online at http://www.mdpi.com/2073-4441/12/10/2917/s1, Figure S1: The entire study catchments (black line) and carbon study parts (MP—Mlynický potok Creek catchment red cross-hatched, VP—Velický potok Creek catchment blue cross-hatched, SP—Slavkovský potok Creek catchment green cross-hatched), Figure S2: Light response curves for selected (a) grass, herb and shrub species; (b) tree species. Table S1: Average 2-week GPP_{max} [μmol C m^{-2} s^{-1}] and α parameters estimated from light response curves for the key species in studied catchments.

Author Contributions: P.F.J., P.F., L.H. designed the experiment, wrote the manuscript, analyzed the data, discussed results, S.C. performed GIS analysis, interpreted data, L.Č., J.R., P.Š. and L.O. sampled and interpreted field data and discussed the results. All authors have read and agreed to the published version of the manuscript.

Funding: This study was carried out through funds made available under the Slovak Agency for Science and Research grant no. APVV 17-0644, the Scientific Grant Agency of the Ministry of Education, Science, Research and Sport of the Slovak Republic under contract VEGA2/0049/18 and State Forest of Tatra National Park.

Acknowledgments: Authors are grateful to Zuzana Kyselová, Peter Michelčík for vegetation sampling, Luboš Slameň for respiration measurement and measuring instruments maintenance, State Forest of TANAP for providing aerial ortophotomaps and forest growth data and Svetlana Bičárová and Joanna Pociask-Karteczka for providing precipitation data from Skalnaté Pleso and Kasprowy wierch. Last but not least, we would like to acknowledge the work of many students who participated in field and laboratory works.

Conflicts of Interest: The authors declare no conflict of interest.

References

1. Thom, D.; Seidl, R. Natural disturbance impacts on ecosystem services and biodiversity in temperate and boreal forests. *Biol. Rev.* **2015**, *91*, 760–781. [CrossRef] [PubMed]
2. Senf, C.; Seidl, R. Mapping the coupled human and natural disturbance regimes of Europe's forests. *BioRxiv* **2020**. [CrossRef]
3. Seidl, R.; Schelhaas, M.J.; Lexer, M.J. Unraveling the drivers of intensifying forest disturbance regimes in Europe. *Glob. Chang. Biol.* **2011**, *17*, 2842–2852. [CrossRef]
4. Anderegg, W.R.L.; Kane, J.M.; Anderegg, L.D.L. Consequences of widespread tree mortality triggered by drought and temperature stress. *Nat. Clim. Chang.* **2013**, *3*, 30–36. [CrossRef]
5. Malhi, Y.; Franklin, J.; Seddon, N.; Solan, M.; Turner, M.G.; Field, C.B.; Knowlton, N. Climate change and ecosystems: Threats, opportunities and solutions. *Philos. Trans. R. Soc. B* **2020**, *375*. [CrossRef]
6. Weiskopf, S.R.; Rubenstein, M.A.; Crozier, L.G.; Gaichas, S.; Griffis, R.; Halofsky, J.E.; Hyde, K.J.W.; Morelli, T.L.; Morisette, J.T.; Muñoz, R.C.; et al. Climate change effects on biodiversity, ecosystems, ecosystem services, and natural resource management in the United States. *Sci. Total Environ.* **2020**, *733*. [CrossRef]
7. IPCC. Climate Change 2013: The Physical Science Basis. In *Contribution of Working Group I to the Fifth Assessment Report of the Intergovernmental Panel on Climate Change*; IPCC: Geneva, Switzerland, 2013.
8. Perdrial, J.; Brooks, P.D.; Swetnam, T.; Lohse, K.A.; Rasmussen, C.; Litvak, M.; Harpold, A.A.; Zapata-Rios, X.; Broxton, P.; Mitra, B.; et al. A net ecosystem carbon budget for snow dominated forested headwater catchments: Linking water and carbon fluxes to critical zone carbon storage. *Biogeochemistry* **2018**, *138*, 225–243. [CrossRef]
9. Bellassen, V.; Luyssaert, S. Managing forests in uncertain times. *Nature* **2014**, *506*, 153–155. [CrossRef]
10. Nabuurs, G.; Schelhaas, M. Carbon profiles of typical forest types across Europe assessed with CO2FIX. *Ecol. Indic* **2002**, *1*, 213–223. [CrossRef]
11. Harmon, M.E.; Bond-Lamberty, B.; Tang, J.; Vargas, R. Heterotrophic respiration in disturbed forests: A review with examples from North America. *J. Geophys. Res. Biogeosci.* **2011**, *116*, 1–17. [CrossRef]
12. Amiro, B.D.; Barr, A.G.; Barr, J.G.; Black, T.A.; Bracho, R.; Brown, M.; Chen, J.; Clark, K.L.; Davis, K.J.; Desai, A.R.; et al. Ecosystem carbon dioxide fluxes after disturbance in forests of North America. *J. Geophys. Res.* **2010**, *115*, G00K02. [CrossRef]
13. Kurz, W.A.; Dymond, C.G.; Stinson, G.J.; Rampley, G.J.; Neilson, E.T.; Carroll, A.L.; Ebata, T.; Safranyik, L. Mountain pine beetle and forest carbon feedback to climate change. *Nature* **2008**, *452*, 987–990. [CrossRef] [PubMed]
14. Niu, S.; Fu, Z.; Luo, Y.; Stoy, P.C.; Keenan, T.F.; Poulter, B.; Zhang, L.; Piao, S.; Zhou, X.; Zheng, H.; et al. Interannual variability of ecosystem carbon exchange: From observation to prediction. *Glob. Ecol. Biogeogr.* **2017**, *26*, 1225–1237. [CrossRef]

15. Calderón-Loor, M.; Cuesta, F.; Pinto, E.; Gosling, W.D. Carbon sequestration rates indicate ecosystem recovery following human disturbance in the equatorial Andes. *PLoS ONE* **2020**, *15*, e0230612. [CrossRef] [PubMed]

16. Anderegg, W.R.L.; Martinez-Vilalta, J.; Cailleret, M.; Camarero, J.J.; Ewers, B.E.; Galbraith, D.; Gessler, A.; Grote, R.; Huang, C.; Levick, S.R.; et al. When a Tree Dies in the Forest: Scaling Climate-Driven Tree Mortality to Ecosystem Water and Carbon Fluxes. *Ecosystems* **2016**, *19*, 1133–1147. [CrossRef]

17. Levy-Varon, J.H.; Schuster, W.S.F.; Griffin, K.L. Rapid rebound of soil respiration following partial stand disturbance by tree girdling in a temperate deciduous forest. *Oecologia* **2014**, *174*, 1415–1424. [CrossRef]

18. Seidl, R.; Schelhaas, M.; Rammer, W.; Verkerk, P. Increasing forest disturbances in Europe and their impact on carbon storage. *Nat. Clim. Chang.* **2014**, *4*, 806–810. [CrossRef]

19. Reichstein, M.; Bahn, M.; Ciais, P.; Frank, D.; Mahecha, M.D.; Senevirante, S.I.; Zscheischler, J.; Beer, C.; Buchmann, N.; Frank, D.C.; et al. Climate extremes and the carbon cycle. *Nature* **2013**, *500*, 287–295. [CrossRef]

20. Hlásny, T.; Barcza, Z.; Fabrika, M.; Balász, B.; Churkina, G.; Pajtík, J.; Sedmák, R.; Turčáni, M. Climate change impacts on growth and carbon balance of forests in Central Europe. *Clim. Res.* **2011**, *47*, 219–236. [CrossRef]

21. Pugh, T.A.M.; Arneth, A.; Kautz, M.; Poulter, B.; Smith, B. Important role of forest disturbances in the global biomass turnover and carbon sinks. *Nat. Geosci.* **2019**, *12*, 730–735. [CrossRef]

22. Klein, D.; Höllerl, S.; Blaschke, M.; Schulz, C. The contribution of managed and unmanaged forests to climate change mitigation-A model approach at stand level for the main tree species in Bavaria. *Forests* **2013**, *4*, 43–69. [CrossRef]

23. Moore, P.T.; DeRose, R.J.; Long, J.N.; van Miegroet, H. Using Silviculture to Influence Carbon Sequestration in Southern Appalachian Spruce-Fir Forests. *Forests* **2012**, *3*, 300–316. [CrossRef]

24. Baldocchi, D.D. Assessing the eddy covariance technique for evaluating carbon dioxide exchange rates of ecosystems: Past, present and future. *Glob. Chang. Biol.* **2003**, *9*, 479–492. [CrossRef]

25. Noe, S.M.; Kimmel, V.; Hüve, K.; Copolovici, L.; Portillo-Estrada, M.; Püttsepp, Ü.; Jõgiste, K.; Niinemets, Ü.; Hörtnagl, L.; Wohlfahrt, G. Ecosystem-scale biosphere-atmosphere interactions of a hemiboreal mixed forest stand at Järvselja, Estonia. *For. Ecol. Manag.* **2011**, *262*, 71–81. [CrossRef]

26. Lindauer, M.; Schmid, H.P.; Grote, R.; Mauder, M.; Steinbrecher, R.; Wolpert, B. Net ecosystem exchange over a non-cleared wind-throw-disturbed upland spruce forest—Measurements and simulations. *Agric. For. Meteorol.* **2014**, *197*, 219–234. [CrossRef]

27. Cleary, M.B.; Naithani, K.J.; Ewers, B.E.; Pendall, E. Upscaling CO_2 fluxes using leaf, soil and chamber measurements across successional growth stages in a sagebrush steppe ecosystem. *J. Arid. Environ.* **2015**, *121*, 43–51. [CrossRef]

28. Wang, K.; Liu, C.; Zheng, X.; Pihlatie, M.; Li, B.; Haapanala, S.; Vesala, T.; Liu, H.; Wang, Y.; Liu, G.; et al. Comparison between eddy covariance and automatic chamber techniques for measuring net ecosystem exchange of carbon dioxide in cotton and wheat fields. *Biogeosciences* **2013**, *10*, 6865–6877. [CrossRef]

29. Luo, X.; Chen, J.M.; Liu, J.; Black, T.A.; Croft, H.; Staebler, R.; McCaughey, H. Comparison of big-leaf, two-big-leaf, and two-leaf upscaling schemes for evapotranspiration estimation using coupled carbon-water modeling. *J. Geophys. Res. Biogeosci.* **2018**, *123*, 207–225. [CrossRef]

30. Kostka, Z.; Holko, L. Role of forest in hydrological cycle—Forest and runoff. *Meteorol. J.* **2006**, *9*, 143–148.

31. Andreassian, V. Waters and forests: From historical controversy to scientific debate. *J. Hydrol.* **2004**, *291*, 1–27. [CrossRef]

32. Ben-Hur, M.; Fernandez, C.; Sakkola, S.; Santamarta-Cerezal, J.C. Overland flow, soil erosion and stream water quality in forest under diferent pertubations and climate conditions. In *Forest Management and Water Cycle*; Bredemeier, M., Cohen, S., Godbold, D., Lode, E., Pichler, V., Eds.; Springer: Berlin/Heidelberg, Germany, 2011; pp. 263–289.

33. Bartík, M.; Holko, L.; Jančo, M.; Škvarenina, J.; Danko, M.; Kostka, Z. Influence of mountain spruce forest dieback on snow accumulation and melt. *J. Hydrol. Hydromech.* **2019**, *67*, 59–69. [CrossRef]

34. Bosch, J.M.; Hewlett, J.D. A review of catchment ex-periments to determine the effect of vegetation changes on water yield and evapotranspiration. *J. Hydrol.* **1982**, *55*, 3–23. [CrossRef]

35. Best, A.; Zhang, L.; McMahon, T.; Western, A.; Vertessy, R. A Critical Review of Paired Catchment Studies with Reference to Seasonal Flows and Climatic Variability. In *CSIRO Land and Water Technical Report 25/03*; Murray Darling Basin Commission: Canberra, Australia, 2003; p. 56. ISBN 1 876 830 57 3.

36. Jones, J.A.; Post, D.A. Seasonal and successional streamflow response to forest cutting and regrowth in the northwestern and eastern United States. *Water Res.* **2004**, *40*, W05203. [CrossRef]

37. Stednick, J.D. Monitoring the effects of timber harvest on annual water yield. *J. Hydrol.* **1996**, *176*, 79–95. [CrossRef]

38. Alila, Y.; Kuraś, P.K.; Schnorbus, M.; Hudson, R. Forests and floods: A new paradigm sheds light on age-old controversies. *Water Resour. Res.* **2009**, *45*, W08416. [CrossRef]

39. The Nature Conservancy. Indicators of Hydrologic Alteration Version 7.1 User's Manual. *Nat. Conserv.* **2009**. Available online: https://www.conservationgateway.org/Files/Pages/indicators-hydrologic-altaspx47.aspx (accessed on 1 March 2014).

40. Olden, J.D.; Poff, N.L. Redundancy and the choice of hydrologic indices for characterizing streamflow regimes. *River Res. Appl.* **2003**, *19*, 101–121. [CrossRef]

41. Gao, Y.; Vogel, R.M.; Kroll, C.N.; Olden, J.D. Development of representative indicators of hydrologic alteration. *J. Hydrol.* **2009**, *374*, 136–147. [CrossRef]

42. Vogel, R.M.; Sieber, J.; Archfield, S.A.; Smith, M.P.; Apse, C.D.; Huber-Lee, A. Relations among storage, yield, and instream flow. *Water Resour. Res.* **2007**, *43*, W05403. [CrossRef]

43. Holeksa, J.; Zielonka, T.; Zywiec, M.; Fleischer, P. Identifying the disturbance history over a large area of larch-spruce mountain forest in Central Europe. *For. Ecol. Manag.* **2016**, *361*, 318–327. [CrossRef]

44. Fleischer, P.; Homolová, Z. Long-term ecological research in larch-spruce forest community after natural disturbances in the Tatra Mts. *Lesn. Cas. For. J.* **2011**, *57*, 237–250.

45. Holko, L.; Fleischer, P.; Novák, V.; Kostka, Z.; Bičárová, S.; Novák, J. Hydrological Effects of a Large Scale Windfall Degradation in the High Tatra Mountains, Slovakia. In *Management of Mountain Watersheds*; Křeček, J., Haigh, M., Hofer, T., Kubin, E., Eds.; Springer: Berlin/Heidelberg, Germany, 2012; pp. 164–179.

46. Holko, L.; Hlavatá, H.; Kostka, Z.; Novák, J. Hydrological regimes of small catchments in the High Tatra Mountains before and after extraordinary wind-induced deforestation. *Folia Geogr. Ser. Geogr. Phys.* **2009**, *15*, 33–44.

47. Holko, L.; Kostka, Z.; Novák, J. Estimation of groundwater recharge, water balance of small catchments in the High Tatra Mountains in hydrological year 2008. In *Sustainable Development and Bioclimate: Reviewed Conference Proceedings*; Pribullová, A., Bičárová, S., Eds.; Geophysical Institute of the Slovak Academy of Sciences: Bratislava, Slovakia, 2009; pp. 93–94.

48. Holko, L.; Škoda, P. Assessment of runoff changes in selected catchments of the High Tatra Mountains ten years after the windthrow. *Acta Hydrolog. Slovaca* **2016**, *17*, 43–50. (In Slovak)

49. Rasband, W.S. Image J. U.S. National Institutes of Health, Bethesda, Maryland, USA. Available online: https://imagej.nih.gov/ij/ (accessed on 30 September 2015).

50. Pajtík, J.; Konôpka, B.; Lukac, M. Biomass functions and expansion factors in young Norway spruce (*Picea abies* [L.] Karst) trees. *For. Ecol. Manag.* **2008**, *256*, 1096–1103. [CrossRef]

51. Pajtík, J.; Konôpka, B.; Bošela, M.; Šebeň, V.; Kaštier, P. Modelling forage potential for red deer: A case study in post-disturbance young stands of rowan. *Ann. For. Res.* **2015**, *58*, 91–107. [CrossRef]

52. Pajtík, J.; Konôpka, B.; Šebeň, V.; Michelčík, P.; Fleischer, P. Biomass allocation of common larch in the first age class in the High Tatra Mts. *Stud. Tatra Natl. Park* **2015**, *44*, 237–249.

53. Pajtík, J.; Konôpka, B. Quantifying edible biomass on young Salix caprea and Sorbus aucuparia trees for Cervus elaphus: Estimates by regression models. *Austrian J. For. Sci.* **2015**, *132*, 61–80.

54. Fleischer, P.; Fleischer, P., Jr.; Gömöryová, E.; Celer, S. Upscaling carbon fluxes from chamber measurement to the landscape-scale in spruce forest disturbed by windthrow in the Tatra Mts. In *Landscape and Landscape Ecology, Proceedings of the 17th int symposium on Landscape Ecology*; Halada, L., Bača, A., Boltižiar, M., Eds.; Institute of Landscape Ecology, Slovak Academy of Sciences: Nitra, Slovakia, 2016; pp. 227–235.

55. Drewitt, G.B.; Black, T.A.; Nesic, Z.; Humphreys, E.R.; Jork, E.M.; Swanson, R.; Ethier, G.J.; Griffis, T.; Morgenstern, K. Measuring forest floor CO_2 fluxes in a Douglas-fir forest. *Agric. For. Meteorol.* **2002**, *110*, 299–317. [CrossRef]

56. Fleischer, P.; Fleischer, P., Jr.; Celer, S. Carbon balance in spruce forest affected by wind and insect disturbances in the Tatra Mts: Methodological approach and preliminary results. In *Aktuálne Problémy v Ochrane Lesa*; Kunca, A., Ed.; National Forest Center: Zvolen, Slovakia, 2013; pp. 113–120. (In Slovak)

57. Marek, M.V. *Carbon in Ecosystems under Changing Climate in Czech Republic*; Academia Praha: Prague, Czech Repulic, 2011. (In Czech)

58. Ryan, M.G.; Lavigne, M.B.; Gower, S.T. Annual carbon cost of autotrophic respiration in boreal forest ecosystems in relation to species and climate. *J. Geophys. Res.* **1997**, *102*, 28871–28883. [CrossRef]

59. Buchmann, N. Biotic and abiotic factors controlling soil respiration rates in *Picea abies* stands. *Soil. Biol. Biochem.* **2000**, *32*, 1625–1635. [CrossRef]

60. Knohl, A.; Søe, A.R.B.; Kutsch, W.L.; Göckede, M.; Buchmann, N. Representative estimates of soil and ecosystem respiration in an old beech forest. *Plant Soil* **2008**, *302*, 189–202. [CrossRef]

61. Searsy, J.K. *Flow-Duration Curves*; Geological Survey Water-Supply-Papet 1542-a: Washington, DC, USA, 1959; p. 33.

62. Curtis, P.S.; Gough, C.M. Forest aging, disturbance and the carbon cycle. *N. Phytol.* **2018**, *219*, 1188–1193. [CrossRef] [PubMed]

63. Yuan, J.; Jose, S.; Hu, Z.; Pang, J.; Hou, L.; Zhang, S. Biometric and eddy covariance methods for examining the carbon balance of a larix principis-rupprechtii forest in the Qinling Mountains, China. *Forests* **2018**, *9*, 67. [CrossRef]

64. Rebane, S.; Jõgiste, K.; Kiviste, A.; Stanturf, J.A.; Metslaid, M. Patterns of carbon sequestration in a young forest ecosystem after clear-cutting. *Forests* **2020**, *11*, 126. [CrossRef]

65. Amiro, B.D.; Barr, A.G.; Black, T.A.; Iwashita, H.; Kljun, N.; McCaughey, H.; Morgenstern, K.; Murayama, S.; Nesic, Z.; Orchansky, A.L.; et al. Carbon, energy and water fluxes at mature and disturbed forest sites, Saskatchewan, Canada. *Agric. For. Meteorol.* **2006**, *136*, 237–251. [CrossRef]

66. Lindroth, A.; Lagergren, F.; Grelle, A.; Klemedtsson, L.; Langvall, O.; Weslien, P.; Tuulik, J. Storms can cause Europe-wide reduction in forest carbon sink. *Glob. Chang. Biol.* **2009**, *15*, 346–355. [CrossRef]

67. Giorgi, S.; Fleischer, P.; Gioli, B.; Kolle, O.; Manca, G.; Matese, A.; Zaldei, A.; Ziegler, W.; Cescatti, A.; Miglietta, F.; et al. Carbon Dioxide Fluxes of a Recent Large Windthrow in the High Tatra. Abstract of CEIP Conference. Sissi-Lassithi, Crete. 2006. Available online: http://carboeurope.org/ceip/conference/abstacts/VU_poster/Giorgi) (accessed on 1 November 2006).

68. Lindroth, A.; Grelle, A.; Morén, A.S. Long-term measurements of boreal forest carbon balance reveal large temperature sensitivity. *Glob. Chang. Biol.* **1998**, *4*, 443–450. [CrossRef]

69. Zhu, K.; Zhang, J.; Niu, S.; Chu, C.; Luo, Y. Limits to growrh of forest biomass carbon sink under climate change. *Nat. Commun.* **2020**. [CrossRef]

70. Michalová, Z.; Morrissey, R.C.; Wohlgemuth, T.; Bače, R.; Fleischer, P.; Svoboda, M. Salvage-Logging after Windstorm Leads to Structural and Functional Homogenization of Understory Layer and Delayed Spruce Tree Recovery in Tatra Mts., Slovakia. *Forests* **2017**, *8*, 88. [CrossRef]

71. Pumpanen, J.; Westman, C.J.; Ilvesniemi, H. Soil CO_2 efflux from a podzolic forest soil before and after forest clear-cutting and site preparation. *Boreal Environ. Res.* **2004**, *9*, 199–212.

72. Koster, K.; Puttsepp, U.; Pumpanen, J. Comparison of soil CO_2 flux between uncleared and cleared windthrow areas in Estonia and Latvia. *For. Ecol. Manag.* **2011**, *262*, 65–70. [CrossRef]

73. Morehouse, K.; Johns, T.; Kaye, J.; Kaye, M. Carbon and nitrogen cycling immediately following bark beetle outbreaks in southwestern ponderosa pine forests. *For. Ecol. Manag.* **2008**, *255*, 2698–2708. [CrossRef]

74. Fleischer, P.; Koreň, M. Selected forest soil properties after the 2004 windfall in the Tatra Mts. In *Sustainable Development and Bioclimate*; Pribullová, A., Bičarová, S., Eds.; Geophysical Institute of the Slovak Academy of Sciences: Bratislava, Slovakia, 2009; pp. 77–78.

75. Moore, D.J.P.; Trahan, N.A.; Wilkes, P.; Quaife, T.; Stephens, B.B.; Elder, K.; Desai, A.R.; Negron, J.; Monson, R.K. Persistent reduced ecosystem respiration after insect disturbance in high elevation forests. *Ecol. Lett.* **2013**, *16*, 731–737. [CrossRef] [PubMed]

76. Hagedorn, F.; Mulder, J.; Jandl, R. Mountain soils under a changing climate and land-use. *Biogeochemistry* **2010**, *97*, 1–5. [CrossRef]

77. Don, A.; Bärwolff, M.; Kalbitz, K.; Andruschkewitsch, R.; Jungkunst, H.F.; Schulze, E.D. No rapid soil carbon loss after a windthrow event in the High Tatra. *For. Ecol. Manag.* **2012**, *276*, 239–246. [CrossRef]

78. Valentini, R. EUROFLUX: An integrated network for studying the long-term responses of biospheric exchanges of carbon, water, and energy of European forests. Fluxes Carbon. *Water Energy Eur. For.* **2003**, *163*, 1–8.

79. Grünwald, T.; Bernhofer, C. A decade of carbon, water and energy flux measurements of an old spruce forest at the Anchor Station Tharandt. *Tellus Ser. B Chem. Phys. Meteorol.* **2007**, *59*, 387–396. [CrossRef]

80. Etzold, S.; Ruehr, N.K.; Zweifel, R.; Dobbertin, M.; Zingg, A.; Pluess, P.; Häsler, R.; Eugster, W.; Buchmann, N. The Carbon Balance of Two Contrasting Mountain Forest Ecosystems in Switzerland: Similar Annual Trends, but Seasonal Differences. *Ecosystems* **2011**, *14*, 1289–1309. [CrossRef]

81. Taylor, A.R.; Seedre, M.; Brassard, B.W.; Chen, H.Y.H. Decline in Net Ecosystem Productivity Following Canopy Transition to Late-Succession Forests. *Ecosystems* **2014**, *17*, 778–791. [CrossRef]

82. Luyssaert, S.; Schulze, E.D.; Börner, A.; Knohl, A.; Hessenmöller, D.; Law, B.E.; Ciais, P.; Grace, J. Old-growth forests as global carbon sinks. *Nature* **2008**, *455*, 213–215. [CrossRef]

83. Yi, C.; Li, R.; Wolbeck, J.; Xu, X.; Nilsson, M.; Aires, L.; Albertson, J.D.; Ammann, C.; Arain, M.A.; De Araujo, A.C.; et al. Climate control of terrestrial carbon exchange across biomes and continents. *Environ. Res. Lett.* **2010**, *5*. [CrossRef]

84. Kulmala, L.; Pumpane, J.; Hari, P.; Vesala, T. Photosynthesis of ground vegetation in different aged pine forests: Effect of environmental factors predicted with a process-based model. *J. Veg. Sci.* **2011**, *22*, 96–110. [CrossRef]

85. Górnik, M.; Holko, L.; Pociask-Karteczka, J.; Bičárová, S. Varibility of precipitation and runoff in the entire High Tatra Mountains in the period 1961–2010. *Prace Geograficzne* **2017**, *151*, 53–74. [CrossRef]

86. Holko, L.; Sleziak, P.; Danko, M.; Bičárová, S.; Pociask-Karteczka, J. Analysis of changes in hydrological cycle of a pristine mountain catchment. 1. Water balance components and snow cover. *J. Hydrol. Hydromech.* **2020**, *68*, 180–191. [CrossRef]

Publisher's Note: MDPI stays neutral with regard to jurisdictional claims in published maps and institutional affiliations.

Article

Long-Term Temporal Changes of Precipitation Quality in Slovak Mountain Forests

Jozef Mind'aš [1,2,*], Miriam Hanzelová [3], Jana Škvareninová [4], Jaroslav Škvarenina [3,*], Ján Ďurský [2] and Slávka Tóthová [5]

[1]   Institute of Ecology and Environmental Sciences, University of Central Europe, Kráľovská 11, 90901 Skalica, Slovakia

[2]   Ecological & Forestry Research Agency EFRA, T. G. Masaryka 8041, 96053 Zvolen, Slovakia; jan.dursky@azet.sk

[3]   Department of Natural Environment, Faculty of Forestry, Technical University in Zvolen, T. G. Masaryka 24, 96000 Zvolen, Slovakia; mirowka@gmail.com

[4]   Department of Applied Ecology, Faculty of Ecology and Environmental Sciences, Technical University in Zvolen, T. G. Masaryka 24, 96000 Zvolen, Slovakia; skvareninova@tuzvo.sk

[5]   National Forest Center, T. G. Masaryka 22, 960 01 Zvolen, Slovakia; tothova@nlcsk.org

*   Correspondence: jozefmindassk@gmail.com (J.M.); skvarenina@tuzvo.sk (J.Š.); Tel.: +421 915 552 744 (J.M.); Tel.: +421-455-206-209 (J.Š.)

Received: 14 September 2020; Accepted: 14 October 2020; Published: 19 October 2020

Abstract: The paper is focused on the evaluation of long-term changes in the chemical composition of precipitation in the mountain forests of Slovakia. Two stations with long-term measurements of precipitation quality were selected, namely the station of the EMEP (European Monitoring and Evaluation Programme) network Chopok (2008 m a.s.l.) and the station of the ICP Forests (International Co-operative Programme on Assessment and Monitoring of Air Pollution Effects on Forests) network Poľana-Hukavský grúň (850 m a.s.l.). All basic chemical components were analyzed, namely sulfur (S-SO4), nitrogen (N-NH4, N-NO3), and base cations (Ca, Mg, and K) contained in precipitation. The time changes of the individual components were statistically evaluated by the Mann–Kendall test and Kruskal–Wallis test. The results showed significant declining trends for almost all components, which can significantly affect element cycles in mountain forest ecosystems. The evaluated forty one-year period (1987 to 2018) is characterized by significant changes in the precipitation regime in Slovakia and the obtained results indicate possible directions in which the quantity and quality of precipitation in the mountainous areas of Slovakia will develop with ongoing climate change.

Keywords: precipitation quality; mountain forests; long-term changes; acid components; base cations

---

1. Introduction

The Slovak Republic is a party of the UN ECE (United Nations Economic Commission for Europe) Convention on Long-range Transboundary Air Pollution. The Implementing Protocols have been progressively adopted to this Convention which, among other things, has been designated by the parties to the Convention to reduce the anthropogenic emissions of pollutants involved in global environmental problems. The Global Environment Report in Europe [1], published by the European Environment Agency, indicates that emissions of acidifying substances have declined significantly since 1990, mainly in Central and Eastern Europe due to economic restructuring. The reduction in Western Europe is mainly related to changes in fuel use, desulfurization, and denitrification of combustion gases and the introduction of three-way catalytic converters in cars. Due to significant emission reductions in most European ecosystems, there is no further acidification, but there are a number of risk areas, especially in Central Europe [2–4].

The need of evaluation of the extent and distribution of pollutants over Europe initiated "The co-operative programme for monitoring and evaluation of the long-range transmission of air pollutants in Europe" (unofficially "European Monitoring and Evaluation Programme" = EMEP). The main objective is to provide governments/decision makers with information of the deposition and concentration of air pollutants, as well as the quantity and the significance of the long-range transmission of air pollutants and their fluxes across boundaries [5,6].

The main source of SOx emissions are large point sources from combustion in energy and transformation industries (56%), but its share decreased by 4% compared to 2012 [7]. Within the last two decades a distinct decrease in emissions in Czech Republic, Slovakia, and Poland has been achieved due to improvement of heat and power production technology, implementation of measures aimed at reducing sulfur emissions from power stations that use brown coal, increased use of low-sulfur fuels in private households and emission control at national and European level [8,9]. Some studies show that precipitation containing a higher proportion of sulfates are more toxic than precipitation at the same pH level but containing a higher proportion of nitrates [10,11].

In Europe, nitrogen oxides (NOx-N) are emitted mainly from stationary combustion sources (power plants and industrial processes) and transportation sources (road, off-road, and ship traffic). NOx-N emissions increased in the 1980s due to increased traffic and started to decrease in the mid-1990s [2,11]. The road transport sector represents the largest source of NOx emissions, accounting for 39% of total EU-28 emissions in 2013 [1,11].

More than 95% of all NH3 emissions come from agriculture, livestock production, and animal waste management. NH3 emissions from the use of artificial nitrogen fertilizers are also significant. NH3 emissions from energy/industry and transport are less significant. NH3 emissions from industry mainly come from the production of nitric acid. NH3 emissions from transport come mainly from road transport. In terms of long-term development, the overall NH3 emissions decrease. The decline in NH3 emissions in the "Agriculture" sector is due to the combined effect of reduced livestock numbers across Europe (especially cattle), changes in the handling and management of organic manures and the abatement of nitrogenous fertilizers [1,12].

The main goal of the article was to analyze the significance of temporal changes of selected parameters of the chemical composition of rainwater in two mountain localities in Slovakia. The aim was to determine the statistical significance of the decrease/increase in the concentrations of selected chemical components in precipitation during the evaluated time period as well as to determine the significance of the differences between the two evaluated localities.

2. Materials and Methods

2.1. Site Description

Chopok is the SHMI (Slovak HydroMeteorological Institute) meteorological observatory located on the ridge of the Low Tatras Mts., at an altitude of 2008 m, (48°56′ N, 19°35′ E) (Figure 1). Measurements started in 1977. Since 1978 it has been part of the EMEP network and the GAW(Global Atmosphere Watch)/BAPMoN(Background Air Pollution Monitoring Network)/WMO(World Meteorological Organisation) network. Chopok is located in a cold climatic zone, C3—cold mountain (July mean temperature below 10 °C). The long-term mean annual precipitation total (1951–1980) is 1142 mm and of which 667 mm is the summer half-year. The mean annual temperature (1951–1980) is −1.2 °C, in the vegetation period (April–September) 3.6 °C. A more detailed climatic characteristic is given by the climate chart (Figure 2) [13].

Research plot Polana Hukavský grúň is located in the territory of the Polana biosphere reserve at an altitude of 850 m (48°38′ N, 19°29′ E) (Figure 1). The mean annual precipitation total (1951–1980) is 853 mm and of which 494 mm in the summer half-year. The mean annual temperature (1951–1980) is

5.8 °C, during the vegetation period (April–September) 11.9 °C. A more detailed climatic characteristic is given by the climate chart (Figure 2).

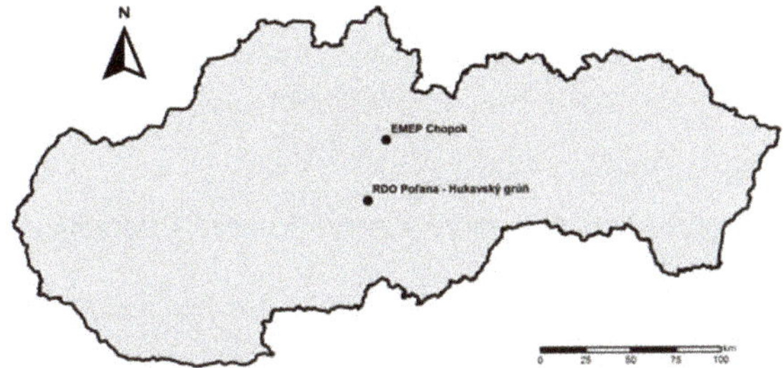

Figure 1. Location of the mountain monitoring stations EMEP (European Monitoring and Evaluation Programme) Chopok and RDO (Research Demonstration Object) Poľana.

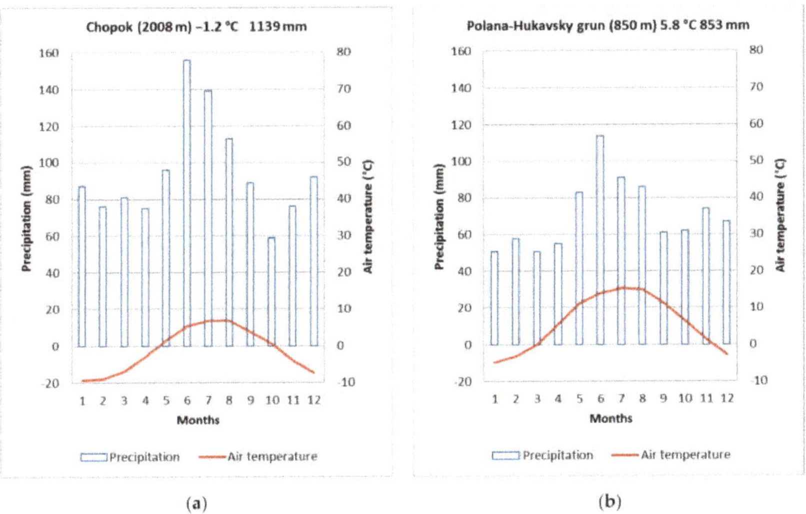

(a)   (b)

Figure 2. Climate diagrams for (**a**) Chopok and (**b**) Polana-Hukavsky grun.

2.2. Precipitation Quality Data

Source data for this article has been obtained for Chopok public data available in the EMEP database. For Polana, the data obtained from the Polana-Hukavsky Grúň Research and Demonstration Facility (VDO) was established by the Forest Research Institute in Zvolen in the spring of 1991. Since 1992, the basic components of precipitation quality have been monitored. It takes place in the weekly measurement and sampling cycle during the summer and the two-week cycle in the winter period is observed in 80% of cases.

The design of precipitation water sampling and the implementation of chemical analyzes of precipitation were performed in accordance with the EMEP manual [14] for the Chopok station and the ICP Forests manual [15] for the Polana-Hukavsky grun station. Chemical analyzes were performed in accredited laboratories of SHMI (Chopok) and Forest Research Institute in Zvolen (Polana) in accordance with the cited manuals. Both laboratories have been involved in international ring testing programs and related quality programs (quality assurance/quality control).

At the Chopok station, total precipitation was measured with a standard METRA rain gauge with wind protection with a collected area of 500 cm^2, samples for chemical analyzes were collected into plastic PET containers with a collected area of 500 cm^2, in a wet-only version for summer period and bulk version for winter period. At the Polana site, precipitation totals were measured with a Hellman's rain gauge with a collected area of 200 cm^2, samples for chemical analyzes were collected into PET bulk collectors (3 pieces with an individual collected area of 200 cm^2). The precipitation quality measurements at Polana site have been changed due to organizational reasons and therefore the statistical analyzes have been performed up to 2014 year.

2.3. Data Analysis and Statistical Methods

The sulfates, nitrates, and ammonium have been recalculated according to the atomic weights to sulfur and nitrogen concentrations.

In the evaluation, individual statistical characteristics are evaluated for annual mean values at monitored stations. Annual weighted means of concentrations shall be assessed. For detecting and estimating trends of annual means have been used the non-parametric Mann–Kendall Test. It is the test of a monotonic increasing or decreasing trend. It is standard method when occur missing values and when data are not normally distributed. Sen's method can be used in cases where the trend can be assumed to be linear. It is an estimator used to quantify the magnitude of potential trends. Thus, Sen's slope is used to estimate the percent reduction in the concentration level while the Mann–Kendall test is used to indicate the significance level of the trend [5,6,16,17]. This test has been used for various studies on long-term environmental and economic impacts [18–21].

Decrease of concentration was calculated from differences of the first three and the last three years averages (depending on availability of measured data) [8]. The rate of decrease of the concentration of the analyzed element in precipitation was expressed on the basis of the regression coefficient "a" of the linear trend from the equation y = a.x + b.

The non-parametric Kruskal–Wallis test was performed by comparing the weighted averages of the stated values from the open areas between the stations Chopok and Poľana.

3. Results

3.1. Precipitation

In the monitored period 1978–2018, the annual precipitation total was recorded at the Chopok regional station, which varied in the range of 840–1590 mm, as pointed out in Figure 3. The maximum was recorded in 2014. From the statistical assessment by the Mann–Kendall test and the graphical representation of the values, it is clear that in the long run there is a slight increase in precipitation totals at a rate of 8.8 mm of atmospheric precipitation per year.

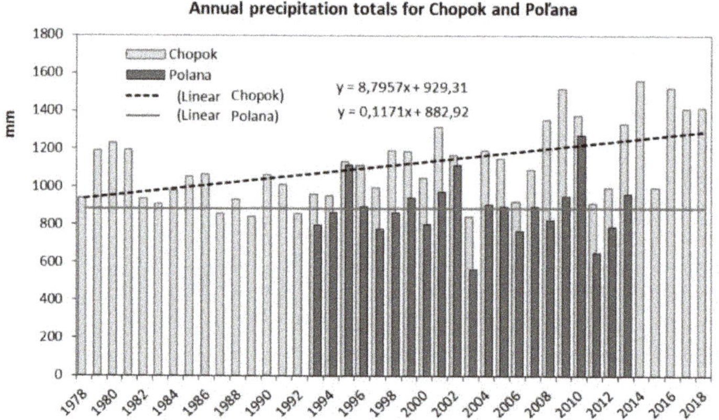

Figure 3. Annual precipitation totals for Chopok and Polana.

Total precipitation has been recorded in Pol'ana since 1993. During the analyzed 21-year period, precipitation totals did not show large fluctuations. The maximum total precipitation was recorded in 2010 at the level of 1270 mm. On average, precipitation at a given locality reaches 886 mm per year. Statistical analysis of precipitation total data obtained from this locality did not confirm the significance of the time trend. Using the Kruskal–Wallis test, we proved a statistically significant difference between these two localities in the years from 1993 to 2018 (see Table 1).

Table 1. Statistical differences between stations Chopok and Pol'ana based on concentrations of elements in precipitation.

| | Kruskal–Wallis Test | | Chopok–Pol'ana | |
Component	Time Period	Count	p-Value	Significance
Precipitation	1993–2013	21	<0.0001	***
pH	1995–2013	21	0.649	NS
H+	1993–2013	21	0.2807	NS
S-SO$_4$	1993–2013	21	0.3022	NS
N-NO$_3$	1993–2013	21	0.252	NS
N-NH$_4$	1995–2013	19	0.3205	NS
Ca	1993–2013	21	0.0019	**
Mg	1995–2013	21	<0.0001	***
K	1995–2015	21	0.0056	**

*** for $p < 0.001$, ** for $p < 0.01$, * for $p < 0.05$, NS for $p \geq 0.05$.

3.2. pH Values and H$^+$ Concentrations

Evaluation of temporal changes of pH values can be quantified only in the case of long-term stable observations performed at the same site. Figure 4 shows changes in pH and H$^+$ values during the observed period for Chopok 1978–2018 and Pol'ana 1993–2013. The Mann–Kendall test confirmed that the acidity of atmospheric precipitation in the observed time period shows a significant trend (see Table 1). At both studied localities, the pH value gradually increases slightly. As the pH increases, the concentration of hydrogen ions H$^+$ decreases naturally. This trend is mainly due to a decrease in the concentrations of major acidifying ions such as sulfate ions, nitrate ions. The decrease of these ions is evident from the following graphical representations (Figure 4).

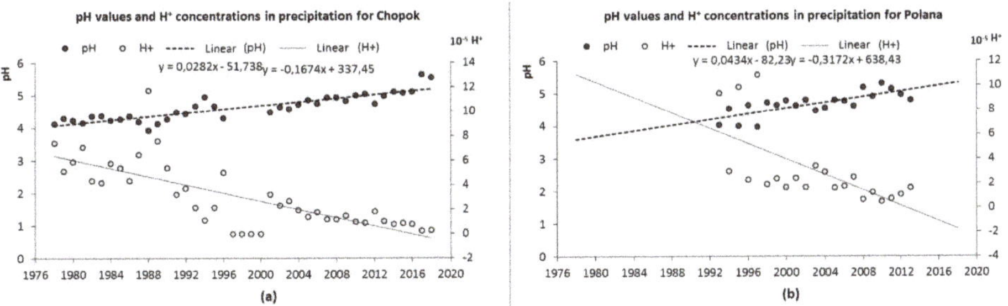

Figure 4. Annual mean weighted pH values and H$^+$ concentrations in precipitation: (**a**) Chopok, (**b**) Polana.

At Chopok, the pH value in annual averages ranges from 3.93 to 5.63. The lowest value is from 1988 and the highest from 2017. The recorded pH values of precipitation in Pol'ana reached their minimum of 3.96 in 1997 and the maximum of 5.4 in 1993. In Pol'ana the statistical significance of the increase in pH values is slightly lower than in Chopok, which can be caused by a shorter reference period. The given localities do not differ statistically significantly from each other in terms of comparison of pH values.

The time series and pH trend over a longer period indicate a decrease in acidity not only in Slovakia, but also in Europe [12,18,22]. The pH values correspond well with the pH values according to the EMEP maps (SHMÚ 2015). At the 22 sites within EMEP with long term pH measurements from 1980 to 2009 the average decrease in H+ concentration was 74% [5]. The increase of precipitation pH value in recent years in bulk precipitation and throughfall has been explained by a greater decrease in acidic anion concentrations [23].

The lowest pH in Europe is observed in the Eastern part of the continent which has relatively high sulfate deposition and a low base cation deposition [5]. Sicard et al. [24] published the pH values in precipitation with a significant decreasing trend of -0.025 ± 0.02 unit pH year^{-1} in France.

3.3. Sulfur and Nitrogen

A basic overview of the weighted annual concentrations of acidic elements (S-SO$_4$, N-NO$_3$, and N-NH$_4$) separately for the station Chopok and Pol'ana is shown in Figure 5. At both monitored stations, sulfate ions dominate in the rainwater. This is confirmed by work [25], that they obtained much higher mean values of sulfur deposition compared to the nitrate-nitrogen deposition—by more than 4-times in case of the open area in mountains in central part of Slovakia.

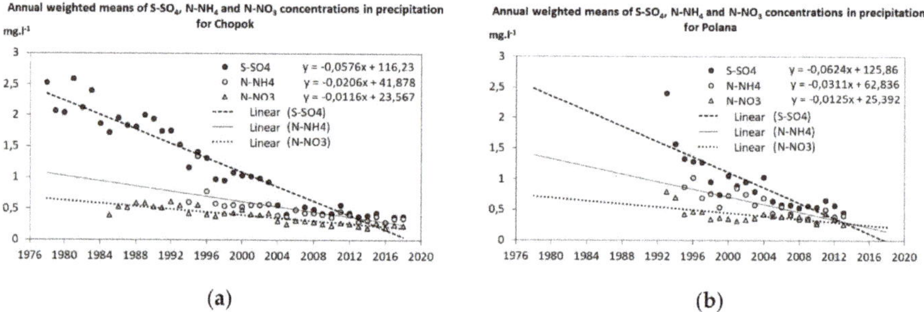

(a) (b)

Figure 5. Annual mean weighted concentrations of acidic elements (S-SO$_4^{2-}$, N-NH$_4^+$, and N-NO$_3^-$) in precipitation: (**a**) Chopok, (**b**) Polana.

For the main air pollutants, the largest reductions across the EU-28 (in percentage terms) since 1990 have been achieved for SO$_x$ emissions (which decreased by 87%), followed by CO (−66%), NMVOCs (−59%), NO$_x$ (−54%), and NH$_3$ (−27%) [1].

Sulfate concentrations together with precipitation pH values have been monitored at the EMEP Chopok station for the longest of all monitored elements since 1978. At the Chopok station, the weighted annual mean sulfur concentrations represented the range 0.38–2.59 mg·L^{-1} in the period 1978–2018. The highest value of the concentration of sulfur in sulfates in rainwater was recorded in 1981. The minimum value of the concentration was reached in 2016. In Pol'ana, the weighted annual averages of S-SO$_4$ concentrations ranged from 0.28 to 2.4 mg·l^{-1}. The values of sulfur concentration in SO$_4^{2-}$ in Pol'ana have been slightly higher in the last decade than the values at the Chopok station. The values of sulfate concentrations are characterized by considerable variability, and it is noticeable to see a gradual decrease in their dispersion. Over the last decade, the weighted annual average concentration of S-SO$_4$ has not exceeded 1 mg·l^{-1}.

Sulfate anions contribute the most to the acidity of precipitation, in Chopok for the whole monitored period from 1978–2018 their concentration decreased at a rate −0.0576 mg·l^{-1} per year and in Pol'ana from 1993–2013 by −0.0624 mg·l^{-1} annually. This decreasing trend of S-SO$_4$ concentration is significantly significant in both cases ($p < 0.0001$) (Table 2).

Table 2. Mann–Kendal test analyze and decrease of concentration elements in precipitation.

	Variable	Time Period	Min	Max	Mean	Median	Std. Deviation	Sen's Slope	*p*-Value	Significance	Decrease %
Chopok	Precipitation	1978–2018	840.5	1560.0	1118.4	1064.1	198.002	6.046	0.0079	**	−29.72
	pH	1978–2018	3.930	5.630	4.631	4.640	0.397	0.025	< 0.0001	***	−28.77
	H^+	1978–2018	2.34×10^{-5}	11.75×10^{-5}	3.32×10^{-5}	2.29×10^{-5}	2.58×10^{-5}	0.000	< 0.0001	***	92.99
	$S\text{-}SO_4$	1978–2018	0.280	2.590	1.201	1.030	0.709	−0.063	< 0.0001	***	84.49
	$N\text{-}NO_3$	1985–2018	0.190	0.610	0.381	0.395	0.128	−0.013	< 0.0001	***	53.15
	$N\text{-}NH_4$	1994–2018	0.290	1.340	0.492	0.430	0.208	−0.017	< 0.0001	***	63.97
	Ca	1992–2018	0.090	1.830	0.384	0.200	0.402	−0.032	0.0001	***	77.95
	Mg	1992–2018	0.020	0.330	0.062	0.040	0.068	−0.005	< 0.0001	***	78.95
	K	1992–2018	0.040	0.700	0.156	0.090	0.147	−0.011	< 0.0001	***	79.49
Polana	Precipitation	1993–2013	561.2	1270.4	886.0	892.7	152.4	−0.009	0.0742	NS	13.35
	pH	1993–2013	3.960	5.310	4.680	4.720	0.350	0.023	< 0.0001	***	−18.75
	H^+	1993–2013	0.49×10^{-5}	10.87×10^{-5}	3.00×10^{-5}	1.91×10^{-5}	2.97×10^{-5}	0.000	0.0003	***	84.83
	$S\text{-}SO_4$	1993–2013	0.454	2.403	0.923	0.798	0.462	−0.053	< 0.0001	***	68.45
	$N\text{-}NO_3$	1993–2013	0.264	0.792	0.409	0.385	0.125	−0.008	0.0134	*	45.01
	$N\text{-}NH_4$	1995–2013	0.282	1.022	0.588	0.550	0.207	−0.032	< 0.0001	***	49.99
	Ca	1993–2013	0.318	1.159	0.638	0.600	0.256	−0.028	0.0011	**	55.87
	Mg	1993–2013	0.050	0.295	0.135	0.124	0.064	−0.008	< 0.0001	***	73.64
	K	1993–2013	0.066	1.223	0.410	0.252	0.332	−0.015	0.0571	NS	57.75

*** for $p < 0.001$, ** for $p < 0.01$, * for $p < 0.05$, NS for $p \geq 0.05$; decrease was calculated from differences of the first three and last three year's averages.

The overall decrease in sulfate concentrations in the long-term time series corresponds to the decrease in SO_2 emissions since 1980 (SHMÚ 2018). From 1978 to 2018, sulfate concentrations in rainwater decreased at the Chopok station by 85.0% and at the Poľana station by 68.5% (1993–2013). This fact is in line with results from the monitoring made within EMEP. It shows large reductions in ambient concentrations and deposition of sulfur species during the last decades. Reductions are in the order of 70–90% since the year 1980 and correspond well with reported emission changes. As a result of the large reductions in sulfur concentrations, the acidity of precipitation has decreased across Europe (TØRSETH et al. 2012). According to [1] between 1990 and 2013, SO_x emissions dropped in the Slovakia by 90% and in the EU-28 by 87% [1]. In 1990 the highest sulfur deposition areas were found in the central-east European areas in countries such as Germany, Poland, Czech Republic, and Slovakia. At year 2006 the highest load, although lower than previously, are found in eastern European countries such as Bulgaria, Romania, Serbia, and Bosnia and Herzegovina [4,26].

In France [24] the concentrations of SO_4^{2-} and $nss\text{-}SO_4^{2-}$ (nss—non see salt) concentrations in precipitation have a significant decreasing trend, -3.0 ± 1.6 and $-3.3 \pm 0.6\%$ year^{-1}, respectively, corresponding with the downward trends in SO_2 emissions in France (-3.3% year^{-1}). A good correlation ($R^2 = 0.84$) between SO_2 emissions and $nss\text{-}SO_4^{2-}$ concentrations was obtained. The decreasing trend of NH_4^+ was more significant ($-5.4 \pm 5.2\%$ year^{-1}) than that of NO_3^- ($-1.3 \pm 2.4\%$ year^{-1}). In Latvia, the $SO_4\text{-}S$ ion concentrations in bulk precipitation also showed a significant negative linear trend [23].

Nitrates are involved in precipitation acidity to a lesser extent than sulfates [11,27]. With the measurement and determination of nitrate concentration in precipitation, Chopok began in October 1985 and Polana in 1993. The annual weighted means of nitrate concentrations converted to nitrogen at the Chopok precipitation ranged from 0.19 to 0.61 mg·l^{-1}. The minimum value is from 2014 and the maximum from 1992. In the monthly values, greater variability of nitrate concentrations was recorded. The N-NO3 concentration was 0.79 mg·l^{-1} in the first year of measurement and the minimum value was 0.19 mg·l^{-1} in the year 2014. The weighted annual concentrations of N-NO3 in Polana in the last decade exceeded the monitored values at Chopok.

The results of the analysis of time changes of nitrates concentrations in precipitation showed a statistically significant trend with decreasing tendency, as is evident from Figure 5 and Table 2. The rate of decrease in N-NO3 concentration in both stations is approximately the same, at Chopk is 0.0116 mg·l^{-1} per year and at Poľana -0.0125 mg·l^{-1} per year. The precipitation content of nitrates does not change as fast as the sulfate content.

At Chopok, nitrate concentrations decreased by 41.0% between 1985 and 2018 and 45.01% at Poľana for 1993–2013. Between 1990 and 2013, NO_x emissions dropped in the Slovakia by 65% and in the EU-28 by 54% [1]. The reduction of NO_x emissions in Europe from 1990 to 2009 were mainly caused by a change from burning of coal and gas to nuclear power. In Eastern Europe increased NO_x

emissions from road traffic after 2000. On the other hand, NO_X emissions from traffic in Western European decreased, even though fuel consumption increased [5].

Monitoring of Ammonium ion in rainfall water Chopok started in 1994 and one year later at Pol'ana. At Chopok, the annual weighted mean of $N-NH_4$ concentrations ranged from 0.29 to 1.34 mg·l^{-1} to the maximum value recorded in 1995. The maximum at Pol'ana was measured in 1996 with a value of 1.02 mg·l^{-1} and a minimum in the year 2010 0.28 mg·l^{-1}. A statistically significant trend ($p < 0.0001$) (Table 2) with a decreasing trend for Chopok −0.0206 mg·l^{-1} nitrogen per annum and Polana −0.0311 mg·l^{-1} is observed from the analysis of time changes of ammonia nitrogen concentrations in precipitation at both stations.

3.4. Base Cations (Mg^{2+}, Ca^{2+}, K^+)

A basic overview of the weighted annual concentrations of base cations (Ca, Mg, and K) separately for the station Chopok and Pol'ana is shown in Figure 6. In Chopok, the measurement and determination of the concentration of basic cations in precipitation began in 1992, and a year later the measurement of the concentrations of these cations also began in Pol'ana.

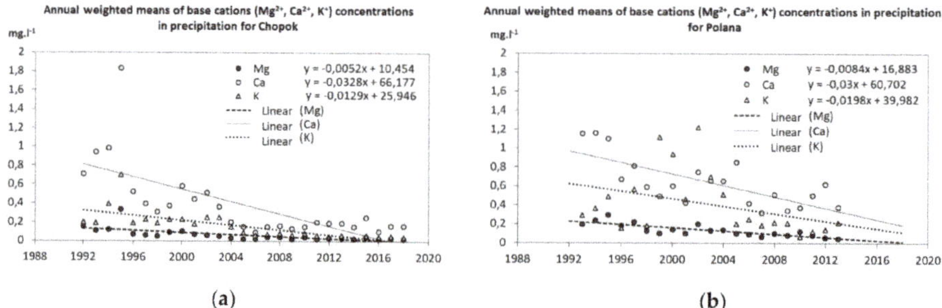

(a)	(b)

Figure 6. Annual mean weighted concentrations of base cations (Mg^{2+}, Ca^{2+}, K^+) in precipitation: (a) Chopok, (b) Polana.

At the Chopok station, for all the basic cations mentioned, the maximum value of the weighted annual concentration was recorded in 1995 for Ca 1.83 mg·l^{-1}, Mg 0.33 mg·l^{-1}, and K 0.7 mg·l^{-1}. A comparison of the concentration of basic cations in precipitation between stations shows mostly higher values at Pol'ana compared to EMEP station Chopok. In this case, since basic cations are the main component of terrestrial dust, the difference in the methods of measuring "wet-only" (rain gauges) vs. "bulk" (rainwater collectors) [28] is evident. In addition, basic cations are produced mainly in the production of building materials, as well as from local dust (roads, stone processing, etc.), which are more localized at lower altitudes [22,29].

From 1992 to 2018, there was an obvious decrease in basic cations in precipitation waters. A more significant decrease is at the Chopok station where the weighted annual average concentration of potassium decreased by 80.0%, magnesium by 86.7%, and calcium by 77.5%. In Pol'ana, within the concentration of the monitored basic cations, magnesium decreased the most by 73.6%, although the decrease of potassium by 57.8% and calcium by 55.9% is not negligible either.

Ca cations have the highest concentration from base cations observed in precipitation. The weighted annual means of calcium concentration in precipitation (Figure 6) at the Chopok station during the observed period varied in the range of 0.09–1.83 mg·l^{-1} and at Pol'ana 0.318–1.159 mg·l^{-1}. The studied localities differ statistically significantly from each other (Table 2). Statistical analysis of the data confirmed a trend that is significantly significant (Table 1). In the case of both stations there is a decreasing tendency of calcium concentrations in precipitation, while at the station Chopok by −0.0328 mg·l^{-1} per year and at Pol'ana by −0.03 mg·l^{-1} per year.

A majority of the EMEP sites showed a decreasing trend of calcium in precipitation with an average decrease of 47% from 1980 to 2009 and 26% from 1990 to 2009. In the early 1990s, the closing of many lignite-fired power stations, iron, and steel smelters as well the implementation of effective abatement technologies for sulfur caused a reduction also in the emissions of base cations [5].

Magnesium concentrations are several times lower than calcium and potassium. When comparing the weighted annual averages of magnesium concentrations between the Poľana and Chopok stations in Figure 6. See the big differences among them. Precipitation from the Poľana station reaches significantly higher magnesium concentrations in almost all monitored years than at Chopok. We confirmed the statistically significant difference ($p < 0.0001$) between the given localities by the Kruskal–Wallis test. Concentrations in the range of 0.05–0.295 mg·l^{-1} were recorded at Poľana and 0.02–0.33 mg·l^{-1} at Chopok. At both stations we can confirm a statistically significant trend ($p < 0.0001$) with a decreasing tendency, in the case of Poľana it is a decrease of -0.0084 mg·l^{-1} and at the station Chopok -0.0052 mg·l^{-1}.

At Chopok station, potassium concentrations change only at a very narrow interval. With regard to the values determined from the precipitation at Poľana, this statement is valid only in the last evaluated years. Values in Poľana stabilized relatively after 2005. Weighted annual potassium concentrations ranged from 0.03 to 0.7 mg·l^{-1} for Chopok and 0.07 to 1.22 mg·l^{-1} for Poľana throughout the period under review. The Mann–Kendal test confirmed a statistically significant trend with a declining character at the Chopok station. In the long run, the trend shows a decrease in potassium concentration by -0.0129 mg·l^{-1} per year. The weighted annual averages of K concentrations in Poľana did not show a statistically significant trend. From Figure 6, however, it can be seen that the concentrations of K reach lower values than in the first half of the observed period. Using the Kruskal–Wallis test, we proved a statistically significant difference between the Chopok and Poľana stations.

4. Discussion

The long-term changes of chemical composition of atmospheric precipitation depends on many factors, but the decisive factors are changes in air pollutant emissions as well as changes in meteorological processes affecting their transformation into a liquid phase in clouds and precipitation [30,31]. The resulting changes are thus an integrated indicator of complex physico-chemical transformations of pollutants from the emission source to the captured precipitation at a specific location [30,32–34]. The results of many observations, especially in the northern hemisphere, point to significant changes in precipitation chemistry, especially in pH values, concentrations of sulfur, nitrogen, but also basic cations. Most notably, these changes are visible at pH values and sulfate concentrations in precipitation water [2,11,24,27,34,35].

Several studies confirm that the primary cause of the decrease of pollutants in atmospheric precipitation is a decrease in sulfur and nitrogen emissions due to the application of measures under the Convention on Long-Range Transboundary Air Pollution—CLRTAP [1,2,7], although the decrease in emissions does not fully explain all observed pollutant trends [3]. Changes in other factors such as changes in air temperature and precipitation, changes in atmospheric circulation, or the occurrence of extreme weather events must also be taken into account [30,32,33].

Trends in decreasing sulfur and nitrogen concentrations in precipitation over the last 2–3 decades have been recorded mainly in Europe and North America, although the intensity of the decrease has been regionally different [2,4,5,23,27,34]. The most significant changes have been identified in Central and Eastern Europe, mainly due to structural changes in energy and industry sectors [1,2,4,5]. Recently studies have emerged on the possible impact of climate change on the further development of acidic and basic components contained in atmospheric precipitation, in particular the impact of changes in precipitation (decrease or increase in precipitation totals), air temperature and atmospheric circulation changes on long-range transmission, and chemical transformation of pollutants in gas and liquid phases [11,12,32,33]. For this reason, it is therefore necessary to maintain the monitoring of rainwater quality as much as possible, especially in localities with their long-term measurement.

Changes in the chemical composition of precipitation also significantly affect the biogeochemical cycles in forest and aquatic ecosystems and can contribute to changes in habitat diversity and biocenoses as well as changes in biomass production (e.g., changes in nitrogen content in atmospheric precipitation). In areas with sufficient precipitation, long-term changes in the chemical composition of precipitation will gradually be reflected in the chemical regime of surface and groundwater [35–37]. The issue of the economic effects of air pollution and precipitation quality, which may take on a different dimension in climate change, must not be forgotten either [38].

5. Conclusions

The presented work deals with the issue of long-term changes in the quality of precipitation in mountain areas of Slovakia at the station EMEP Chopok and VDO Poľana-Hukavský grúň. A 41-year time series of measured data of most of the analyzed elements of precipitation chemistry was available for the Chopok locality. At the Poľana station, the concentrations of elements in precipitation are monitored and are available for a 21-year period. In this work we evaluated the long-term development of concentrations of elements in precipitation (S-SO$_4$, N-NO$_3$, N-NH$_4$, Ca, and Mg, K).

The concentration of sulfur in sulfates in precipitation water decreases significantly at both monitoring stations. At Chopok the decrease represents up to 84.5% decrease compared to the first three years of measurement and at Poľana 68.5% decrease. Nitrogen concentrations in nitrate and ammonium ions decreased with a significant trend at both monitored mountain stations. In both cases, its deposition decreased by almost 50%. Nitrogen concentrations in both nitrate and ammonium ions decreased at approximately the same rate per year at both stations.

For all basic cations, a decrease in their concentration in atmospheric precipitation is observed, with the exception of the potassium cation at the Poľana station, where the trend was not statistically confirmed.

Author Contributions: Conceptualization, J.M. and J.Š.(Jaroslav Škvarenina); methodology, M.H. and J.M.; software and statistics, J.Ď. and M.H.; data management, S.T.; writing—original draft preparation, J.M. and M.H.; writing—review and editing, J.Š.(Jana Škvareninová). All authors have read and agreed to the published version of the manuscript.

Funding: This research received no external funding.

Acknowledgments: This work was accomplished as a part of VEGA projects No.: 1/0500/19, 1/0111/18 of the Ministry of Education, Science, Research, and Sport of the Slovak Republic and the Slovak Academy of Science; and the projects of the Slovak Research and Development Agency No.: APVV-18-0347; APVV-15-0425 and APVV-19-0340. The authors thank the agencies for the support.

Conflicts of Interest: The authors declare no conflict of interest.

References

1. EEA (European Environment Agency) 2015. *European Union Emission Inventory Report 1990-2013 under the UNECE Convention on Long-Range Transboundary Air Pollution (LRTAP). Copenhagen: European Environment Agency, No 8/2015*; European Environment Agency: København K, Denmark, 2015; 125p. [CrossRef]

2. Colette, A.; Aas, W.; Banin, L.; Braban, C.F.; Ferm, M.; Ortiz, A.G.; Ilyin, I.; Mar, K.; Pandolfi, M.; Putaud, J.-P.; et al. *Air Pollution Trends in the EMEP Region between 1990 and 2012. Joint Report of the EMEP Task Force on Measurements and Modelling (TFMM), Chemical Co-Ordinating Centre (CCC), Meteorological Synthesizing Centre-East (MSC-E), Meteorological Synthesizing Centre-West (MSC-W), EMEP: CCC-Report 1/2016*; Norwegian Institute for Air Research: Kjeller, Norway, 2016; ISBN 978-425-2834-6.

3. Fazekasova, D.; Boguská, Z.; Fazekas, J.; Skvareninova, J.; Chovancová, J. Contamination of vegetation growing on soils and substrates in the unhygienic region of Central Spis (Slovakia) polluted by heavy metals. *J. Environ. Biol.* **2016**, *37*, 1335–1340.

4. Keresztesi, Á.; Birsan, M.; Nita, I.A.; Bodor, Z.; Szép, R. Assessing the neutralisation, wet deposition and source contributions of the precipitation chemistry over Europe during 2000–2017. *Environ. Sci. Eur.* **2019**, *31*, 50. [CrossRef]

5. Tørseth, K.; Aas, W.; Breivik, K.; Fjæraa, A.M.; Fiebig, M.; Hjellbrekke, A.G.; Lund Myhre, C.; Solberg, S.; Yttri, K.E. Introduction to the European Monitoring and Evaluation Programme (EMEP) and observed atmospheric composition change during 1972–2009. *Atmos. Chem. Phys.* **2012**, *12*, 5447–5481. [CrossRef]

6. Oulehle, F.; Kopacek, J.; Chuman, T.; Cernohous, V.; Hunová, I.; Hruska, J.; Krám, P.; Lachmanová, Z.; Navrátil, T.; Stepánek, P.; et al. Predicting sulphur and nitrogen deposition using a simple statistical method. *Atmos. Environ.* **2016**, *140*, 456–468. [CrossRef]

7. EMEP 2015. *Transboundary Particulate Matter, Photo-Oxidants, Acidifying and Eutrophying Components, Status Report 1/2015*; Norwegian Meteorological Institute: Oslo, Norway, 2015; ISSN 1504-6192. Available online: http://emep.int/publ/reports/2015/EMEP_Status_Report_1_2015.pdf (accessed on 25 November 2019).

8. Drápela, K.; Drápelová, I. Application of Mann-Kendall test and the Sen´s slope estimates for trend detection in deposition data from Bílý Kříž (Beskydy Mts., the Czech Republic) 1997–2010. *Beskydy* **2011**, *4*, 133–146.

9. Kopáček, J.; Hejzlar, J.; Krám, P.; Oulehle, F.; Posch, M. Effect of industrial dust on precipitation chemistry in the Czech Republic (Central Europe) from 1850 to 2013. *Water Res.* **2016**, *103*, 30–37. [CrossRef] [PubMed]

10. Ashenden, T.W. Efects of Wet Deposited Acidity. In *Air Pollution and Plant Life*; Bell, J.N.B., Treshow, M., Eds.; John Wiley and Sons: Chichester, UK, 2002; pp. 237–250.

11. Liu, Z.; Yang, J.; Zhang, J.; Xiang, H.; Wei, H. A Bibliometric Analysis of Research on Acid Rain. *Sustainability* **2019**, *11*, 3077. [CrossRef]

12. Trentman, M.T. The impact of long-term regional air mass patterns on nutrient precipitation chemistry and nutrient deposition within a United States grassland ecosystem. *J. Atmos. Chem.* **2018**, *75*, 399–410. [CrossRef]

13. Sitár, M.; Hanzelová, M.; Mind'áš, J.; Škvarenina, J. Long-Term Temporal Changes of Precipitation Quality in the Mountainous Region of Chopok (Low Tatras, Slovakia). In *Mendel and Bioclimatology*; Brzezina, J., Ed.; Mendel Unversity and Czech Bioclimatological Society: Brno, Czech, 2016; pp. 306–320.

14. EMEP Manual for Sampling and Chemical Analysis. EMEP/CCC-Report 1/95 Revision 1/2001. EMEP Co-operative Programme for Monitoring and Evaluation of the Long-range Transmission of Air Pollutants in Europe EMEP Manual for Sampling and Chemical Analysis. Norwegian Institute for Air Research, P.O. Box 100, N-2007 Kjeller, Norway. Available online: http://www.nilu.no/projects/ccc/manual/index.html (accessed on 23 August 2020).

15. Ulrich, E.; Mosello, R.; Derome, J.; Derome, K.; Clarke, N.; König, N.; Lövblad, G.; Draaijers, G.P.J.; Zlindra, D. Manual on Methods and Criteria for Harmonized Sampling, Assessment, Monitoring and Analysis of the Effects of air Pollution on Forests. Part VI—Sampling and Analysis of Deposition—Updated 2009. United Nations Economic Commission for Europe, Convention on long-range transboundary air pollution, International Co-operative Programme on Assessment and Monitoring of Air Pollution Effects on Forests (ICP Forests). Available online: https://www.icp-forests.org/pdf/manual/2000/Chapt6_2009-compl.pdf (accessed on 25 August 2020).

16. Gilbert, R.O. *Statistical Methods for Environmental Pollution Monitoring*; Van Nostrand Reinhold: New York, NY, USA, 1987.

17. Salmi, T.; Määttä, A.; Anttila, P.; Ruoho-Airola, T.; Amnell, T. *Detecting Trends of Annual Values of Atmospheric Pollutants by the Mann-Kendall Test and Sen´s Slope Estimates—The Excel Template Application MAKESENS*; Publications of air quality Finnish Meteorological Institute. No 31; Ilmatieteen laitos: Helsinki, Finland, 2002; ISSN 1456-789X. ISBN 951-697-563-1.

18. Chugunkova, A.V.; Pyzhev, A.I. Impacts of Global Climate Change on Duration of Logging Season in Siberian Boreal Forests. *Forests* **2020**, *11*, 756. [CrossRef]

19. Hrvol', J.; Horecká, V.; Škvarenina, J.; Střelcová, K.; Škvareninová, J. Long-term results of evaporation rate in xerothermic Oak altitudinal vegetation stage in Southern Slovakia. *Biologia* **2009**, *64*, 605–609.

20. Vaničková, M.; Stehnová, E.; Středová, H. Long-term development and prediction of climate extremity and heat waves occurrence: Case study for agricultural land. *Contrib. Geophys. Geod.* **2017**, *47*, 247–260. [CrossRef]

21. Zeleňáková, M.; Vido, J.; Portela, M.M.; Purcz, P.; Blištán, P.; Hlavatá, H.; Hluštík, P. Precipitation Trends over Slovakia in the Period 1981–2013. *Water* **2017**, *9*, 922.

22. Dadashazar, H.; Ma, L.; Sorooshian, A. Sources of pollution and interrelationships between aerosol and precipitation chemistry at a central California site. *Sci. Total Environ.* **2019**, *651*, 1776–1787. [CrossRef]

23. Terauda, E.; Nikodemus, O. Sulphate and nitrate in precipitation and soil water in pine forests in Latvia. *Water Air Soil Pollut. Focus* **2007**, *7*, 77–84. [CrossRef]
24. Sicard, P.; Coddeville, P.; Sauvage, S.; Galloo, J.C. Trends in chemical composition of wet-only precipitation at rural French monitoring stations over the 1990–2003 period. *Water Air Soil Pollut. Focus* **2007**, *7*, 49–58. [CrossRef]
25. Dubová, M.; Bublinec, E. Evaluating of sulphur and nitrate-nitrogen deposition to forest ecosystems. *Ekol. Bratisl.* **2006**, *25*, 366–376.
26. Fagerli, H.; Spranger, T.; Posch, M. Chapter 3: Acidification and Eutrophication—Progress Towards the Gothenburg Protocol Target Year 2010. In *Transboundary Acidification, Eutrophication and Ground Level Ozone in Europe since 1990 to 2004*; EMEP Status Report 1/06; Norwegian Meteorological Institute: Oslo, Norway, 2006; Available online: http://emep.int/publ/reports/2006/status_report_1_2006_ch.pdf (accessed on 28 August 2020).
27. Perikhanyan, Y.; Shahnazaryan, G.; Gabrielyan, A. Long term trends of wet deposition and atmospheric concentrations of nitrogen and sulfur compounds at EMEP site in Armenia. *J. Atmos. Chem.* **2020**, *77*, 101–116. [CrossRef]
28. Mind'áš, J. Characteristics of the precipitation and soil water quality in the mature mixed forest stand (fir-spruce-beech) of midmountain region Pol'ana—II. base cations. *Lesn. Čas.–For. J.* **2006**, *52*, 159–173.
29. Mihalíková, K.; Škvarenina, J.; Strelcová, K.; Gömöryová, E. Throughfall chemistry and atmospheric deposition in a Norway spruce–subalpine climax forest in the Pol'ana Biosphere reserve, Slovakia. *Folia Oecol.* **2008**, *35*, 30–39.
30. Szép, R.; Mateescu, E.; Niţă, I.A.; Birsan, M.-V.; Bodor, Z.; Keresztesi, Á. Effects of the Eastern Carpathians on atmospheric circulations and precipitation chemistry from 2006 to 2016 at four monitoring stations (Eastern Carpathians, Romania). *Atmos. Res.* **2018**, *214*, 311–328. [CrossRef]
31. Mind'áš, J.; Škvarenina, J. Chemical composition of fog cloud and rain snow water in Biosphere Reserve Pol'ana. *Ekol. Bratisl.* **1995**, *14*, 125–137.
32. Ballard, T.C.; Sinha, E.; Michalak, A.M. Changes in Precipitation and Temperature Have Already Impacted Nitrogen Loading. *Environ. Sci. Technol.* **2019**, *53*, 5080–5090. [CrossRef]
33. Hou, P.; Wu, S. Long-term Changes in Extreme Air Pollution Meteorology and the Implications for Air Quality. *Sci. Rep.* **2016**, *6*, 23792. [CrossRef] [PubMed]
34. Johansen, A.M.; Duncan, C.; Reddy, A.; Swain, N.; Sorey, M.; Nieber, A.; Agren, J.; Lenington, M.; Bolstad, D.; Samora, B.; et al. Precipitation chemistry and deposition at a high-elevation site in the Pacific Northwest United States (1989–2015). *Atmos. Environ.* **2019**, *212*, 221–230. [CrossRef]
35. Moiseenko, T.I.; Dinu, M.I.; Bazova, M.M.; de Wit, H.A. Long-Term Changes in the Water Chemistry of Arctic Lakes as a Response to Reduction of Air Pollution: Case Study in the Kola, Russia. *Water Air Soil Pollut.* **2015**, *226*, 98. [CrossRef]
36. Báliková, K.; Dobšinská, Z.; Paletto, A.; Sarvašová, Z.; Korená Hillayová, M.; Štěrbová, M.; Výbošťok, J.; Šálka, J. The Design of the Payments for Water-Related Ecosystem Services: What Should the Ideal Payment in Slovakia Look Like? *Water* **2020**, *12*, 1583.
37. Bartík, M.; Jančo, M.; Střelcová, K.; Škvareninová, J.; Škvarenina, J.; Mikloš, M.; Vido, J.; Waldhauserová, P.D. Rainfall interception in a disturbed montane spruce (*Picea abies*) stand in the West Tatra Mountains. *Biologia* **2016**, *71*, 1002–1008. [CrossRef]
38. Zhang, Y.; Li, Q.; Zhang, F.; Xie, G. Estimates of Economic Loss of Materials Caused by Acid Deposition in China. *Sustainability* **2017**, *9*, 488. [CrossRef]

Publisher's Note: MDPI stays neutral with regard to jurisdictional claims in published maps and institutional affiliations.

Article

Density of Seasonal Snow in the Mountainous Environment of Five Slovak Ski Centers

Michal Mikloš [1,*], Jaroslav Skvarenina [1,*], Martin Jančo [2] and Jana Skvareninova [3]

[1]   Department of Natural Environment, Faculty of Forestry, Technical University in Zvolen,
      Ul. T.G. Masaryka 24, 960 53 Zvolen, Slovakia

[2]   Institute of Hydrology, Slovak Academy of Sciences, Dúbravská cesta 9, 841 04 Bratislava, Slovakia;
      martinjanco11@gmail.com

[3]   Faculty of Ecology and Environmental Sciences, Technical University in Zvolen, Ul. T.G. Masaryka 24,
      960 53 Zvolen, Slovakia; skvareninova@is.tuzvo.sk

*   Correspondence: miklosmiso@gmail.com (M.M.); skvarenina@tuzvo.sk (J.S.); Tel.: +421-455-206-209 (J.S.)

Received: 17 August 2020; Accepted: 15 December 2020; Published: 18 December 2020

Abstract: Climate change affects snowpack properties indirectly through the greater need for artificial snow production for ski centers. The seasonal snowpacks at five ski centers in Central Slovakia were examined over the course of three winter seasons to identify and compare the seasonal development and inter-seasonal and spatial variability of depth average snow density of ski piste snow and uncompacted natural snow. The spatial variability in the ski piste snow density was analyzed in relation to the snow depth and snow lances at the Košútka ski center using GIS. A special snow tube for high-density snowpack sampling was developed (named the MM snow tube) and tested against the commonly used VS-43 snow tube. Measurements showed that the MM snow tube was constructed appropriately and had comparable precision. Significant differences in mean snow density were identified for the studied snow types. The similar rates of increase for the densities of the ski piste snow and uncompacted natural snow suggested that the key density differences stem from the artificial (machine-made) versus natural snow versus processes after and not densification due to snow grooming machines and skiers, which was relevant only for ski piste snow. The ski piste snow density increased on slope with decreasing snow depth (18 kg/m^3 per each 10 cm), while snow depth decreased 2 cm per each meter from the center of snow lances. Mean three seasons maximal measured density of ski piste snow was 917 ± 58 kg/m^3 the density of ice. This study increases the understanding of the snowpack development processes in a manipulated mountainous environment through examinations of temporal and spatial variability in snow densities and an investigation into the development of natural and ski piste snow densities over the winter season.

Keywords: snow density; snow tube; technology; artificial snow; ski slope; piste; VS-43; snow depth; snow lances; water balance

1. Introduction

A seasonal snow cover that is deposited and melts annually is a major component of regional and global hydrologic balances due to its effects on energy and moisture budgets [1,2]. To calculate snowmelt, a major source of water from mountainous areas in temperate zones [3], the development of snow water equivalent (SWE) over the winter season shall be monitored [4]. The SWE is possible to identify from the extracted snow core directly by a calibrated scale or indirectly by calculation from the snow depth and density (weight) measurements [5]. However, both methods are time consuming due to snow core extraction and weighing. Mathematical models of snow density development over the winter season and/or models of relationship between snow depth and density could potentially save lot of time and effort by allowing SWE calculations to be made on the basis of snow depth measurements and estimated snow density only [6,7]. Such models have not yet been developed for a ski piste snowpack; thus, doing so is a goal of this study. For this purpose, a series of snow density measurements must be performed via snow core extraction and weight measurements. To extract a gravimetric snow core from a snowpack, a snow tube is used [8]. The snow tube was first popularized by Church [5,9] and is still utilized by snow surveyors in forested, remote, or hilly watersheds [10]. Manual measurements utilizing a snow tube frequently underestimate depth-average snow density and the SWE of a snowpack due to sampling difficulties associated with the high density of ground ice [11]. It is impossible to extract an ice sample or sample through an ice layer when a conventional snow tube is used. Keller et al. [12], when dealing with impact of ski-slope grooming on the snowpack and soil properties, mention that after mid-December, the snow on the ski slope they were investigating was too hard for manual snow-density measurements to be obtained with a snow tube designed for uncompacted natural snow. Therefore, Rixen et al. [13] sampled such snow with a motor-driven SIPRE corer [14] to compare the physical properties of artificial and natural snow; however, this approach is not practical due to construction and operation costs. In addition, motor-driven devices for snow core extraction are much more complicated to construct and operate than snow tubes. However, snow tubes suitable for a groomed ski piste snowpack or a natural snowpack of high density have not yet been developed (an aim of the present paper). Another approach is to use a PICO coring auger [15] designed for ice sampling for the manual extraction of snow samples from a high-density snowpack [16]. The snow density of a ski piste snowpack is essential for the assessment of the environmental impacts of ski piste management [17], for the calculation of the SWE and water balance [18], for defining the snow type [19], and mainly for a quality assessment of the ski piste snowpack for winter sports [20]. Snow metamorphoses, depending on the meteorological conditions, the physical properties of the overlying snow, time on the ground, and mechanical disturbances, can change the snowpack's density [21]. Compared to natural snow cover, the snow on the ski pistes is increased mechanically via snow grooming machines and artificial snow [22]. Federolf et al. [23] observed that groomed ski slopes covered by natural snow have densities ranging from 330 to 660 kg/m^3, while Rixen et al. [24] identified a similar density range for freshly-produced artificial snow. A whole season's worth of systematic measurements of the depth-average density of a groomed ski piste snowpack with artificial snow added has not yet been published (another aim of the present paper).

Ongoing and future changes in forest composition and climatic conditions have and will modify the seasonal distribution of both precipitation and runoff in the pilot region [25,26]. As a consequence, the operability of Central Slovakian ski slopes up until 1000 m a.s.l. is highly dependent on snow production, as shown in our previous study [27]. Winter precipitation and water availability during the skiing season is decreasing, in general [28,29]. Therefore, a high volume of artificial snow is produced at the beginning of each season for the groomed ski pistes of Central Slovakia. However, the snow densities and microstructures, which are used to define snow types [19], of natural and artificial snow differ markedly, both in the case of freshly fallen/produced snow and in the case of older, metamorphic snow on the ground [30,31]. The following snow types were examined in the present article: (i) new natural snow (max. 2-day-old snowpack), (ii) new artificial snow (max. 2-day-old machine-made snowpack), (iii) uncompacted natural snow, and iv) ski piste snow. The main goal of this study was to identify models of temporal and spatial development of ski piste snow density that could allow density estimation to be made on the basis of the term in winter season and / or snow depth measurements. The spatial development of ski piste snow density was analyzed only in one ski center at the end of the winter season; thus, this model can be generalized only for the late snow ablation period. To reach the main goal and fill gaps in the research on the groomed and snowed ski piste snowpack described above, the following sub-objectives were set:

A. Snow tube construction:

- Design and construct a snow tube suitable for depth-average snow density measurements of a groomed ski piste snowpack with artificial snow added.
- Identify the precision of the designed snow tube on ski piste and off-piste sites through comparison of snow density measurements with commonly used VS-43 snow tube.

B. Density of snow at five ski centers:

- Identify and compare the mean, minimal, and maximal densities of ski piste snowpack with uncompacted natural snowpack on a seasonal and monthly scale and do the same for new artificial snow and new natural snow measured at the beginning of the winter season.
- Identify and compare the seasonal and inter-seasonal course densities and variabilities in the densities of ski piste and uncompacted natural snow and describe them with linear mathematical models if applicable to identify slope of increase over the season.

C. Snow density versus snow depth relationship on the example of Košútka ski center:

- Identify the spatial distribution of the snow depth and the spatial variability of the snow density in the ski piste area and analyze the correlations between these two variables and the positions of fixed snow-making lances at the end of the winter season.

2. Materials and Methods

2.1. Study Sites

The study was conducted at five ski centers located in Central Slovakia (Figure 1). The elevation of the ski slopes varied from 510 to 1402 m a.s.l., while all study sites at the ski centers were located at elevations lower than 1000 m a.s.l. (Figure 1). According to the geomorphological classification [32,33], all the ski centers are located in the Inner Western Carpathian sub-province. According to Slovakia's climate classifications [34], the higher ski resorts (Krahule and Donovaly) belong to subregion C1 (moderately cool), while the lower resorts (Košútka and Jasenská) are included in the climate subregions M7 (moderately warm, very humid, highlands) and M6 (moderately warm, humid, highlands), respectively. Based on the annual temperature amplitudes, the lower research plots belong to the transitional maritime climate to moderately continental climate, while, for the higher ski resorts, the temperature amplitude decreases and the oceanity of the climate increases [35]. Further detailed characteristics of the resorts' climatic and natural conditions are given in Table 1, which has been processed according to the Climate Atlas of Slovakia [36] and Hrvol' et al. [37]. As we can see from the Table 1, the number of days with snow cover and the average snow depth increase with increasing altitude [38–40]. The potential natural vegetation is described in terms of Skvarenina et al. [41].

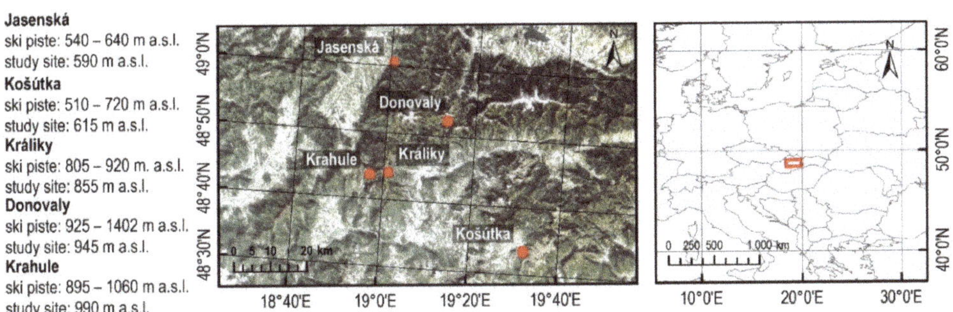

Figure 1. Ski centers' elevations and locations in Central Slovakia and Central Europe.

Table 1. The basic environmental characteristics of five ski centers in Central Slovakia at which the snow density was examined, according to the Climate Atlas of Slovakia [36] where data from 1961 to 2010 were processed.

	Jasenská	Košútka	Králiky	Krahule	Donovaly
Altitude (m a. s. l.)	590	615	855	990	945
GPS	N 49.00976° E 19.00959°	N 48.55909° E 19.53484°	N 48.73626° E 19.01706°	N 48.72824° E 18.94727°	N 48.87313° E 19.22425°
Geomorphological units	Veľká Fatra	Vepor Mountains	Kremnica Mountains	Kremnica Mountains	Staré Hory Mountains
Aspect	N	NW	NE	SW	N
Average annual air temperature (°C)	6.9	6.7	5.2	4.8	4.5
Average winter air temperature (°C)	−2.8	−3.1	−3.9	−4.1	−4.4
Average annual precipitation total (mm)	850	755	1080	1025	1180
Average winter precipitation total (mm)	160	130	240	220	250
Average number of day with snow cover (≥10 cm)	55	48	70	68	87
Average number of day with snow cover (≥20 cm)	35	27	45	47	66
Average number of day with snow cover (≥50 cm)	9	6	10	12	27
Average snow cover depth (cm)	40	37	53	56	65
Climatic sub region	M7	M6	C1	C1	C1
Watercourse/ Drainage basin	Beliansky potok/Váh	Slanec/Hron	Tajovský potok/Hron	Krahulský potok/Hron	Korytnica/Váh
Potential natural vegetation	3th oak–beech stage	4th beech stage	5th fir–beech stage	5th fir–beech stage	6th spruce-beech-fir stage

2.2. Snow Tube

Snow tubes are used to extract gravimetric samples of snow (snow cores) in order to measure the depth-averaged snow density and SWE [8]. The snow tube used on the groomed ski pistes with high-density snow was developed (Table 2, see Figure 2b in Section 3.1) and named the MM snow tube due to the highly used man-made (artificial) snow on the groomed ski slopes and the initials of its designer (first author of this article). To extract the snow core with the designed tube, it was necessary to use a heavy hammer (approx. 3.5 kg). Before removing the tube from the snowpack for weighing, the snow core had to be pressed inside to increase sample adhesiveness. To make it easier to pull it from the snowpack and taking into consideration the possible use of a hanging weight, a stainless-steel hose clamp with hanging rope was placed on the top of the tube. After weighing, the snow core was pushed out of the tube with an aluminum rod with a penny washer (M8 × 30 mm) on one side.

Table 2. Main characteristics of the snow tubes used in this study.

Snow Tube	Cap	Material	Length (mm)	Weight (Cap) (kg)	Inner Ø (mm)	Sampling Time (min.) *
MM	+	Stainless Steel	800	4.18 (0.76)	40	4 ± 2
VS-43	+	Aluminum	600	1.25 (0.17)	84	15 ± 5

* Average sampling time of the 30 paired measurements (mean ± standard deviation).

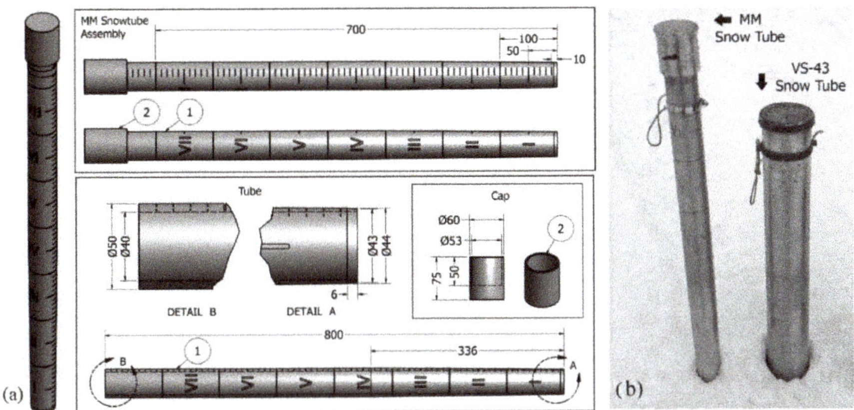

Figure 2. (**a**) Assembly of the MM snow tube, which consisted of a tube and cup (dimensions in mm); (**b**) Commonly used VS-43 snow tube in comparison with the designed tube.

To assess precision and reliability of the MM tube on the groomed ski piste, 30 paired measurements were performed in January 2017 at Králiky. A snow tube that is popular in Europe and Russia named the VS-43 was used for comparison (Table 2) [42]. The assembly of the VS-43 snow tube with its original mechanical scales is illustrated in Figure A1 in Appendix A. The VS-43 snow tube was destroyed during the 30 paired measurements, because the aluminum did not stand up to the hammering. The precision of the new snow tube was assessed on uncompacted natural snow, as well. In this case, 30 paired measurements were taken at Donovaly in March 2017.

2.3. Snow Density

Depth average density of the four following snow types were identified at the five ski centres (Jasenská, Košútka, Králiky, Krahule, Donovaly) across three winter seasons from 2014/15 to 2016/17: (i) new artificial snow (max. two-day-old machine-made snow), (ii) new natural snow (max. two-day-old snowpack), (iii) ski piste snow (groomed snow with artificial snow added), and (iv) uncompacted natural snow on off-piste sites. The densities of all four types of snow were identified via series of five measurements. If natural snow occurred on the off-piste site, then the measurements of the ski piste and natural snowpacks were always paired. Measurements began with the first occurrence of natural snow for the season and carried on over the course of irregular time intervals, which were 17 days long, on average. The density measurements for the ski piste and natural snowpacks were identified on the horizontal transect that intersected groomed ski piste and adjacent off-piste sites with uncompacted natural snow. A series of five measurements were made at groomed ski piste and off-piste sites that were paired according to site properties, such as elevation, slope, aspect, curvature of relief, and obscuration. The densities of uncompacted and new natural snow were sampled with the VS-43 snow tube, while new artificial snow and ski piste snow were sampled with the MM snow tube.

The samples were compared via paired or unpaired Student's t-tests at an alpha level of 0.05 and $n-1$ degrees of freedom. In each case, the proper test was chosen according to the assumptions required for the samples. Both tests assumed that the variables were normally distributed, the observations were sampled independently, and the variables were from two related groups or matched pairs; the unpaired t-test assumed that the variables were from two independent groups with the same variance. The type of t-test used is identified when the results of a case are discussed. Mean values mentioned in results are accompanied with the standard deviation and number of samples (N/n) used for calculation. The total number of measured samples (N) or number of mean values (n) can be found.

2.4. Snow Density Versus Snow Depth

At the end of the winter season on 15 March 2015, when the natural snow on the adjacent off-piste sites has melted away, the snow depth and density of the ski piste snowpack were measured along the whole ski piste at Košútka to analyze the correlations between the snow density and depth variables and the distance from the snow lances. The snow depth was measured at 96 sampling points, while the snow density was measured at 72 of these points. To analyze the spatial distribution of the snow and spatial variability of the snow density, the snow depth, and density raster layers were interpolated from the point measurements in ArcGIS 10.3.1. The raster layers, which consist of matrices of cells organized into rows and columns, where each cell contains a value representing information, such as snow depth or density, as in this paper, were created to provide pictures of these real-world phenomena. The interpolation spline technique was used to create raster layers with 1×1 m cell sizes. The spline technique was chosen, because it guarantees that the resulting surface will pass through the data points exactly, making it ideal for generating gently varying surfaces, such as snowpack surfaces or elevation. The correlations between the snow depth and density raster layers were computed with the Band Collection Statistics tool in ArcGIS 10.3.1. Concentric circles around snow lances were used to analyze the relationship between the snow depth and density [16]. The mean snow density and depth were calculated on concentric circles of radius R = 5, 7.5, 10 … 30 m around each of the 17 snow lances. The maximum radius of 30 m was equal to the maximum distance that water/ice particles emitted from the used snow lances in Košútka ski centre could travel without wind support. This distance depends on the type and technology of used snowmaking machines.

3. Results

3.1. Snow Tube

A 800 mm-long snow tube with a 50-mm outside diameter and 5-mm-thick walls that began tapering 336 mm from the cutting end was constructed (Figure 2a). The wall thickness of the sharpened cutting end was 1.5 mm. The weight of the tube without its cap was 4.2 kg, while the 100 mm of the tube used for construction weighed 0.6 kg. An essential part of the tube was its removable cap. EN 1.4310 (X10CrNi18-8) stainless steel was selected as the construction material due to its hardness and corrosion resistance. The measurement scale was engraved on the outside wall of the tube that was used to measure the snow depth or identify how deep the tube had penetrated.

The paired t-test showed no significant difference (p-value = 0.23) between the samples collected by the MM tube and the commonly used VS-43 snow tube (Figure 3). The average difference between the observations in pairs was 22.2 ± 18.1 kg/m^3, while the MM tube underestimated the snow density compared to the VS-43 tube by about 2.1 ± 3.9%, on average. The VS-43 snow tube (Figure 2b) has a design that is similar to that of the MM tube, but, when it was used on a ski slope, the following disadvantages were noted: (i) long sampling time, (ii) it had difficulty penetrating the snowpack, (iii) it was hard to remove from the snowpack (the sampler usually stuck in the snowpack due to the bigger diameter of the cutting end compared to the diameter of the tube), and (iv) inadequate construction (after a few samples, the sampler was deformed by the hammering). The tendency of the MM snow tube to underestimate the snow density was detected for natural snow as well, and it underestimated the density by about 6.9 ± 2.9% (Figure 3). In the case of uncompacted natural snow, the paired t-test detected a difference between the samples taken by the two snow tubes. Therefore, the MM tube is not recommended for use mainly when the snowpack is shallow and has a low density (new natural snow), possibly due to low sample weights (small tube diameter), which are difficult to detect precisely in the field with a hanging weight.

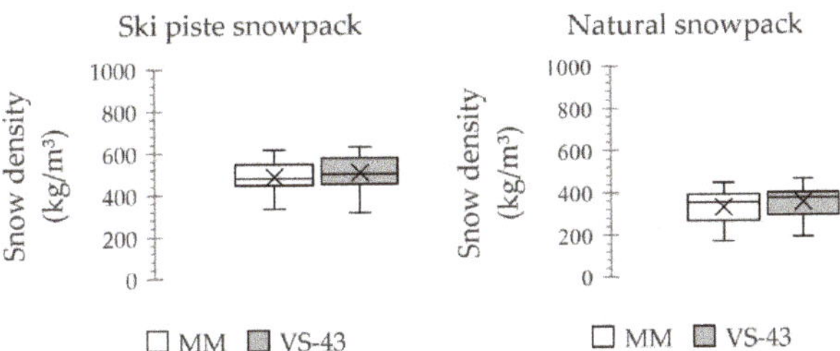

Figure 3. Comparison of 30 paired measurements of the snow densities of ski piste and uncompacted natural snow measured by designed MM snow tube versus the VS-43 snow tube.

3.2. Snow Density

The mean densities of the types of snow were calculated for three seasons with data from the five ski centres. These densities indicated that the mean density of new artificial snow was 2.3 times higher than the mean density of new natural snow (409.6 ± 66.2 kg/m³ versus 175.3 ± 71.6 kg/m³; N = 209 versus 199 samples) and that the mean density of ski piste snow was 2.5 times higher than the density of uncompacted natural snow on the off-piste sites (595.9 ± 101.0 kg/m³ versus 238.5 ± 95.3 kg/m³; N = 722 versus 400 samples; Figure 4). The following is a list of snow types according to their mean densities (in descending order): ski piste snow (mixture of natural and artificial snow), new artificial snow (max. two-day-old machine-made snow), uncompacted natural snow, and new natural snow. The data over the three seasons showed significant differences (unpaired *t*-test: $p < 0.5$) across all four studied snow types. The ski piste snow had greater seasonal variability than the uncompacted natural snow (Figure 4). Mean difference between the maximal and minimal density (except for outliers; Figure 4b) identified in each of five ski centers over the three winter seasons for the types of snow were as follows: (i) 420 ± 87 kg/m³ for ski piste snow, (ii) 328 ± 80 kg/m³ for uncompacted natural snow, (iii) 273 ± 66 kg/m³ for new artificial snow, and (iv) 211 ± 59 kg/m³ for new natural snow. A paired comparison of the mean differences between the maximal and minimal densities of snow (excluding outliers) at the ski centers showed no significant difference (paired *t*-test: $p > 0.5$) between ski piste and uncompacted natural snow (comparable differences, not maximal and minimal values) but showed a significant difference (paired *t*-test: $p < 0.5$) between new artificial and new natural snow. Outliers were identified mainly for ski piste snow at all five ski centers. The mean density of the ski piste snowpack calculated from the peak outliers identified at the ski centers was 917 ± 58 kg/m³ ($n = 5$; peak outlier is highest measured value displyed in Figure 4a), which is the density of ice. The peak outliers of ski piste snow were not significantly different (paired *t*-test: $p > 0.5$) from the density of ice.

The highest three seasons mean density of new artificial snow (435.8 ± 55.7 kg/m³, N = 34 samples) and ski piste snow (621.1 ± 111.3 kg/m³, N = 139 samples) were identified at a low-elevation ski center, namely, Košútka (Figure 4). The lowest and highest three seasons mean densities of new natural snow were identified at high-elevation ski centers, namely, Krahule (134.6 ± 37.4 kg/m³; N = 30 samples) and Donovaly (210.4 ± 68.2 kg/m³, N = 49 samples), respectively. The mean density of uncompacted natural snow was lowest at Košútka (181.1 ± 88.9 kg/m³, N = 38 samples) and highest at Donovaly (270.0 ± 96.6 kg/m³, N = 103 samples).

The mean density of ski piste snow in December calculated from all measured data was 538.9 ± 135.2 kg/m³, while its increase per month was 13.2 kg/m³ (Figure 5). The mean monthly density of ski piste snow had a tendency to increase over the whole winter season, except at the end of the season in April, when the mean monthly density was equal to the mean density in March. The mean density of uncompacted natural snow in December was 190.8 ± 116.5 kg/m³, while its increase per month was 28.2 kg/m³. The mean monthly density of uncompacted natural snow had a tendency to increase over the whole season from December to April, except for the month of March. The mean monthly densities of ski piste snow and uncompacted natural snow were significantly different (paired *t*-test: $p < 0.5$). The average seasonal difference between the mean monthly densities of ski piste snow and uncompacted natural snow (displayed in Figure 5) was 324.6 ± 46.0 kg/m^3. The low standard deviation of this difference the mean monthly densities of both types of snow showed comparable increases. The highest difference in the mean monthly densities of ski piste snow and uncompacted natural snow was found in March (372 kg/m^3), which also had the highest maximal (980 kg/m^3) and mean (635 kg/m^3) densities of ski piste snowpack (Figure 5). The mean monthly density of uncompacted natural snow exceeded the mean monthly density of ski piste snow in January and April only by about 60 and 35 kg/m^3, respectively. In contrast, the minimal mean monthly density of ski piste snow never dropped below the mean monthly density of uncompacted natural snow.

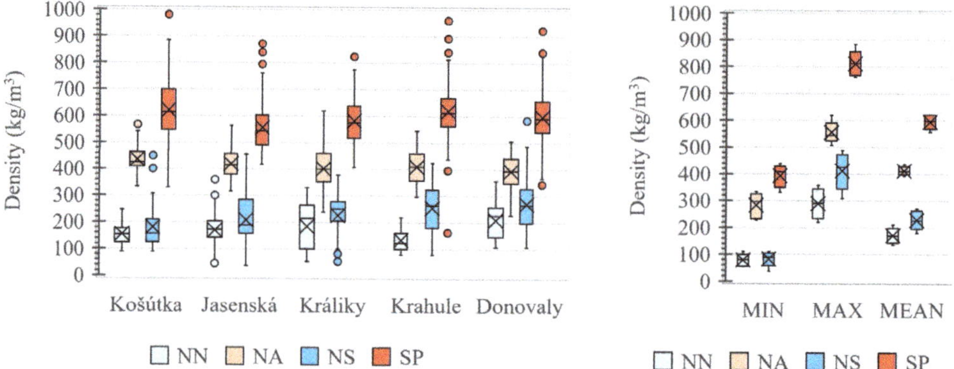

Figure 4. (**a**) Comparison of the data for three seasons with regard to the snow density of new natural snow (NN), new artificial snow (NA), uncompacted natural snow (NS), and ski piste snow (SP) for the five ski centers. The lowest number of measured samples (N = 30) was in the case of NA in Kraíliky and NN in Krahule. Description of boxplot: cross = mean, horizontal line in the box = median, box = 50% of values, box with whiskers = 99% of values, points = outliers. (**b**) Boxplots of the minimal, maximal, and mean density identified in each of five ski centers (excluding outliers) over the three winter seasons for four snow types (*n* = 5).

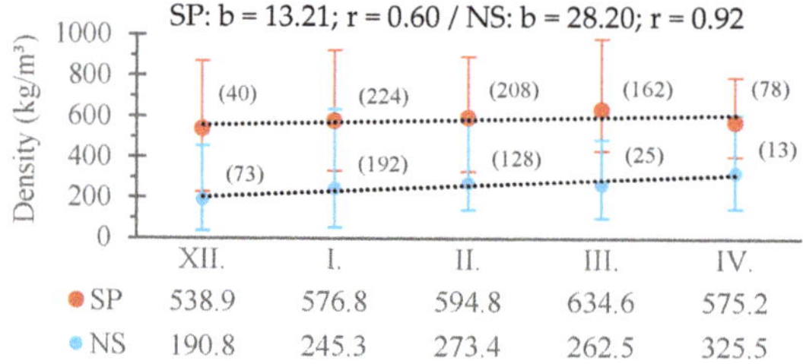

Figure 5. Increases in the mean monthly density of ski piste snow (SP) and uncompacted natural snow (NS) over the winter season (mean ± min/max). The means from three seasons calculated from all data measured at the five ski centers are displayed. b = slope of regression line, r = correlation coefficient, (x) = the total number of samples.

The densities of the ski piste snow and uncompacted natural snow showed an increasing seasonal trend and inter-seasonal variability in all three winter seasons across all five ski centers, except for seasons with shortages of natural snow (Figure 6). The slopes of the regression models in Figure 6 indicate no significant difference between ski piste snow and uncompacted natural snow when compared with the paired *t*-test ($p < 0.5$). The mean values of these slopes were comparable when calculated from data with moderate or higher correlations (r ≥ 50; Figure 6). While the increase in the density of the ski piste snow over a season was 1.9 ± 1.2 kg/m^3 per day (mean ± standard deviation), on average, the increase in the density of uncompacted natural snow over the season was 2.3 ± 1.3 kg/m^3 per day, on average. The similar rates of increase for the densities of the two snow types suggest that the key density differences stem from the artificial (machine-made) versus natural snow versus processes after and not densification due to snow grooming machines and skiers, which was relevant only for ski piste snow. The difference (paired *t*-test: $p < 0.5$) between the linear models for the ski piste snow and natural snow was in the starting density. At the beginning of the winter season, the density of the ski piste snow was 434.5 ± 63.9 kg/m^3, while the density of the uncompacted natural snow was 169.5 ± 63.9 kg/m^3, on average (mean values for three seasons calculated from the means of first surveys of the seasons; $n = 15$). Consistent relationships between time and mean density are illustrated by the trend lines and supported by the correlation coefficients, which are high in most of the seasons, although the correlation coefficients may not be reliable measures given the few data points (Figure 6). The correlation between these variables was reduced by the large outlier values, mainly in case of the ski piste snow. The outliers, for the ski piste snow were identified in the following surveys: 27 January 2015 (Králiky); 18 January 2015 (Krahule); and 21 December 2017 (Donovaly). When the density of the ski piste snow increased to outliers in these surveys, the density of natural snow did not follow same pattern. Therefore, such exceptional growth in the ski piste snow density should be connected with management activities regarding the ski piste snowpack (snowmaking, grooming, and so on).

The identification of occurrences or durations of these snowpacks was not the aim of the present paper; however, in most of the paired measurements, the measurements for natural snow were missing approximately from the middle of February until the end of the winter season (Figure 6). The reason for these missing values was the absence of natural snow. Therefore, the density of ski piste snow increased for a longer period and reached higher mean seasonal values compared to the density of natural snow, which melted earlier.

3.3. Snow Density Versus Snow Depth on the Ski Piste of Košútka Ski Center

The snow depth and density were measured manually on 5 March 2015, from 72 sampling points at Košútka, and a moderately strong negative linear correlation was calculated with the data (Figure 7). According to this correlation, the density increased 18 kg/m^3 per 10 cm of decreasing snow depth. The average deviation of this linear model was calculated as ± 72 kg/m^3 and represents the standard error of estimation (RMSE with two degrees of freedom). The snow density ranged from 457 to 969 kg/m^3 (Figure 7), with a mean of 632 ± 85 kg/m^3 (mean ± standard deviation). The minimal average density was found at a maximal depth of 135 cm, while the maximal density was associated with nearly the minimal depth of snow (969 kg/m^3 for 13 cm versus 640 kg/m^3 for 5 cm). The Snow profile was saturated with meltwater during measurements; thus, the maximal density of the snow was higher than the density of ice.

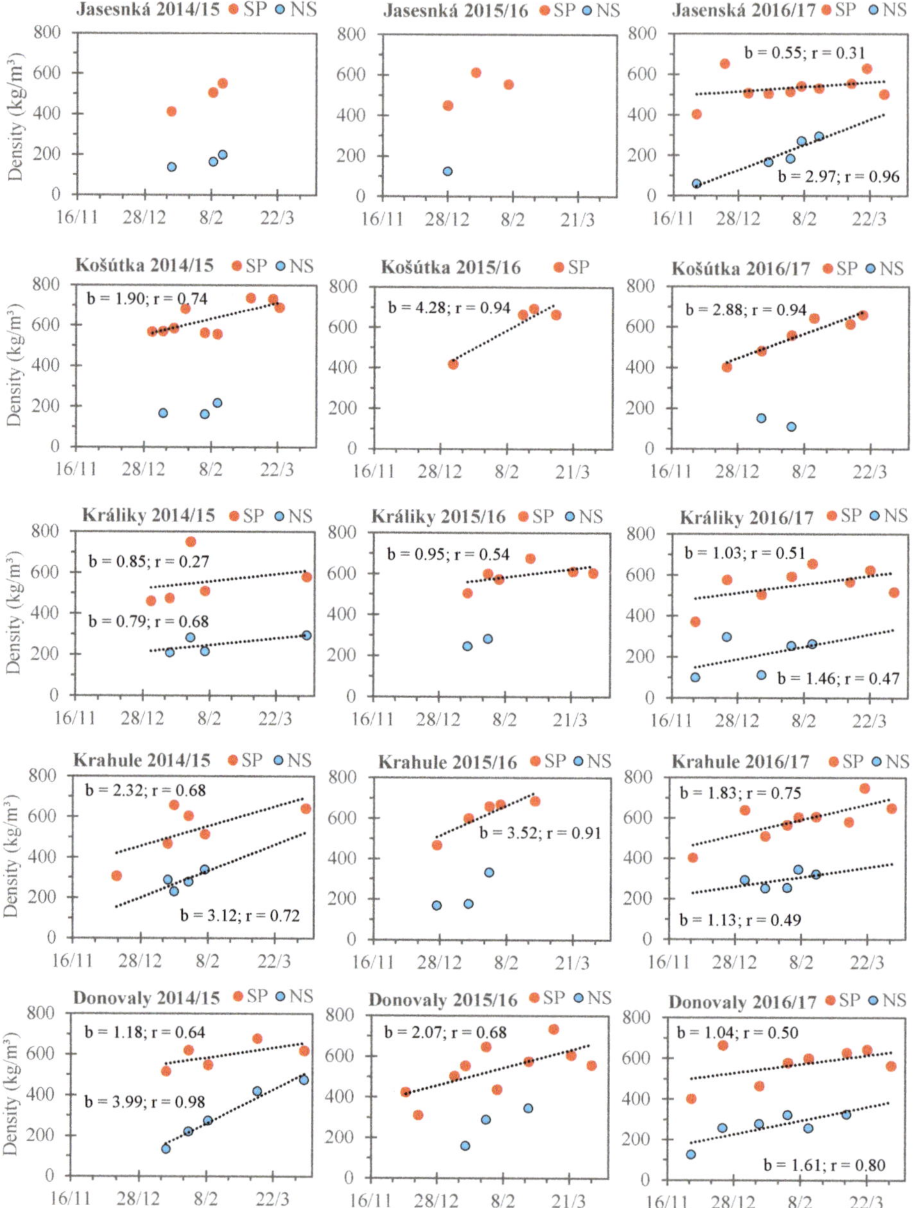

Figure 6. The mean densities of ski piste snow (SP) and uncompacted natural snow (NS) and their increases over three winter seasons (from 2014/15 to 2016/17) at five ski centers. The mean density was calculated from at least five measurements/samples. The slopes of the regression lines (b) indicate the increase in the mean densities of the snow per day, and the correlation coefficients (r) indicates weak (+30), moderate (+50), or strong (+70) linear models. The snow density measurements for SP and NS were paired. The absence of measurements for NS means the absence of natural snowpack on these dates.

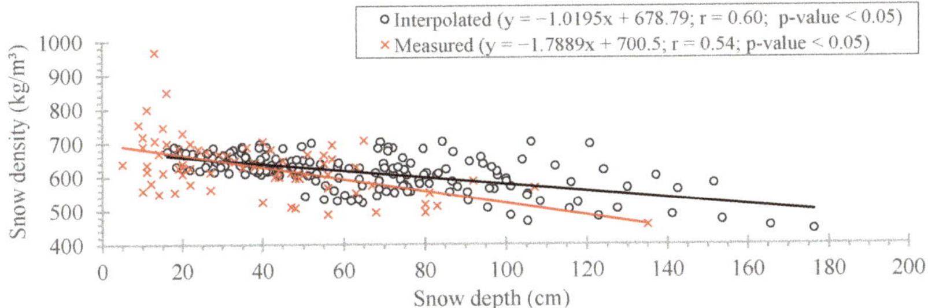

Figure 7. Correlation between measured values on 5 March 2015, (red line) and correlation between the mean depth and density of the interpolated ski piste snowpack on the basis of 181 concentric circles around snow lances. Correlation coefficients (r) and *p*-values from analysis of variance (ANOVA) are displayed.

The snow depth and snow density raster layer were interpolated from the 72 sampling points (Figure 8) to analyze the spatial distribution of these variables on the ski piste. Both snow density and depth showed similar, but reverse, geometric patterns along the entire ski piste, as indicated by the negative correlation. The correlation between these two raster layers was lower than that between the 72 paired, manually measured values (r = 0.31 versus r = 0.54). Intensive snow making at the ski center resulted in the occurrence of snow piles with maximum measured and interpolated depths of 198 and 211 cm, respectively (Figure 8). Each of the 17 snow lances produced a different volume of snow during the season; thus, the depths of the snow piles differed significantly. According to the interpolated snow depth raster (Figure 8), the centers of snow piles were located 12.5 ± 4.9 m from the closest snow lance, on average. These centers were situated down the slope, except for first two snow lances at the foot of the slope. The average snow depth and snow density in the centers of the snow piles were 110 ± 52 cm and 573 ± 70 kg/m^3, respectively. Strength and slope of correlation between the mean depth and density of the interpolated values on the basis of 181 concentric circles around snow lances was comparable as correlation between measured values (Figure 7). The mean snow depth and density calculated on such circles decreased 2 cm and 3 kg/m^3 per each meter from the center of snow lance, respectively. These ratios were identified from slopes of the linear relationships between a series of raddi around snowmaking lances (5, 7.5, 10 ... 30 m) and (i) snow depth (y = −1.9648x + 99.019; r = 0.97; n = 181), and (ii) depth average snow density (y = 2.8164x + 563.41; r = 1.00; n = 181).

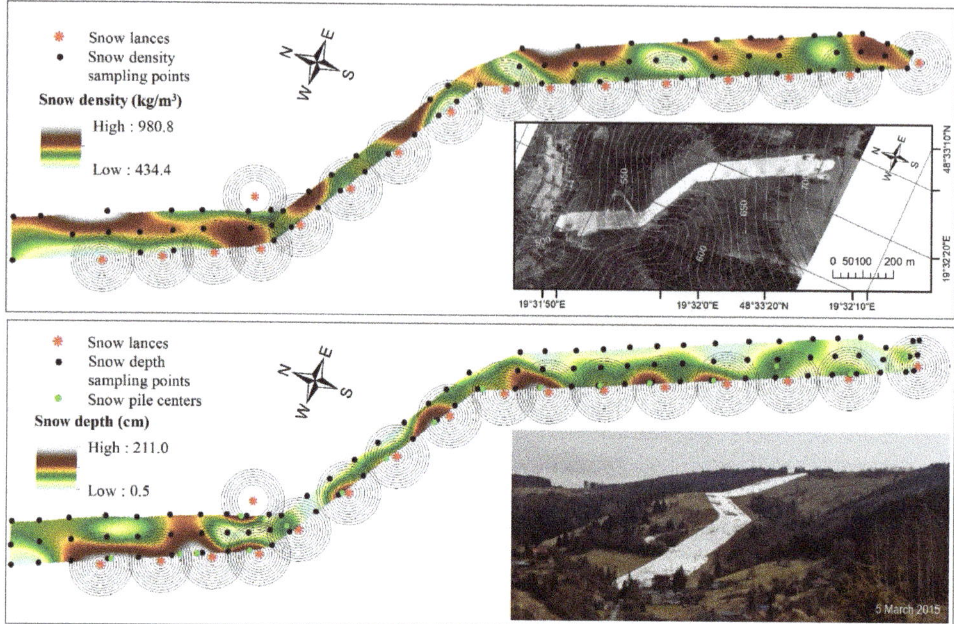

Figure 8. Snow densities and depth mapping of the ski piste snowpack for 5 March 2015, after the disappearance of natural snow from the off-piste sites. Displayed are snow pile centers (points of maximum snow depth within a 20 m radius around snow lances) and concentric circles around snow-making lances with 5 m spacing and maximum radii of 30 m.

4. Discussion

4.1. Snow Tube

The difference in construction of designed MM snow tube resulted in a different method for snow sample extraction. While commonly used snow tubes are pushed and turned through the snowpack [5], the MM snow tube is hammered. Compared to most of the of commonly used snow tubes (VS-43, Standard Federal, SnowHydro) characterized by aluminum or non-opaque plastic bodies, the presence/absence of slots in the tube, and serrated cutting ends with teeth [43,44], the MM snow tube was made of stainless steel and designed with a keen cutting end and no slots in the body of the tube. This design reduces significant errors resulting from the presence of slots in the tube, which increase the snow density and the presence of a cutter with teeth, which underestimates the snowpack [8,45,46]. The MM snow tube slightly underestimated the density of the ski piste and natural snowpacks compared to the VS-43 snow tube, most likely due to its sharp and non-serrated cutting end. According to Bindon [47] and Beaumont [45], the sharper tube allows the cleaner extraction of snow samples from the snowpack and reduces possible overestimation by up to 50%. The serrated cutting end of the VS-43 snow tube had difficulty penetrating through the ski piste snowpack when hammered, resulting in snow compression and higher density readings. These results coincide with the findings of Turčan and Loijens [48], who showed that the average snow density at depth can be artificially increased during snow tube penetration through the snowpack as the result of snow sample compression. The lower densities of the snow samples extracted by the MM tube were not the result of missed ice-soil plugs, as described by Dixon et al. [49] when comparing three snow tubes used in Canada, because the snow inside the examined tubes was pressed before removing it from the snowpack, and therefore ice-soil plugs were not missed. The underestimation of the MM snow tube with its small cutting end area (13 cm^2) is in contrast to the findings of Peterson and

Brown [46] and Farnes et al. [50], who found that snow tubes with small cutting end areas overestimate. These authors also found that snow tubes with cutting end areas >20 cm^2 had the least error. The results of abovementioned studies could differ due to different snowpack characteristics during snow surveys, mainly in terms of the hardness of the snow/ice layers in the snow profile and the snow grain size, which could influence penetration and consequently the accuracy of the measurements.

4.2. Snow Density

The mean minimal and mean maximal densities (excluding outliers) of uncompacted natural snow in the ski centers of Central Slovakia were 83 kg/m^3 and 411 kg/m^3, respectively. Mössner et al. [20] found comparable densities for seasonal natural snow, which varied from 100 kg/m^3 to 500 kg/m^3. Fassnacht [21] observed that prior to melt, the snow can attain a density of 300 to 500 kg/m^3, depending on the time period, meteorological conditions, and depth of the snow. As the present article shows, the density of new natural snowpack, maximally two days old, can vary between 78 to 289 kg/m^3. Singh [51] explained that the density of natural snow increased from 80 to 250 kg/m^3 when freshly fallen to a density of 300 kg/m^3 over 100 days due to equi-temperature snow metamorphism, during which the snow strength increases and compression occurs. This explanation coincides with our findings that the average density of the February snow was 263 kg/m^3 when this snow cover started to form in December. López-Moreno et al. [7], who analyzed snowpack characteristics in the Spanish Pyrenees (1517–3015 m a.s.l.), discovered an increase in snow density from 300 kg/m^3 in February to 455 kg/m^3 in April. These values are slightly higher than the mean values identified in the present paper, which could be explained by the longer durability of old, dense snow in the higher elevations of the Pyrenees or by the wider range of snow density values at lower elevations due to the occurrence of low-density new snow on previously melted areas. The current paper found a wide range of snow densities in April, which varied from 149 to 611 kg/m^3, that could be explained by new snow from the beginning of winter metamorphosing into snow from late winter. Jonas et al. [6] identified a comparable range of values at the end of the winter season for elevations below 1400 m a.s.l. (Swiss Alps). These authors also found a lower range of snow density values at higher elevations.

The present paper has demonstrated a classification of four snow types according to their mean densities (in descending order): ski piste snow (snowed/groomed snowpack), new artificial snow, uncompacted natural snow, and new natural snow. The findings of Rixen et al. [13,30] confirm the higher density of ski piste snow compared to uncompacted natural snow due to the intensive production of artificial snow and compaction of snow on the ski pistes; however, new artificial and natural snow were not examined in these studies. In the present article, the mean density of March snow on the snowed/groomed ski pistes was 2.4 times higher than the mean density of uncompacted natural snow beside pistes (635 versus 263 kg/m^3). Rixen et al. [24] found lower ratio in Swiss ski centers (elevation above 1500 m a.s.l.) in March (1.3 times; approximately 570 versus 430 kg/m^3; [30]), because of the lower density of ski piste snow and the higher density of natural uncompacted snow beside the piste, compared to present study. This difference can be explained by the lower elevation of the Slovakian ski pistes, where artificial snow has to be produced even at higher air temperatures [27] and where the continuous natural snowpack is present for a shorter duration compared to conditions at elevations above 1000 m a.s.l. [27]. A shorter continuous duration of the snowpack means a shorter time for snow metamorphism, during which the snow density increases [52]. The mean minimal and maximal densities of snow on the snowed and groomed ski pistes of Central Slovakian were 392 and 812 kg/m^3. Fauve et al. [53] and Federolf et al. [23] identified comparable minimums (400 and 430 kg/m^3), while their maximums were considerably lower (600 and 660 kg/m^3). The all-season high maximal density of the ski piste snowpack in the studied ski centers could have resulted from a high density of artificial snow produced at the beginning of each winter season in all studied ski centers. The mean density of new artificial snow at the ski centers was 410 kg/m^3, while it ranged between 282 and 556 kg/m^3, on average. Melanie and Rixen [31] found a comparable range for the density of new artificial snow (350 to 600 kg/m^3). The present study shows that the mean density of new artificial snow was 2.3 times higher than that of new natural snow. The higher density of artificial snow compared to natural snow results from its small grain size and subsequently higher degree of compaction [54].

4.3. Snow Density versus Snow Depth

High snow production at Košútka resulted in snow piles with average depths and distances from snow lances of 110 cm and 13 m down the slope, respectively. Spandre et al. [16], who examined ski piste snow from November 2015 to January 2016 near the Les 2 Alpes ski resort (Oisans Range, French Alps; elevation 1680 m a.s.l.), identified lower snow depths in the centers of snow piles at approximately half the distance from snow lances than found in this article, probably due an earlier winter season and the lower slopes of the ski piste (5° versus 20°). The snow lances used in the studies were comparable, while a low mean hourly wind speed was observed during months with snow production, even at Košútka [27]. The presented article shows that snow depth decreases away from snow lances, on average. Spandre et al. [16] showed a decrease in the snow depth of piles that were approximately 7 m away from the snow lances.

The current article found a moderately strong negative correlation between the snow depth and density of the ski piste snowpack on 15 March, 2015, at Košútka. No one else has published anything on this relationship yet, in contrast to studies dealing with uncompacted natural snow [55]. Numerous studies cited by Lunberg et al. [55] that analyzed data on a regional or continental scale identified a positive correlation between the depth and density of natural snow. For example, a long-term survey done in Yakutia (USSR; [56]) showed an increase in the snow density of about 180 kg/m^3 per each meter of increased snow depth. The present article showed a decrease in ski piste snow density. Lopez Moreno et al. [7], who performed measurements at approximately 100 m intervals using a local scale in the Spain Pyrenees, found an inconsistent relation between snow depth and density (strong/weak and negative/positive correlations) and pointed out that snow depth alone explained as much variability in the snow density as any other variable (terrain characteristics). In any comparison with the present article, it is important to point out that the natural snowpack does not have such variability on a local scale [7], thereby bolstering the consistent relationship between snow depth and density in the case of the ski piste snowpack. The negative correlation between the snow depth and density on the ski piste could be explained by the occurrence of basal ice layers on the bottom of the snowpack, as found for Košútka by Mikloš et al. [2]. These authors found, that with lower snow depth, the higher thickness of basal ice layer or bare ice (ice instead of snow) can be expected at the end of winter when warm and frosty days alternates [2]. Thus, it could be assumed that as the depth of the snow above the basal ice layer decreases during melting, the depth average density of snowpack increases in Košútka until snow melted away and pure ice remained.

5. Conclusions

Compared to the commonly used VS-43 snow tube, the designed MM snow tube has a smaller diameter, higher wall thickness, and sharpened cutting end and is resistant to damage during snow sample extraction. The MM snow tube has proved to be suitable for sampling high-density snow, such as the snow found on snowed/groomed ski pistes, due to its precision, rugged construction, and short sampling time. The MM tube is not recommended for sampling snow of low density and depth due to the small diameter of the tube and hardly detectable sample weights. Four snow types occurring in at ski centers were classified according to their mean seasonal densities and are listed in descending order: ski piste snow, new artificial snow, uncompacted natural snow, and new natural snow. While the mean seasonal densities of these four snow types differ significantly, the ranges were similar between new natural and new artificial snow and between uncompacted natural snow and ski piste snow. The lowest density recorded for freshly fallen natural snow was slightly above zero, while the density of ski piste snow can reach the density of ice or higher if it is saturated with meltwater. The density of the seasonal snow increases at a comparable rate over the season at piste and off-piste sites. Therefore, snow metamorphosis changes are major factors driving the snow density increases, because densification by snow-grooming machines and skiers was relevant only for ski pistes. The increase in the snow density on pistes is less rapid due to the high initial density caused by artificial snow and snow compaction via snow-grooming vehicles and skiers. At the end of the winter season, the range of the snow density on the pistes is comparable with the wide seasonal range. At this time of melting period, the spatial variability of the snow density on the piste changes with snow depth and distance from the snow lances. The snow density increases as the snow depth decreases, while the snow depth decreases with greater distances from the snow lances. The basal ice layers increase depth average density of snowpack when snow above is melting, while the snow piles at almost half the distance of the snow lances' range occur on the piste until the end of season as result of high artificial snow production.

Author Contributions: M.M. designed the MM snow tube and conducted the main data collection, analyses, and writing; M.J. helped with data collection; J.S. (Jaroslav Skvarenina) and J.S. (Jana Skvareninova) provided overall guidance and supervised the study. All authors have read and agreed to the published version of the manuscript.

Funding: This work was accomplished as a part of VEGA projects No. 1/0500/19, 1/0111/18 of the Ministry of Education, Science, Research, and Sport of the Slovak Republic and the Slovak Academy of Science and the projects No. APVV-15-0425, APVV-18-0347 of the Slovak Research and Development Agency. The authors thank the agencies for their support.

Acknowledgments: Authors would like to extend their gratitude to Igor Mikloš for his help with the MM snow tube construction and production.

Conflicts of Interest: The authors declare no conflict of interest.

Appendix A

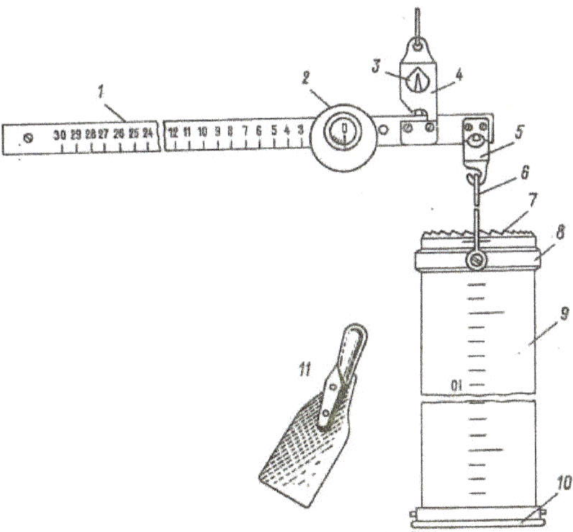

Figure A1. Assembly of the VS-43 snow tube with original mechanical scales and shovel. Scales: 1-metal ruler, 2-movable rider, 3-pointer, 4-suspension, 5-hook; Snow tube: 6-handle, 7-cutting end, 8-movable ring, 9-aluminum tube, 10-tube cover (cap), 11-a shovel. Source: [42,57]. Digital Kern HDB10K10N scales were used in the presented article instead of mechanical scales.

References

1. Kumar, A. Seasonal snow cover. In *Encyklopedia of Snow, Ice and Glaciers*, 1st ed.; Singh, V.P., Singh, P., Haritashya, U.K., Eds.; Springer: Dordrecht, The Netherlands, 2011; pp. 974–975.
2. Mikloš, M.; Igaz, D.; Šinka, K.; Skvareninova, J.; Jančo, M.; Vyskot, I.; Skvarenina, J. Ski piste snow ablation versus potential infiltration (Veporic Unit, Western Carpathians). *J. Hydrol. Hydromech.* **2020**, *68*, 28–37. [CrossRef]
3. Barrett, T.P.; Adam, J.C.; Lettenmaier, D.P. Potential impacts of warming climate on water availability in snow-dominated regions. *Nature* **2005**, *438*, 303–309. [CrossRef] [PubMed]
4. Mikloš, M.; Vyskot, I.; Šatala, T.; Korísteková, K.; Jančo, M.; Skvarenina, J. Effect of forest ecosystems on the snow water equivalent in relation to aspect and elevation in the Hučava river watershed, Poľana Biosphere Reserve (Slovakia). *Ekológia* **2017**, *36*, 268–280.
5. Kinar, N.J.; Pomeroy, J.W. Measurement of the physical properties of the snowpack. *Rev. Geophys.* **2015**, *53*, 481–544. [CrossRef]
6. Jonas, T.; Marty, C.; Magnusson, J. Estimating the snow water equivalent from snow depth measurements in the Swiss Alps. *J. Hydrol.* **2009**, *378*, 161–167. [CrossRef]
7. López-Moreno, J.I.; Fassnacht, S.R.; Heath, J.T.; Musselman, K.N.; Revuelto, J.; Latron, J.; Morán-Tejeda, E.; Jonas, T. Small scale spatial variability of snow density and depth over complex alpine terrain: Implications for estimating snow water equivalent. *Adv. Water Resour.* **2013**, *55*, 40–52.
8. Goodison, B.; Ferguson, H.; McKay, G. Measurement and data analysis. In *Handbook of Snow: Principles, Processes, Management and Use*; Gray, D.M., Male, D.H., Eds.; Pergamon Press: Toronto, ON, Canada, 1981; pp. 191–274.
9. Church, J.E. Snow surveying: Its principles and possibilities. *Geogr. Rev.* **1933**, *23*, 529–563. [CrossRef]
10. Bartík, M.; Sitko, R.; Oreňák, M.; Slovik, J.; Skvarenina, J. Snow accumulation and ablation in disturbed mountain spruce forest in West Tatra Mts. *Biologia* **2014**, *69*, 1492–1501. [CrossRef]
11. Singh, A.K. Snow layer. In *Encyklopedia of Snow, Ice and Glaciers*, 1st ed.; Singh, V.P., Singh, P., Haritashya, U.K., Eds.; Springer: Dordrecht, The Netherlands, 2011; pp. 1059–1060.
12. Keller, T.; Pielmeier, C.; Rixen, C.; Gadient, F.; Gustafsson, D.; Stähli, M. Impact of artificial snow and ski-slope grooming on snowpack properties and soil thermal regime in a sub-alpine ski area. *Ann. Glaciol.* **2004**, *38*, 314–318. [CrossRef]
13. Rixen, C.; Haeberli, W.; Stoeckli, V. Ground temperatures under ski pistes with artificial and natural snow. *Arct. Antarct. Alp. Res.* **2004**, *36*, 419–427. [CrossRef]
14. Horner, R.A. Techniques for sampling sea-ice algae. In *Polar Marine Diatoms*; Medlin, L.K., Priddle, J., Eds.; British Antarctic Survey, Natural Environment Research Council: Cambridge, UK, 1990; pp. 19–23.
15. Koci, B.R.; Kuivinen, K.C. The PICO lightweight coring auger. *J. Glaciol.* **1984**, *30*, 244–245. [CrossRef]
16. Spandre, P.; François, H.; Thibert, E.; Moris, S.; George-Marcelpoil, E. Determination of snowmaking efficiency on a ski slope from observations and modelling of snowmaking events and seasonal snow accumulation. *Cryosphere* **2017**, *11*, 891–909. [CrossRef]
17. Meijer zu Schlochtern, M.P.; Rixen, C.; Wipf, S.; Cornelissen, J.H. Management, winter climate and plant–soil feedbacks on ski slopes: A synthesis. *Ecol. Res.* **2014**, *29*, 583–592. [CrossRef]
18. De Jong, C. Challenges for mountain hydrology in the third millennium. *Front. Environ. Sci.* **2015**, *3*, 38. [CrossRef]
19. Fierz, C.; Armstrong, R.L.; Durand, Y.; Etchevers, P.; Greene, E.; McClung, D.M.; Nishimura, K.; Satyawali, P.K.; Sokratov, S.A. *The International Classification for Seasonal Snow on the Ground*; IHP-VII Technical Documents in Hydrology No. 83, IACS Contribution No. 1; UNESCO-IHP: Paris, France, 2009; pp. 3, 63.
20. Mössner, M.; Innerhofer, G.; Schindelwig, K.; Kaps, P.; Schretter, H.; Nachbauer, W. Measurement of mechanical properties of snow for simulation of skiing. *J. Glaciol.* **2013**, *59*, 1170–1178. [CrossRef]
21. Fassnacht, S. Snow density. In *Encyklopedia of Snow, Ice and Glaciers*, 1st ed.; Singh, V.P., Singh, P., Haritashya, U.K., Eds.; Springer: Dordrecht, The Netherlands, 2011; 1045p, ISBN 978-90-481-2642-2.
22. Spandre, P.; François, H.; George-Marcelpoil, E.; Morin, S. Panel based assessment of snow management operations in French ski resorts. *J. Outdoor Recreat. Tour.* **2016**, *16*, 24–36. [CrossRef]
23. Federolf, P.; JeanRichard, F.; Fauve, M.; Lüthi, A.; Rhyner, H.-U.; Dual, J. Deformation of snow during a carved ski turn. *Cold Reg. Sci. Technol.* **2006**, *46*, 69–77. [CrossRef]

24. Rixen, C.; Stoeckli, V.; Ammann, W. Does artificial snow production affect soil and vegetation of ski pistes? A review. *Perspect. Plant. Ecol.* **2003**, *5*, 219–230. [CrossRef]

25. Vido, J.; Tadesse, T.; Šustek, Z.; Kandrík, R.; Hanzelová, M.; Skvarenina, J.; Skvareninova, J.; Hayes, M. Drought occurrence in central european mountainous region (Tatra National Park, Slovakia) within the period 1961–2010. *Adv. Meteorol.* **2015**, *2015*, 1–8. [CrossRef]

26. Kohnová, S.; Rončák, P.; Hlavčová, K.; Szolgay, J.; Rutkowska, A. Future impacts of land use and climate change on extreme runoff values in selected catchments of Slovakia. *Meteorol. Hydrol. Water Manag.* **2019**, *7*, 47–55. [CrossRef]

27. Mikloš, M.; Jančo, M.; Korísteková, K.; Skvareninova, J.; Skvarenina, J. The Suitability of Snow and Meteorological Conditions of South-Central Slovakia for Ski Slope Operation at Low Elevation—A Case Study of the Košútka Ski Centre. *Water* **2018**, *10*, 907. [CrossRef]

28. Ďurigová, M.; Ballová, D.; Hlavčová, K. Analyses of Monthly Discharges in Slovakia Using Hydrological Exploratory Methods and Statistical Methods. *Slovak J. Civ. Eng.* **2019**, *27*, 36–43. [CrossRef]

29. Minďaš, J.; Bartík, M.; Skvareninova, J.; Repiský, R. Functional effects of forest ecosystems on water cycle–Slovakia case study. *J. For. Sci.* **2018**, *64*, 331–339.

30. Rixen, C.; Freppaz, M.; Stoeckli, V.; Huovinen, C.; Huovinen, K.; Wipf, S. Altered snow density and chemistry change soil nitrogen mineralization and plant growth. *Arct. Antarct. Alp. Res.* **2008**, *40*, 568–575. [CrossRef]

31. Melanie, P.; Rixen, C. Management, winter climate and plant-soil feedbacks on ski slopes: A synthesis. *Ecol. Res.* **2014**, *29*, 583–592.

32. Mazúr, E.; Lukniš, M. *Geomorfologické členenie SSR a ČSSR. Časť Slovensko*; Slovenská kartografia: Bratislava, Slovakia, 1986.

33. Vilček, J.; Koco, Š. Integrated index of agricultural soil quality in Slovakia. *J. Maps* **2018**, *14*, 68–76. [CrossRef]

34. Lapin, M.; Faško, P.; Melo, M.; Šťastný, P.; Tomlain, J. Climatic regions. In *Landscape Atlas of the Slovak Republic*, 1st ed.; Miklós, L., Hrnčiarová, T., Eds.; Slovak Environmental Agency: Bratislava, Slovakia, 2002; p. 99.

35. Vilček, J.; Skvarenina, J.; Vido, J.; Nalevanková, P.; Kandrík, R.; Skvareninova, J. Minimal change of thermal continentality in Slovakia within the period 1961–2013. *Earth Syst. Dyn.* **2016**, *7*, 735–744. [CrossRef]

36. SHMI (Slovak Hydrometeorologic Institute). *Climate Atlas of Slovakia*; Slovak Hydrometeorologic Institute: Banská Bystrica, Slovakia, 2015; 228p.

37. Hrvoľ, J.; Horecká, V.; Skvarenina, J.; Střelcová, K.; Skvareninova, J. Long-term results of evaporation rate in xerothermic Oak altitudinal vegetation stage in Southern Slovakia. *Biologia* **2009**, *64*, 605–609.

38. Hríbik, M.; Vida, T.; Skvarenina, J.; Skvareninova, J.; Ivan, L. Hydrological effects of Norway spruce and European beech on snow cover in a mid-mountain region of the Poľana Mts. *J. Hydrol. Hydromech.* **2012**, *60*, 319–332. [CrossRef]

39. Bartík, M.; Jančo, M.; Střelcová, K.; Skvareninova, J.; Skvarenina, J.; Mikloš, M.; Vido, J.; Waldhauserová, P.D. Rainfall interception in a disturbed montane spruce (*Picea abies*) stand in the West Tatra Mountains. *Biologia* **2016**, *71*, 1002–1008. [CrossRef]

40. Šatala, T.; Tesař, M.; Hanzelová, M.; Bartík, M.; Šípek, V.; Skvarenina, J.; Minďáš, J.; Waldhauserová, P.D. Influence of beech and spruce sub-montane forests on snow cover in Poľana Biosphere Reserve. *Biologia* **2017**, *72*, 854–861. [CrossRef]

41. Skvarenina, J.; Tomlain, J.; Hrvoľ, J.; Skvareninova, J. Occurrece of dry and wet periods in altitudinal vegetation stages of West Carpathians in Slovakia: Time-Series Analysis 1951–2005. In *Bioclimatology and Natural Hazards*, 1st ed.; Střelcová, K., Matyas, C., Kleidon, A., Lapin, M., Matejka, F., Blazenec, M., Skvarenina, J., Holecy, J., Eds.; Springer: Dordrecht, The Netherlands, 2009; pp. 97–106.

42. Haberkorn, A. *European Snow Booklet*; COST Association: Brussels, Belgium, 2019; p. 363.

43. Crook, A.; Freeman, T. A comparison of techniques of sampling the arctic-subarctic snowpack in Alaska. In *Proceedings of the 41st Annual Western Snow Conference, Grand Junction, CO, USA, 17–19 April 1973*; Crook, A.G., Freeman, T.G., Eds.; Western Snow Conference: Brush Prairie, WA, USA, 1973; pp. 62–68.

44. McKay, G.A.; Blackwell, S. Plains snowpack water equivalent from climatological records. In *Proceedings of the 29th Annual Western Snow Conference, Spokane, WA, 11–13 April 1961*; Western Snow Conference: Brush Prairie, WA, USA, 1961; pp. 27–43.

45. Beaumont RT. Field accuracy of volumetric snow samplers at Mt. Hood, Oregon. In *Proceedings of the Conference on Physics of Snow and Ice, Sapporo, Japan, 14–19 August 1966*; Institute of Low Temperature Science, Hokkaido University: Hokkaido, Japan, 1967; pp. 1007–1013.
46. Peterson, N.R.; Brown, A.J. Accuracy of snow measurements. In *Proceedings of the 43rd Annual Western Snow Conference, Coronado, CA, 23–25 April 1975*; Western Snow Conference Association: Coronado, CA, USA, 1975; Volume 43, pp. 1–9.
47. Bindon, H.H. The design of snow samplers for Canadian snow surveys. In Proceedings of the 21st Annual Eastern Snow Conference, Utica, NY, USA, 13–14 February 1964; pp. 23–28.
48. Turčan, J.; Loijens, H. Accuracy of snow survey data and errors in snow sampler measurements. In Proceedings of the 32nd Annual Eastern Snow Conference, Manchester, NH, USA, 6–7 February 1975; pp. 2–11.
49. Dixon, D.; Boon, S. Comparison of the SnowHydro snow sampler with existing snow tube designs. *Hydrol. Process.* **2012**, *26*, 2555–2562. [CrossRef]
50. Farnes, P.; Peterson, N.; Goodison, B.; Richards, R. Metrification of manual snow sampling equipment. In *Proceedings of the 50th Western Snow Conference, Reno, Nevada, 19–23 April 1982*; Western Snow Conference: Brush Prairie, WA, USA, 1982; pp. 120–132.
51. Singh, A.K. Snow metamorphism. In *Encyklopedia of Snow, Ice and Glaciers*, 1st ed.; Singh, V.P., Singh, P., Haritashya, U.K., Eds.; Springer: Dordrecht, The Netherlands, 2011; pp. 1060–1061.
52. Singh, A.K. Snow course. In *Encyklopedia of Snow, Ice and Glaciers*, 1st ed.; Singh, V.P., Singh, P., Haritashya, U.K., Eds.; Springer: Dordrecht, The Netherlands, 2011; p. 1032.
53. Fauve, M.; Rhyner, H.; Schneebeli, M. *Preparation and Maintenance of Pistes: Handbook for Practitioners*; Swiss Fed. for Snow and Avalanche Research SLF: Davos Dorf, Swizterland, 2002; 134p.
54. Jones, H.G.; Devarennes, G. The chemistry of artificial snow and its influence on the germination of mountain flora. In *Biogeochemistry of Seasonal Snow-Covered Catchments, Proceedings of a Boulder Symposium, Boulder, CO, USA, 12–13 July 1995*; IAHS: Wallingford, UK, 1995; Volume 228, pp. 355–360.
55. Lundberg, A.; Richardson-Näslund, C.; Andersson, C. Snow density variations: Consequences for ground penetrating radar. *Hydrol. Process.* **2006**, *20*, 1483–1495. [CrossRef]
56. Gavrilèv, R.I. *Zavisimost' plotnosti snezhnogo pokrova v IAkutii ot ego vysoty (Dependence of the Density of the Snow Cover in Yakutai upon Its Thickness)*; Akademiia Nauk SSSR, Sibirskoe otdelenie, Institut merzlotovedeniia, Protsessy teplo-i massoobmena v merzlykh gornykh porodakh: Nauka, Moscow, 1965; pp. 45–49.
57. Losev, A.P. Praktikum po agroklimaticheskomu obespecheniyu rastenievodstva. In *Workshop on Agro-Climatic Security Crop*; Gidrometeoizdat: Saint Petersburg, Russia, 1994.

Publisher's Note: MDPI stays neutral with regard to jurisdictional claims in published maps and institutional affiliations.

Article

Influence of Warmer and Drier Environmental Conditions on Species-Specific Stem Circumference Dynamics and Water Status of Conifers in Submontane Zone of Central Slovakia

Adriana Leštianska [1,*], Peter Fleischer Jr. [1,2], Katarína Merganičová [3,4], Peter Fleischer [1] and Katarína Střelcová [1]

1 Faculty of Forestry, Technical University in Zvolen, T.G. Masaryka 24, 96001 Zvolen, Slovakia; p.fleischerjr@gmail.com (P.F.J.); p.fleischersr@gmail.com (P.F.); strelcova@tuzvo.sk (K.S.)
2 Department of Plant Ecophysiology, Institute of Forest Ecology, Slovak Academy of Sciences, Štúrova 2, 96053 Zvolen, Slovakia
3 Faculty of Forestry and Wood Sciences, Czech University of Life Sciences Prague, Kamýcká 129, 16500 Praha 6—Suchdol, Czech Republic; k.merganicova@forim.sk
4 Department of Biodiversity of Ecosystems and Landscape, Slovak Academy of Sciences, Štefánikova 3, P.O.Box 25, 81499 Bratislava, Slovakia
* Correspondence: adriana.lestianska@tuzvo.sk

Received: 14 September 2020; Accepted: 18 October 2020; Published: 21 October 2020

Abstract: The frequency and intensity of droughts and heatwaves in Europe with notable impact on forest growth are expected to increase due to climate change. Coniferous stands planted outside the natural habitats of species belong to the most threatened forests. In this study, we assess stem circumference response of coniferous species (*Larix decidua* and *Abies alba*) to environmental conditions during the years 2015–2019. The study was performed in Arboretum in Zvolen (ca. 300 m a.s.l., Central Slovakia) characterised by a warmer and drier climate when compared to their natural habitats (located above 900 m a.s.l.), where they originated from. Seasonal radial variation, tree water deficit (ΔW), and maximum daily shrinkage (MDS) were derived from the records obtained from band dendrometers installed on five mature trees per species. Monitored species exhibited remarkably different growth patterns under highly above normal temperatures and uneven precipitation distribution. The magnitudes of reversible circumference changes (ΔW, MDS) were species-specific and strongly correlated with environmental factors. The wavelet analysis identified species-specific vulnerability to drought indicated by pronounced diurnal stem variation periodicity in rainless periods. *L. decidua* exhibited more strained stem water status and higher sensitivity to environmental conditions than *A. alba*. Tree water deficit and maximum daily shrinkage were found appropriate characteristics to compare water status of different tree species.

Keywords: climate change; dendrometer; stem water deficit; shrinkage; morlet wavelet

---

1. Introduction

Climate projections in the near future indicate rising temperatures and increase in the frequency and severity of climatic extremes [1,2]. Changes in temperature and precipitation patterns may (dis)favour a given species [3] and modify forest composition. In the light of ongoing and predicted climate change, the growth performance of economically important tree species under climatic extremes, especially extreme drought events, has been frequently discussed [4,5]. In the context of climate change, it is critical to understand whether climatic conditions are becoming more or less favourable for tree growth. The strong need to adapt forests to future climate conditions through changes in tree species composition is frequently in stark contrast to the dearth of information about the suitability of individual species and their provenances for these future conditions [6]. The increase in frequency of extremely dry years, as predicted consequences of climate change [7], may shift the dynamics of growth more or less abruptly in favour of less drought-sensitive tree species [8]. A substantial reduction in growth and increased mortality has been observed for some tree species growing outside their natural habitats—e.g., Norway spruce (*Picea abies* L. Karst) and European larch (*Larix decidua* Mill) [9,10]. European larch is typical for mountainous regions of the Carpathians and the Alps [11]. At lower altitudes, it is considered a non-native species. Owing to its deciduous character, larch has a shorter growing season to reach similar above-ground production rates as adjacent evergreen conifers. To this end, its photosynthetic rates are greater [12]. Silver fir (*Abies alba* Mill.) is a European coniferous species occurring in the Alps and the Carpathians. Although fir was frequently regarded as a species that prefers cold and moist climate [13], recent findings suggest that higher temperatures and more frequent drought events do not seem to have significant negative effects on the growth of this species [6,14,15]. Moreover, recent growth of Silver fir (*Abies alba*) has been accelerated [16] since its recovery from the severe growth decline due to air pollution in 1970s and 1980s [17].

Due to the ongoing climate change, more emphasis has been recently laid upon the monitoring of the changes in spatial and temporal distribution of precipitation, and temperature regime [2]. Expected climate changes, such as rising temperatures, or more frequent drought events [1,2], will have an impact on a wide range of physiological functions and biochemical reactions [18,19], which control tree growth [20,21] and tree water status [22,23]. Daily stem variations are products of irreversible stem size changes due to cambial cell division and cell enlargement processes, and reversible changes caused by contraction and expansion of water storage tissues [24]. A substantial amount of internally stored tree water can be used by a plant [25] for transpiration to balance the differences in water content of roots and shoots [26]. Water in namely elastic tissues of the bark (i.e., cambium, phloem, and parenchyma) and mesophyll of needles [27] is partly depleted and replenished on a daily basis due to changing water potential gradients within the plant [28]. Stem radius variations detrended for growth were named tree water deficit (ΔW) by [27], and this parameter was found to be proportional to water content in the living tissues of the bark [29]. Another stem water indicator derived from stem variation is maximum daily shrinkage (MDS)—e.g., [30–32]. The amplitude of daily shrinkage is a function of water loss through the leaves and water uptake by roots. Hence, daily shrinkage often correlates with evaporative demand [33]. Trees attain their maximum circumference just before dawn—i.e., before they open their stomata and start losing water to the atmosphere more rapidly than they can take up from the soil [34]. Changes in xylem water tension [35] and water storage in the bark and phloem [36] cause tree stems to shrink, reaching minimum circumference in the afternoon, after which stems begin to swell until they reach their maximum before the next dawn. According to mentioned authors, both water status indicators (ΔW and MDS) are closely related to tree drought stress and are mainly determined by a combination of atmospheric and soil conditions.

High temporal resolution measurements of stem diameter variation can provide valuable information on the radial growth process as well as the tree water status [37–39]. Stem radius variation can be continuously monitored with automatic dendrometers from periods of minutes to hours [40,41]. It is an appropriate method to study tree water relations and radial growth over a whole year [42,43]. Therefore, dendrometers are widely used in climate-growth relation studies—e.g., [6,44,45].

Analysis of the climate–growth relationship makes it possible to identify and assess the influence of climatic factors on tree growth. The assessment of species-specific stem size response to climate is challenging because it requires homogeneous site and stand conditions to determine the influence of climate solely. In our study, we compared stem circumference variations of two coniferous species, namely *Abies alba* and *Larix decidua*, growing at the same site located in warmer and drier conditions than in their natural habitats situated at higher elevations (Figure 1, Table 1). The conditions at the current site can represent the likely future climate the species will have to face also in their natural habitats, since the temperature of the current site exceeded the one in their natural habitats by 2 °C or more, and the annual precipitation total was by approximately 100 mm less than in their original habitats (Figure 1).

Table 1. The characteristics of the original location. Air temperature (°C) is long-term annual air temperature representing a period of 1961–1990. Precipitation (mm) is the long-term mean annual precipitation total representing a period of 1961–1990.

| Species | Original Location of Selected Species | | | | |
	Orographic Unit	Locality	Elevation (m a.s.l.)	Air Temperature (°C)	Precipitation (mm)
L. decidua	Spišsko-gemerský kras	Voniaca dolina	900	4.7	835
A. alba	Kremnica mountains	Flochovský back	950	6.5	786

We hypothesise that species-specific ecophysiological traits—e.g., photosynthetic and transpiration rates—lead to significant differences in climate–growth relationships of the two investigated species. Based on band dendrometer records (BDR), we aim to (i) identify species-specific reversible changes in stem circumference under warmer and drier environmental conditions than in their natural environment; (ii) identify periodicity in stem size changes and their relation to environmental conditions; and (iii) derive a set of variables describing tree water status. We expected that drought would limit their stem growth and stimulate diurnal reversible changes. At low elevation, the combination of low precipitation and high temperatures could lead to water deficit and respective stomata closure and sap flow reduction during dry periods. We also hypothesised that differences in water regime patterns are more pronounced between species than between trees of the same species.

2. Material and Methods

2.1. Study Area

The study site is located in Arboretum of Technical University in Zvolen, Central Slovakia (48°35′ N, 19°07′ E, altitude from 290 m a.s.l. to 377 m a.s.l.). The facility serves for preserving a gene pool of the Carpathian dendroflora ex situ [46]. The site represents common upland forest communities in Central Slovakia. Mean annual air temperature is 8.2 °C, and annual precipitation total is 651 mm. During a growing season (April–September), a long-term average temperature is 14.7 °C and precipitation total is 377 mm (calculated from long-term data from a nearby meteorological station of Sliač, 313 m a.s.l. provided by the Slovak Hydrometeorological Institute, representing a period of 1961–1990). Cambisol is a dominant soil type and *Querceto-Fagetum* community represents potential forest vegetation (https://geo.enviroportal.sk/atlassr).

Two research plots, each representing single species (*Larix decidua* and *Abies alba*), were selected at sites with similar environmental conditions, which are generally warmer and drier than their original natural habitats (Figure 1). In the case of *L. decidua*, the long-term (1961–1990) values of temperature and precipitation as well as their values representing individual monitored years of the study site occurred outside the current species range (Figure 1). For *A. alba*, the two last monitored years 2018 and 2019 did not occur inside the current species range (Figure 1). Basic site characteristics of the original location of the selected tree species provenances are in Table 1.

At each plot, five adult trees with similar age and size were selected for the purpose of this study. Tree diameters at breast height and tree heights were measured in the years 2015 and 2016, respectively. The monitored trees of *L. decidua* had an average diameter at breast height ($d_{1.3}$) of 30.9 ± 3.9 cm, height of 27.5. ± 1.1 m, and tree age of 51 years. The monitored trees of *A. alba* had an average diameter at breast height of 31.7 ± 4.7 cm, height of 21.4. ± 1.5 m, and tree age of 46 years.

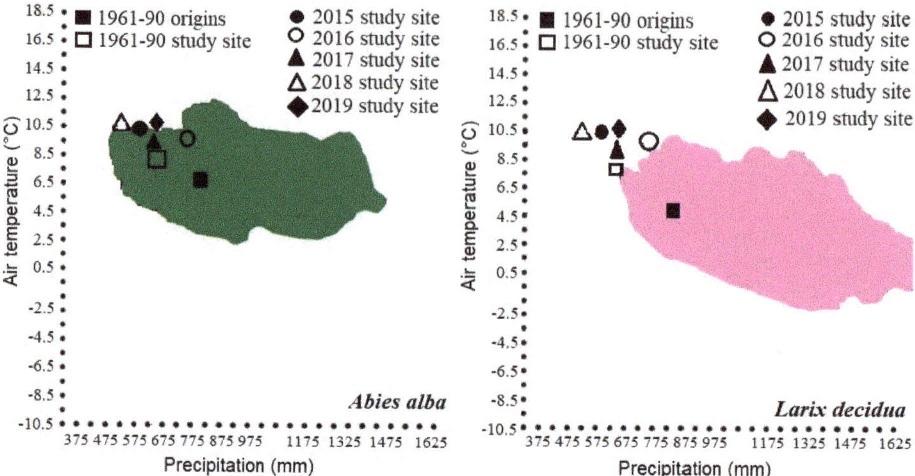

Figure 1. Species-specific distribution with regard to annual values of two main climate parameters (coloured areas) derived from Kölling [47] compared with local long-term values, local values representing the years 2015–2019, and values representing their original habitats.

2.2. Environmental Data

During the study period, meteorological data were recorded with an automatic meteorological station (EMS Brno, CZ) installed at an open area near study plots (at a distance 80–150 m). The meteorological station recorded global radiation (GR, W.m^{-2}), air temperature (AT, °C), relative air humidity (RH, %), and precipitation (P, mm) with automatic sensors every 10 min. From these values, daily mean air temperature, daily mean relative air humidity, daily precipitation totals and daily global radiation sums were calculated. Daily mean vapour pressure deficit in the air (VPD, kPa) was calculated from the daily means of air temperature and relative air humidity. Soil water potential (SWP, bar) in 15, 30, and 50 cm soil depths was measured under forest canopy within the study plots (gypsum blocks and MicroLog SP3, EMS Brno, CZ). Measuring intervals were set to 20 min, and mean daily SWP values per plot calculated from all depths were used for further analyses.

2.3. Band Dendrometer Records (BDR)

Stem circumference variation of 10 sample trees (five trees per species) was recorded with high temporal resolution automatic band dendrometers (model DRL 26, EMS Brno, CZ, accuracy ±1 μm). To ensure a close contact of dendrometer bands with tree stems and to reduce the influence of hygroscopic swelling and shrinkage of the bark, the outermost part of the bark (periderm) was carefully removed before the installation of dendrometers. Circumference measurements were recorded in 20-min intervals.

BDR were processed by applying two distinguished methods: (i) a daily cycle and (ii) a stem cycle approach. Both daily and stem cycles consist of three distinguished phases based on stem dynamics: stem contraction phase (con), expansion phase (exp), and stem circumference increment phase (inc) [37,42] (Figure 2). The daily approach operates at a daily scale (from 0:00 to 0:00). The stem cycle describes stem dynamics regardless of calendar days—i.e., at a scale that can differ from 24 h. With the daily approach we derived the daily mean, maximum, and amplitude of BDR. A contraction phase is a period between BDR maximum and minimum. An expansion phase is a period from BDR minimum to the following maximum value. An increment phase is a part of the expansion phase from the time when the stem size exceeds the previous maximum until it reaches the subsequent maximum

(Figure 2). The duration (h, hours) of each phase was derived from the dendrometer records. Seasonal radial stem increment (cum_inc) was calculated as a sum of increments during the whole season.

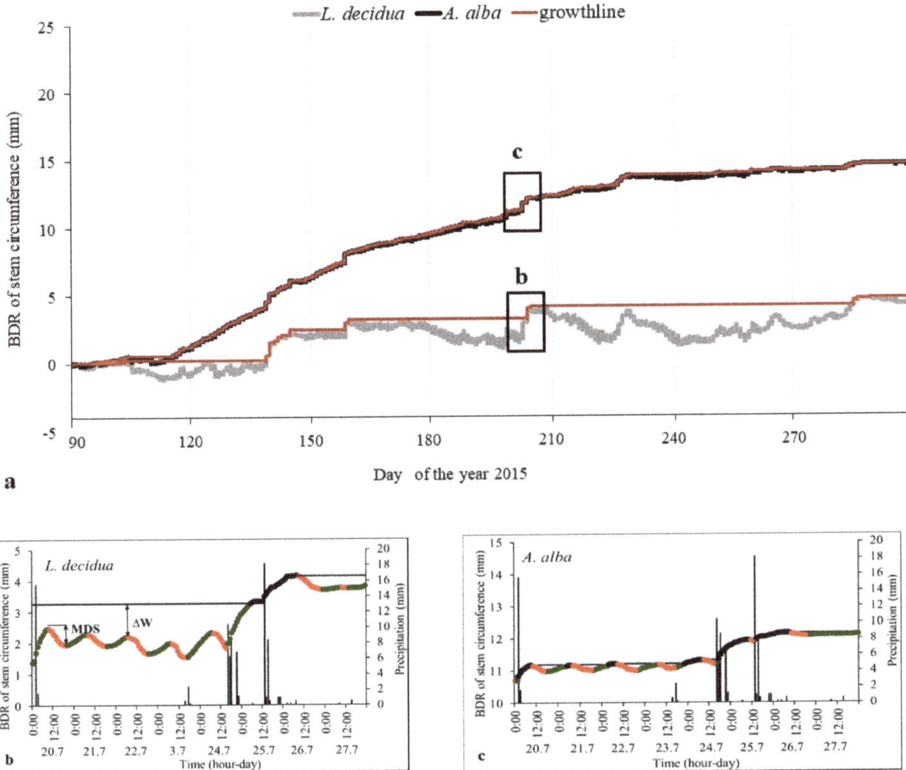

Figure 2. Growth lines (solid black lines) and band dendrometer records (BDR) of stem circumference of *L. decidua* and *A. alba* for the year 2015 (**a**). Detailed figures (**b**) and (**c**) depict rainless (20–23 July 2015) and rainy (24–26 July 2015) periods, respectively, showing distinct phases of a stem cycle: contraction (red), expansion (green), and increment (black points). Individual points represent hourly data. Maximum daily shrinkage (MDS) is a difference between daily maximum and minimum stem size. Stem water deficit (ΔW) is a difference between the actual stem size and the growth line representing the stem size under fully hydrated conditions.

2.4. Tree Water Status

We applied two indicators to quantify tree water status. The first one is tree water deficit (ΔW in mm) that defines the actual tree state in comparison to a fully hydrated state [30,31]. At first, the growth line was derived from BDR using a moving maximum of the current and previous dendrometer readings. Afterwards, tree water deficit was calculated as a difference between the actual BDR value and the respective growth line value of stem size, which represents a tree state under fully hydrated conditions (i.e., when ΔW = 0) [48] (Figure 2). Hence, increasingly negative values of ΔW indicate increasing tree drought stress.

The second characteristic is maximum daily shrinkage (MDS in mm) defined as the difference between daily maximum and minimum stem size (BDR). This indicator quantifies the daily cycle of water uptake at night and water loss from elastic cambial and phloem tissues during a day [43] (Figure 2).

Stem cycles in BDR, duration of each phase, water deficit (ΔW), and maximum daily shrinkage (MDS) were determined from BDR with "DendrometeR" R package [49].

2.5. BDR and Environmental Variables

The relationships of daily environmental variables (precipitation, relative air humidity (RH), vapour pressure deficit (VPD), minimum, maximum and mean air temperature (ATmin, ATmax, ATmean), soil water potential (SWP)) with daily parameters extracted from BDR (inc, ΔW, MDS) within the analysed period of the year (April–October) were quantified with the Spearman rank-correlation coefficients. Correlations were tested with the Statgraphics Centurion XIV Version 6.1.11 software.

2.6. Species-Specific BDR Variability

Besides the mentioned indicators (inc, ΔW, MDS), another 13 variables were derived from BDR. We used the "daily.stats" and "cycle.stats" methods in dendrometerR package to process the data. Daily values were calculated with "daily.stats", while "cycle.stats" were used to identify magnitude of three distinct phases (contraction, expansion and increment, Figure 2) The derived variables are as follows:

1. Number of cycles (n_cyc) (number) in the season—based on the stem cycle approach one cycle always comprises contraction and expansion phases, while the increment phase is optional;
2. Cumulative duration of contraction (cum_d_cont) (hour)—seasonal sum of all contraction time lengths;
3. Average duration of contraction (avg_d_cont) (hour)—cumulative duration of contraction divided by number of stem cycles;
4. Cumulative amplitude of contraction (cum_cont) (mm)—seasonal sum of all contractions;
5. Average amplitude of contraction (avg_cont) (mm)—seasonal sum of contractions divided by number of stem cycles;
6. Cumulative duration of expansion (cum_d_exp) (hour)—seasonal sum of all expansion time lengths;
7. Average duration of expansion (avg_d_exp) (hour)—cumulative duration of expansion divided by number of stem cycles;
8. Cumulative amplitude of expansion (cum_exp) (mm)—seasonal sum of all expansions;
9. Average amplitude of expansion (avg_exp) (mm)—cumulative amplitude of expansion divided by number of stem cycles;
10. Cumulative duration of increment (cum_d_inc) (hour)—seasonal sum of all increment time lengths;
11. Average duration of increment (avg_d_inc) (hour)—cumulative duration of increment divided by number of stem cycles;
12. Cumulative increment (cum_inc) (mm)—seasonal sum of all daily increments;
13. Average daily increment (avg_inc) (mm)—cumulative increment divided by number of stem cycles.

PCA (principal components analysis) was performed in R using the "factoextra" package. The highly correlated variables were removed based on the correlation circle, and hierarchical clustering was performed with 8 variables (Figure 3) to assess differences and similarities between studied trees in the monitored years. To identify groups of trees with similar values of water status indicators, we used hierarchical clustering with principal components. The individual trees and years were clustered in a hierarchical tree using Ward's criterion. The generated clusters were mapped in genuine PCA axes, and K-means clustering was applied to individual trees and years based on the nearest distance between cluster means and individuals. Hierarchical clustering with principal components was performed in R using the "FactoMiner", and data were visualised using "ggplot2". The number of desired clusters was set prior to the analysis.

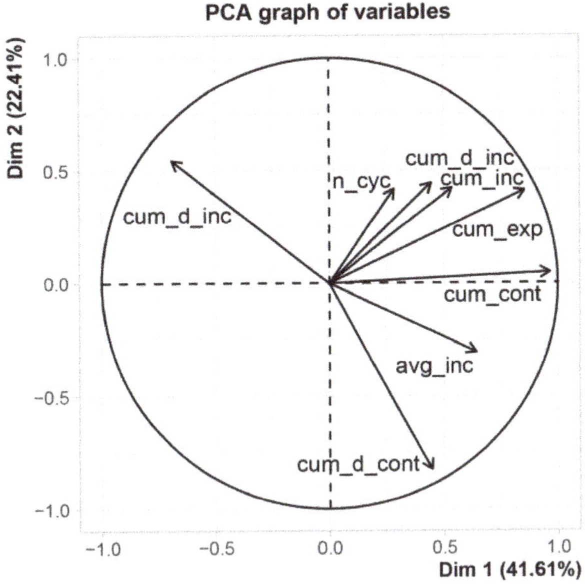

Figure 3. Variables describing radial stem growth pattern and water status derived from band dendrometer records used in principal component analysis (cum_d_inc (h)—cumulative duration of increment, cum_d_cont (h)—cumulative duration of contraction, avg_inc (mm)—average daily increment, cum_cont (mm)—cumulative contraction, cum_exp (mm)—cumulative expansion, cum_inc (mm)—cumulative increment, cum_d_inc (h)—cumulative duration of increment, n_cyc—number of cycles).

2.7. Periodicity of BDR

We performed a wavelet analysis [50] to examine significant periodicities in BDR ranging from hours to weeks. Specifically, we analysed residuals of BDR time series of individual years to respective fitted Weibull functions that were transformed using the Morlet transformation to distinguish random fluctuations from periodic events [50]. Wavelet analysis was performed using WaveletComp R package [51]. The lower period was set to 20 min intervals, while the upper one to 1024 20-min intervals (covering 16 days).

3. Results

3.1. Environmental Conditions During the Studied Periods 2015–2019

The measured meteorological data representing the years 2015–2019 were compared with the long-term normal (1961–1990) calculated from the nearest meteorological station (Sliač, 300 m a.s.l., 3.5 km from the study area of Borová hora) (Table 2). All mean temperatures representing observed periods (April–October) of the years 2015–2019 were more than 1.5 °C higher in comparison to the 30-year-long average (1961–1990) at the Sliač station (Table 2). Monthly average air temperatures were above their respective long-term values in almost all examined months of the studied years (Figure 4). Below-average temperatures were only observed in October 2016 and in May 2019 (Figure 4). The temporal precipitation distribution varied during the study periods. The observed period of the year 2018 was the warmest and driest from all periods between 2015 and 2019, and it was characterised by low monthly precipitation totals during the whole period (Table 2, Figure 4. Below-average precipitation total was also observed in the 2019 period) (Table 2). Precipitation total in the 2015 period was by 5% greater than the long-term average (Table 2) but mainly because of high precipitation in October 2015 (Figure 4). Precipitation totals in 2016 and 2017 exceeded the long-term average due to several rainy months (July, August, and October 2016 and July, September, and October 2017; Figure 4; Table 2). The lowest seasonal mean of VPD was observed in 2016, while the highest value was reached in 2018 (Table 2). Higher air temperature and lower precipitation total in 2018 resulted in higher values of vapour pressure deficit in comparison with other years (Table 2).

Table 2. Climatic characteristics representing the period April–October (A–O) of the years 2015–2019, where N (%) is percentage of long-term normal (long-term average calculated from the years 1961–1990), Mean is seasonal mean air temperature of the respective period, Difference from Normal is difference between mean air temperature and long-term average for the years 1961–1990, vapour pressure deficit (VPD) is seasonal mean vapour pressure deficit.

Month	Precipitation (mm)		Air Temperature (°C)		VPD
	Precipitation Total (mm)	N (%)	Mean (°C)	Difference from Normal (°C)	(kPa)
(A-O) 2015	412.2	105	15.7	2.2	0.552
(A-O) 2016	512.3	125	15.3	1.8	0.455
(A-O) 2017	501.0	124	15.2	1.6	0.500
(A-O) 2018	320.8	75	17.0	3.5	0.557
(A-O) 2019	387.2	92	15.8	2.3	0.480

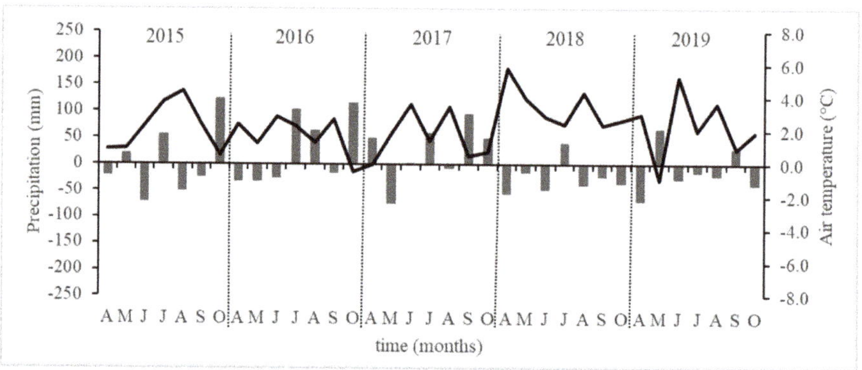

Figure 4. Anomalies of monthly precipitation totals (bars) and monthly mean air temperature values (lines) in the studied period April–October (A–O) of the years 2015–2019 in comparison to long-term mean climate conditions in the period 1961–1990. The 0 line represents the long-term monthly mean air temperature and long-term monthly precipitation total.

3.2. Stem Growth and Tree Water Status Derived from BDR

Stem circumference records and tree species-specific seasonal stem circumference increment characteristics derived from BDR showed pronounced differences between species (Figure 5). More intensive radial growth was found for *A. alba* with low fluctuations over the season, while *L. decidua* showed pronounced seasonal fluctuations in stem increment (Figures 2 and 5). In all investigated years, greater seasonal radial increments were recorded for *A. alba* (Table 3). Dendrometer records showed the greatest annual radial growth of both species in 2017 (Table 3, Figure 5). In 2018, the annual stem circumference increment of *A. alba* was only a half of 2017 (Table 3). For *L. decidua*, the lowest value of radial increment was observed in 2015.

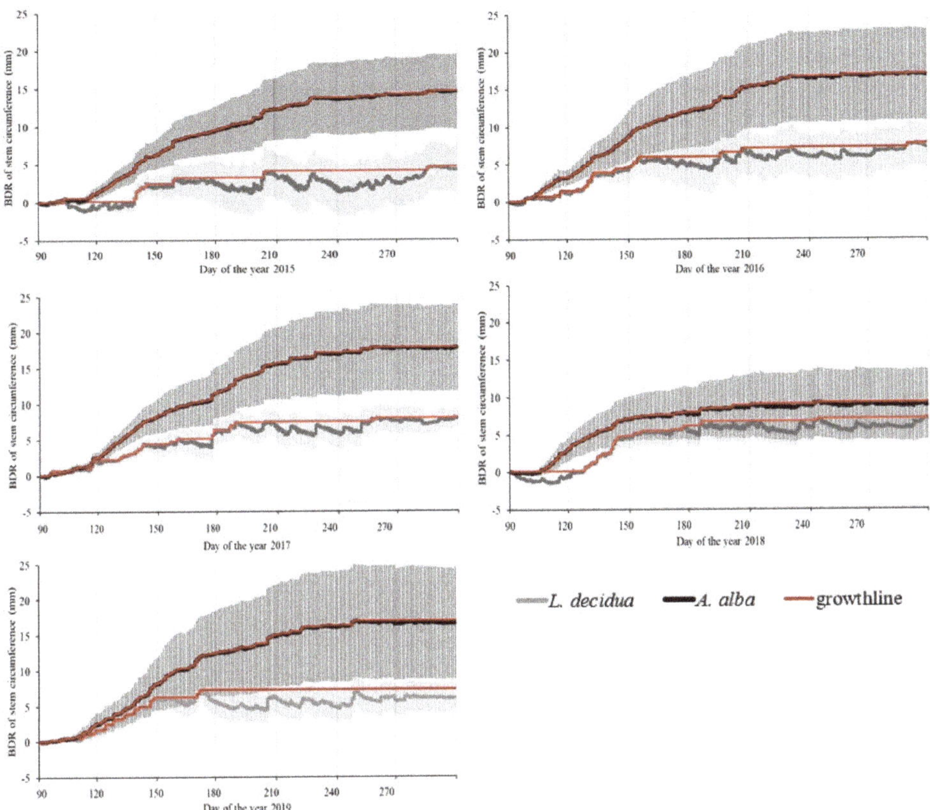

Figure 5. Seasonal courses of band dendrometer records (BDR) of stem circumference of *L. decidua* (dark grey line) and *A. alba* (black line) and growth lines (red lines) in individual years. Each line represents an average from five trees of the same species. Bars (light grey lines for *L. decidua* and dark grey lines for *A. alba*) represent standard deviations of BDR (*n* = 5).

Table 3. Seasonal species-specific characteristics of growth and stem water status dynamics, where cum_inc is average cumulated increment (mm), ΔWcum is average cumulated stem water deficit (mm), MDScum is average cumulated maximum shrinkage (mm), and SE is standard error of the respective characteristics.

	A. Alba			L. Decidua		
	Cum_Inc ± SE	**ΔWcum ± SE**	**MDScum ± SE**	**Cum_Inc ± SE**	**ΔWcum ± SE**	**MDScum ± SE**
2015	14.59 ± 2.18	−17.34 ± 0.54	29.35 ± 3.67	4.79 ± 1.33	−234.78 ± 28.33	68.11 ± 7.34
2016	17.04 ± 2.74	−14.69 ± 1.79	29.26 ± 4.49	7.74 ± 1.28	−146.51 ± 11.97	58.53 ± 4.63
2017	17.88 ± 2.71	−19.17 ± 1.61	32.52 ± 5.39	8.36 ± 0.72	−122.32 ± 11.35	53.29 ± 4.21
2018	9.34 ± 2.01	−49.60 ± 6.75	35.07 ± 6.32	7.41 ± 0.75	−169.86 ± 13.25	56.00 ± 3.60
2019	16.96 ± 3.55	−29.76 ± 3.71	37.31 ± 6.59	7.38 ± 0.82	−241.92 ± 22.03	48.12 ± 1.74

The greatest proportion of larch radial increment was usually created at the beginning of the periods (especially in May) except for the year 2015. In that year, continuous radial growth of larch occurred only within a short period in May (Figure 5). In subsequent months, larch radial circumference increased in a stepwise manner, usually after rain events (Figure 5). Unlike larch, fir continuously grew during the whole period of all years except for the year 2018, when we observed stagnation of fir stem circumference already at the beginning of June (Figure 5).

Significant differences in cumulative MDS between species were revealed for all years (Table 4), while species-specific seasonal cumulative maximum daily shrinkage was significantly higher for *L. decidua* than for *A. alba* (Table 3). We also observed differences in cumulative MDS between years, although these were not always significant. Greater differences in cumulative MDS between individual years were found for *L. decidua*. The two species differed in the years with the greatest and lowest cumulative MDS, since e.g., larch obtained the greatest MDS in 2015, while *A. alba* in 2019 (Table 4).

Table 4. Differences between species in growth and stem water status characteristics dynamics, where cum_inc is average cumulated increment (mm), ΔWcum is average cumulated stem water deficit (mm), MDScum is average cumulated maximum shrinkage (mm), * significant difference between species at 95% confidence level, ** significant difference between species at 99% confidence level, *** significant difference between species at 99.9% confidence level.

| | *p-values* | | |
	Cum_Inc	ΔWcum	MDScum
2015	0.005 **	0.000 ***	0.002 **
2016	0.015 *	0.000 ***	0.002 **
2017	0.009 **	0.000 ***	0.016 *
2018	0.400	0.000 ***	0.021 *
2019	0.030 *	0.000 ***	0.024 *

Extracted stem water deficit values indicated limited water storage relative to fully hydrated stem conditions. Increasingly negative values mean more pronounced lack of water in storage tissues. Over the examined periods, stem water deficit gradually decreased in species with different magnitudes (Figures 5–7). This trend was occasionally disrupted by precipitation events, after which stem water deficit reached values close to zero (Figures 5–7). Cumulative ΔW of *A. alba* reached the highest value in 2016 and the lowest value in 2018. Cumulative ΔW of *L. decidua* reached the highest value in 2017 and the lowest value in 2019 (Table 4). The cumulative values of stem water deficit in larch were in most cases ten times greater than in fir.

3.3. Tree Water Status and Environmental Conditions

We analysed the relationships between environmental factors and the daily changes of stem water status during the growing periods (1 April–30 October) of the years 2015–2019 using Spearman correlation coefficients. All but one correlation of *A. alba* stem water deficit to global radiation were significant (Table 5). The results showed that stem water deficit of both species was negatively correlated to daily mean and maximum air temperatures and VPD, while *L. decidua* was found to be more sensitive to them than *A. alba*. Moreover, *L. decidua* was also negatively correlated to daily minimum air temperature (Table 5). The highest positive correlation was revealed between ΔW and soil water potential for both species. The close positive correlation of stem water deficit and precipitation was detected, too (Table 5).

Table 5. Spearman rank-correlations between daily stem water deficit (ΔW) and maximum daily shrinkage (MDS) of two investigated tree species with daily environmental variables (GR—global radiation; ATmean—daily mean air temperature; ATmin—daily minimum air temperature; ATmax—daily maximum air temperature; P—daily precipitation total; P–1—previous day precipitation; RAH—daily mean relative air humidity; VPD—daily mean vapour pressure deficit; SWP—daily mean soil water potential) of the whole period (from 1 April to 30 October). Significance levels: * 95% significance; ** 99% significance; *** 99.9% significance.

		GR	ATmean	ATmin	ATmax	P	P-1	RAH	VPD	SWP
A. alba	ΔW	−0.04	−0.17 ***	0.08 *	−0.37 ***	0.20 ***	0.27 ***	0.17 ***	−0.24 ***	0.41 ***
	MDS	0.40 ***	0.48 ***	0.42 ***	0.44 ***	0.12 ***	−0.04	−0.25 ***	0.39 ***	−0.24 ***
L. decidua	ΔW	−0.08 **	−0.37 ***	−0.30 ***	−0.38 ***	0.15 ***	0.28 ***	0.14 ***	−0.30 ***	0.48 ***
	MDS	0.34 ***	0.57 ***	0.61 ***	0.49 ***	0.22 ***	0.09 **	−0.11 ***	0.33 ***	−0.09 **

Maximum daily shrinkage of both species was positively correlated with mean, minimum and maximum air temperature, precipitation, and VPD, and negatively correlated with relative air humidity and soil water potential (Table 5).

Morlet wavelet analysis revealed significant daily cycles in BDR ($p < 0.05$) of both tree species, which were notably more pronounced during the rainless periods (not shown here). The wavelet analysis confirmed more distinct diurnal stem variations in *L. decidua* compared to *A. alba*. Revealed daily periodicities correspond with MDS variation, which was greater in *L. decidua* (Figure 6, Table 4). Although MDS values of *A. alba* were small, Morlet analysis still revealed daily periodicities (Figure 7). Even small changes in BDR can indicate periodic events. Rainy events or SWP increase caused disturbances in daily periodicities. While periodicities shorter than one day were not significant in either of tree species (Figures 6 and 7), significant periodicities of several days up to 2 weeks occurred in wavelet spectra of both species. In the case of *L. decidua*, longer periodicities (from 8 days up to 16 days) occurred almost continuously over the studied periods (Figure 6). In the case of fir, longer periodicities were less frequent due to more continuous increase in BDR over time (Figure 5). Abrupt changes of ΔW after rainy events resulted in the co-occurrence of both daily and longer periodicities (Figure 6).

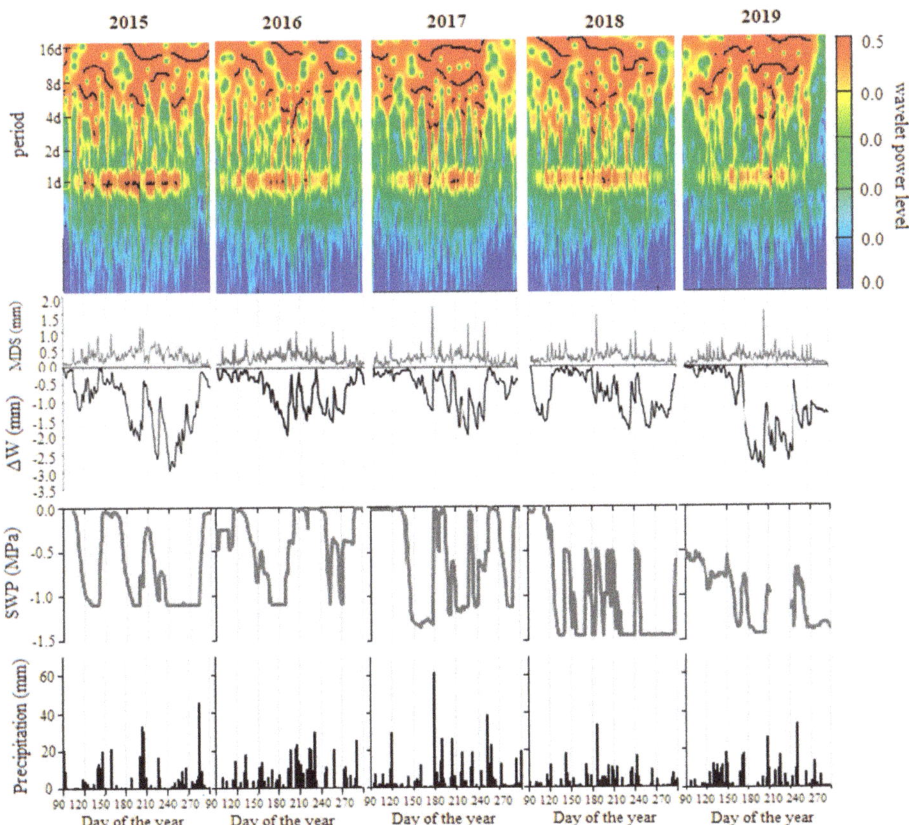

Figure 6. Morlet wavelet spectra of 20-min measured records of stem circumference for *L. decidua*, maximum daily shrinkage (MDS), stem water deficit (ΔW), daily precipitation (bars), and soil water potential (SWP, grey line) in the studied period (April–October) of the years 2015–2019. Dark red and white colours are assigned to the highest wavelet power spectra, whereas the dark blue colour is assigned to the lowest values. Wavelet power levels were set from 0.0 to 1.10^{-1}.

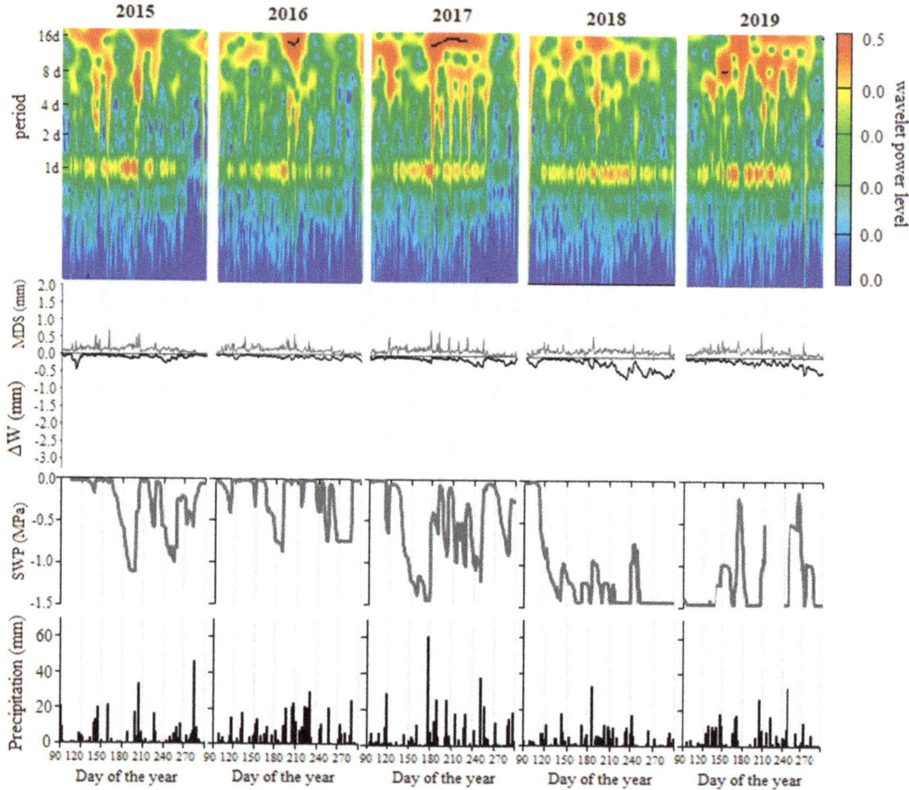

Figure 7. Morlet wavelet spectra of 20-min measured records of stem circumference for *A. alba*, maximum daily shrinkage (MDS), stem water deficit (ΔW), daily precipitation (bars), and soil water potential (SWP, grey line) in the studied period (April–October) of the years 2015–2019. Dark red and white colours are assigned to the highest wavelet power spectra, whereas the dark blue colour is assigned to the lowest values. Wavelet power levels were set from 0.0 to 1.10^{-1}.

The differences in growth responses and water status parameters between trees, species, and years were further analysed by cluster analysis (Figure 8). The two variables derived by principal component analysis describing growth patterns and water status in detail (Figure 3) were used as inputs for cluster analysis. From Figure 8a, we see that in the case of two clusters, the analysis clearly distinguished clusters of individual species. If we pre-defined five clusters for the whole dataset, the specified clusters still grouped trees of the same species, which indicates that the differences between species prevailed over the differences between the years (Figure 8b). Similarly, in the case of 10 clusters, these were still determined within individual tree species, but grouped different trees and different years together (not shown here). This suggests that the impact of genetics and climate on tree water status was mixed, and one did not prevail the other. The same result was revealed by cluster analysis performed for individual species (Figure 8c,d), for which we predefined five clusters to examine if the individual trees or years could be separated. In most cases, different trees and different years were combined into clusters, which indicates that each tree has its own sensitivity to environmental conditions, and its unique response to their various combinations. We interpret the occurrence of the particular tree in different clusters as its ability to react to different environmental conditions in a different way—i.e., its plasticity. Hence, the plasticity of *Abies* tree No. 3 was the lowest, as 4 out of 5 examined years occurred within one cluster of the given tree (Figure 8c).

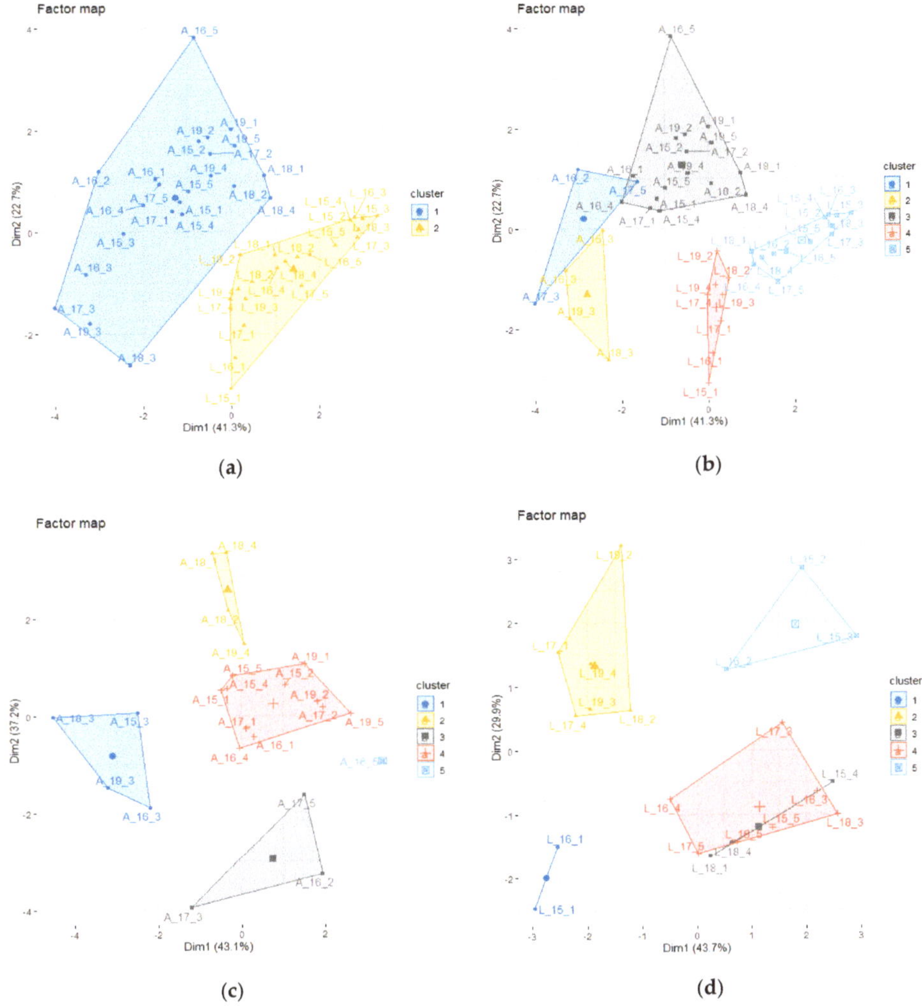

Figure 8. Species-specific responses of stem circumference to environmental conditions analysed by principal component analysis and cluster analysis with predefined two clusters for the whole dataset (**a**), five clusters for the whole dataset (**b**), five clusters for *A. alba* (**c**), and five clusters for *L. decidua* (**d**). Abbreviations in labels: A and L stand for *A. alba* and *L. decidua*, respectively; numbers 15–19 represent years (2015–2019); numbers 1–5 represent individuals of the respective tree species.

4. Discussion

In recent decades, various authors have analysed high temporal resolution stem growth data of a variety of tree species to better understand their growth responses—e.g., [45,52,53]. Tree growth is controlled by a vast array of conditions, out of which climate is considered as one of the most important. Climate envelope models visualise plant species distribution under contemporary climate [47]. Despite some limitations in their prediction capacity, which include their lack of ability to account for biotic interactions and evolutionary changes [54], this approach has a potential to estimate future spatial species distribution and their sensitivity under changing climate [55]. Our study was performed under conditions projected in coming decades [1,2] as the mean air temperature in the analysed growing seasons exceeded the long-term values of the current site (Table 2), as well as the values of their original natural habitats (Figure 1). Higher annual mean air temperature was observed at the study site also in the preceding years (2014: +2.8 °C; 2013: +1.5 °C; 2012: +1.7 °C). The observed reaction of stem radial growth and water status to environmental conditions in 2015–2019 reflected the species-specific distance from their current distribution described by main climate characteristics (Figure 1)—i.e., higher sensitivity was found for *L. decidua*—for which the long-term (1961–1990) and all monitored years climate conditions of the study site occurred outside the current species range (Figure 1). Weaker responses were revealed for *A. alba*, for which both the long-term (1961–1990) and the years 2015, 2016, and 2017 climate conditions occurred inside the current species range (Figure 1).

Seasonal differences in tree circumference growth result from species-specific temperature and/or photoperiod thresholds of cambial activity [56,57]. The increase in stem circumference of both species was mainly favoured by higher precipitation events. In particular, *L. decidua* grew in cascades separated by plateaus representing stagnation periods. The remarkable plateaus in BDR of *L. decidua* (Figure 5) observed during June and July in all years correspond with the prevalence of the contraction phase. The growth of *A. alba* seemed to be less limited by the prevailing dry conditions than *L. decidua* as the persistence of larger increments indicates (Figure 5), which might be due to its more effective stomatal control mechanism. We assume that limited growth of *L. decidua* at the studied site is a consequence of conditions that are beyond the climatic distribution of the species causing its higher stress and stimulating strong individual tree responses. Transpiration drives daily cycles in BDR [58,59], which are strictly dependent on soil water content and microclimatic conditions, and can quickly change according to weather conditions. The stem shrinkage corresponded to the periods when ΔW presented a clear decreasing trend, which can be associated with the exhaustion of the internal water storage. In the summer, soil water content diminishes and day length increases, which decreases recovery. Due to this, transpiration demands cannot be fulfilled, and the stem water deficit gradually increases [60]. Transpiration is controlled by stomatal responses to water availability and atmospheric conditions [59]. The physiological consequences of stomata closure are carbon starvation and secondary growth decline due to the allocation of carbon to higher ranking physiological processes such as root growth [59,61]. As a result, trees reduce their metabolism and enter in quiescence [62,63]. During the summer, trees could not compensate for daily water losses, presenting the most negative ΔW values. If a tree can no longer replenish the water lost by transpiration, then contraction would have to be restrained, which would result in a higher dependence between duration and amplitude.

The differences in drought responses between the species may be explained by their intrinsic differences in morphology and physiology. Silver fir has a longer wood formation period than larch (Figure 5). Similarly to our results, the beginning of fir growth in Slovenian forests was observed in early April and the end in late October [64,65]. In larch, we observed maximum radial increase at the beginning of the study periods, usually in May, after which stem circumference increased only sporadically (Figure 5). Early cessation of radial growth as a result of limited water availability has also been observed by Oberhuber et al. [45]. The functional significance of water storage in individual tree species depends on their water status regulation strategy, hydraulic architecture, and wood density [66]. Although *L. decidua* has been shown to develop a specific drought avoidance strategy by osmotic adjustment resulting from the accumulation of solutes [67], if growing at dry and low elevation sites, this species was found to be sensitive to water stress [68], especially during the summer months, which was also confirmed by our results (Figure 5). Weak adjustability of *L. decidua* to drought is possibly related to its deciduous habit [69] and/or anisohydric strategy [67], when high transpiration rates are maintained under drought finally causing impairment of tree water status [8]. MDS and ΔW of both tree species showed similar responses to all monitored environmental variables (Table 5). The positive correlation found between MDS and air temperature (mean, minimum, and maximum) suggests that elevated transpiration caused water loss and stem contraction [25,40]. During periods when temperatures are high and soil water content is lower, stomatal control on transpiration rates increases [59], and stem contraction is reduced. Coupling of ΔW to atmospheric conditions indicates that increasing temperature due to climate warming can negatively affect plant growth [70] as well as its water status through its effects on VPD, which increases exponentially with increasing temperature [71]. High sensitivity of stem water status to VPD and soil moisture was reported in several experimental studies [1,39,72]. High VPD reduces cell turgor pressure, which subsequently inhibits cell enlargement and growth [30,73]. High temperatures stimulate evaporation rates causing constraints in water availability, which is characterised by low soil water potential. SWP decline during rainless periods explains "plateaus" in seasonal courses of circumference variation (Figure 5). This is valid for *L. decidua*, but it is less pronounced for *A. alba*. The results showed that *A. alba* is a more resistant tree species to changing conditions defined by increased temperature and periodical drought than *L. decidua*. Due to this, fir is often considered as a prospective species under climate change [74], since fir productivity should not be adversely affected by increasing temperature [3,75]. The results of climate-growth relationships reflect the growth of tree species in response to water stress. Positive correlations between stem water deficit and precipitation and negative correlations with temperature suggest that tree water status and tree growth is limited by moisture. Increasingly negative numbers of ΔW indicate increasing drought stress (Table 3, Figures 5–7).

Diurnal pattern of stem circumference variation is also underlined by high power levels and regions of significant periodicities for both studied species. The most straightforward relationship is between SWP and precipitation (Figures 6 and 7). Wavelet analysis confirmed more pronounced daily cycles in *L. decidua* BDR than in *A. alba*. Diurnal periodicities became more distinct when soil drought occurred and SWP values decreased (Figures 6 and 7). This is coupled to MDS temporal course, since MDS represent short-term changes in stem water status. In rainless periods, the values of MDS increase due to water consumption and intensive transport of water from storage tissues to conductive tissues. MDS increases until it reaches a breaking point [76], which we, however, did not observe in our experiment. Wavelet analysis presents absolute values of periodic events, which can be linked to dry or wet periods. Due to this, the interpretation of its results is clearer if linked with ΔW. In the year 2018, fir radial growth was the lowest because the absolute values of ΔW were largest (Table 3), which was reflected in daily periodicity during the whole season (Figure 7). Tree water deficit reflects losses of water over a time longer than one day. Longer periodical cycles occurred either after rain events or as a result of drought, when ΔW gradually increased or decreased. The length of the period of ΔW changes in one direction specifies the periodicity interval. While longer periodicities in *A. alba* were influenced by rain events, periodicities in *L. decidua* were also affected by drought. In larch, we observed more frequent and greater power levels of longer periodicities than in fir, because ΔW of *L. decidua* changed more substantially over time (Figures 6 and 7). This results from the anisotropic character of larch, due to which this species uses more water from its tissues for transpiration than fir.

Cluster analysis identified two groups representing individual investigated species, which confirms the differences between fir and larch in their tree water status (Figure 8). However, the differences between individual trees or years could not be detected, as the identified clusters comprised different years and trees (Figure 8). This indicates that the impact of genetics and environment was mixed and none of them prevailed. Although we assumed that the occurrence of a single tree within multiple clusters may indicate its ability to react to specific environmental conditions, according to Carpenter and Brock [77], increasing variance in respective system processes serves as a decisive symptom of approaching a critical threshold. Hence, more evidence is needed before making a final conclusion.

5. Conclusions

Knowledge on the relationships between climate and growth is essential for assessing future performance of tree species under a warmer and drier climate. Species and communities might strongly differ in their responses to extreme climatic events. Understanding tree growth responses to climatic stresses is an essential element of climate change research because species-specific drought resistance will affect the development of forest ecosystems under changed climate by changing species composition and inducing shifts in forest distribution. Better understanding of species-specific effects of weather conditions on tree growth could help foresters to direct future forest management.

The results of our study confirmed our hypothesis that species-specific ecophysiological and morphological traits of *A. alba* and *L. decidua* led to significant differences in climate-growth relationships. Monitored species exhibited remarkably different growth patterns over the growing seasons characterised by highly above normal temperature and uneven precipitation distribution. More pronounced daily dehydration/rehydration changes were observed for *L. decidua*. Although *A. alba* had greater seasonal radial increments, and lower values of stem water status characteristics, in the year 2018 with extremely above normal temperature and below normal precipitation this species significantly reduced its increment and increased its stem water deficit because of the lowest SWP in that year. In the case of *L. decidua*, the impact of limited moisture conditions was not so straightforward, because the lowest increment was observed in the year 2015, while the worst soil water conditions characterised by the greatest stem water deficit were observed in the year 2019. We assume that this could be caused by longer rainless periods that occurred in these two years, while in the year 2018, rain events were more regular although of lower intensity. More long-term research and information needs to be collected to understand the response of individual tree species to changing conditions thoroughly.

Cluster analysis revealed significant differences between *A. alba* and *L. decidua*, but not between individual trees and years suggesting that the impact of genetics and environment on tree response was mixed and could not be separated.

Both species were able to cope with changing environmental conditions, and continued to grow under the conditions of above average air temperature and limited soil water. Long-term increase in air temperature, more frequent heat waves coupled with more intense and longer drought periods could affect the ability of species to respond to environmental changes. Therefore, further research is needed to contribute to elucidating individual responses of forest trees and individual coniferous species to external factors outside their natural habitat of environmental and climatic conditions.

Author Contributions: Conceptualization, A.L., P.F.J. and K.M.; methodology, A.L., P.F.J. and K.M.; validation, A.L. and P.F.J.; formal analysis, A.L. and P.F.J.; investigation, A.L.; data curation, A.L.; writing—original draft preparation, A.L.; writing—review and editing, A.L., P.F.J., P.F. and K.M.; supervision, K.S.; project administration, K.S.; funding acquisition, K.S. All authors have read and agreed to the published version of the manuscript.

Funding: The study was supported by research grants of the Slovak Research and Development Agency APVV-16-0306, APVV-18-0390, APVV-18-0347, APVV-16-0325, Slovak Grant Agency for Science no. VEGA 2/0049/18, the grant "EVA4.0", No. CZ.02.1.01/0.0/0.0/16_019/0000803 financed by OP RDE, and the project: Scientific support of climate change adaptation in agriculture and mitigation of soil degradation (ITMS2014+ 313011W580) supported by the Integrated Infrastructure Operational Programme funded by the ERDF.

Conflicts of Interest: The authors declare no conflict of interest.

References

1. Will, R.E.; Wilson, S.M.; Zou, C.B.; Hennessey, T.C. Increased vapour pressure deficit due to higher temperature leads to greater transpiration and faster mortality during drought for tree seedlings common to the forest-grassland ecotone. *New Phytol.* **2013**, *200*, 366–374. [CrossRef] [PubMed]

2. IPCC. *Climate Change 2013. The Physical Science Basis. Contribution of Working Group I to the Fifth Assessment Report of the Intergovernmental Panel on Climate Change*; Stocker, T.F., Qin, D., Plattner, G.-K., Tignor, M., Allen, S.K., Boschung, J., Nauels, A., Xia, Y., Bex, V., Midgley, P.M., Eds.; Cambridge University Press: Cambridge, UK, 2013.

3. Usoltsev, V.; Merganičová, K.; Konôpka, B.; Osmirko, A.A.; Tsepordey, I.S.; Chasovskikh, V.P. Fir (*Abies* spp.) stand biomass additive model for Eurasia sensitive to winter temperature and annual precipitation. *Cent. Eur. For. J.* **2019**, *65*, 166–172. [CrossRef]

4. Albert, M.; Hansen, J.; Nagel, J.; Schmidt, M.; Spellmann, H. Assessing risks and uncertainties in forest dynamics under different management scenarios and climate change. *For. Ecosyst.* **2015**, *2*, 14. [CrossRef]

5.   IPCC. *Climate Change 2014: Impacts, Adaptation, and Vulnerability. Part A: Global and Sectoral Aspects. Contribution of Working Group II to the Fifth Assessment Report of the Intergovernmental Panel on Climate Change*; Cambridge University Press: Cambridge, UK; New York, NY, USA, 2014.

6.   Vitali, V.; Büntgen, U.; Bauhus, J. Silver fir and Douglas fir are more tolerant to extreme droughts than Norway spruce in south-western Germany. *Glob. Chang. Biol.* **2017**, *23*, 5108–5119. [CrossRef] [PubMed]

7.   Chauvin, F.; Denvil, S. Changes in severe indices as simulated by two French coupled global climate models. *Global Planet. Chang.* **2007**, *57*, 96–117. [CrossRef]

8.   Bréda, N.; Huc, R.; Granier, A.; Dreyer, E. Temperate forest trees and stands under severe drought: A review of ecophysiological responses, adaptation processes and long-term consequences. *Ann. For. Sci.* **2006**, *63*, 625–644. [CrossRef]

9.   Lebourgeois, F. Climatic signal in annual growth variation of silver fir (*Abies alba* Mill.) and spruce (*Picea abies* Karst.) from the French Permanent Plot Network (RENECOFOR). *Ann. For. Sci.* **2007**, *64*, 333–343. [CrossRef]

10.   Lévesque, M.; Saurer, M.; Siegwolf, R.; Eilmann, B.; Brang, P.; Bugmann, H.; Rigling, A. Drought response of five conifer species under contrasting water availability suggests high vulnerability of Norway spruce and European larch. *Glob. Chang. Biol.* **2013**, *19*, 3184–3199. [CrossRef]

11.   Danek, M.; Chuchro, M.; Walanus, A. Variability in Larch (*Larix Decidua* Mill.) Tree-ring growth response to climate in the polish Carpathian Mountains. *Forests* **2017**, *8*, 354. [CrossRef]

12.   Schulze, E.D.; Čermák, J.; Matyssek, R.; Penka, M.; Zimmermann, R.; Vasícek, F.; Gries, W.; Kučera, J. Canopy transpiration and water fluxes in the xylem of the trunk of *Larix* and *Picea* trees—A comparison of xylem flow, porometer and cuvette measurements. *Oecologia* **1985**, *66*, 475–483. [CrossRef]

13.   Ellenberg, H. *Vegetation Ecology of Central Europe*, 4th ed.; Cambridge University Press: Cambridge, UK, 2009.

14.   van der Maaten-Theunissen, M.; Kahle, H.-P.; van der Maaten, E. Drought sensitivity of Norway spruce is higher than that of silver fir along an altitudinal gradient in southwestern Germany. *Ann. For. Sci.* **2013**, *70*, 185–193. [CrossRef]

15.   Ruosch, M.; Spahni, R.; Joos, F.; Henne, P.D.; van der Knaap, W.O.; Tinner, W. Past and future evolution of *Abies alba* forests in Europe-comparison of a dynamic vegetation model with palaeo data and observations. *Glob. Chang. Biol.* **2016**, *22*, 727–740. [CrossRef]

16.   Bošela, M.; Petráš, R.; Sitková, Z.; Priwitzer, T.; Pajtík, J.; Hlavatá, H.; Sedmák, R.; Tobin, B. Possible causes of the recent rapid increase in the radial increment of silver fir in the Western Carpathians. *Environ. Pollut.* **2014**, *184*, 211–221. [CrossRef]

17.   Büntgen, U.; Tegel, W.; Kaplan, J.O.; Schaub, M.; Hagedorn, F.; Bürgi, M.; Brázdil, R.; Helle, G.; Carrer, M.; Heussner, K.U.; et al. Placing unprecedented recent fir growth in a European-wide and Holocene-long context. *Front. Ecol. Environ.* **2014**, *12*, 100–106. [CrossRef]

18.   Hartmann, H.; Moura, C.F.; Anderegg, W.R.L.; Ruehr, N.K.; Salmon, Y.; Allen, C.D.; Arndt, S.K.; Breshears, D.D.; Davi, H.; Galbraith, D.; et al. Research frontier for improving our understanding of drought-induced tree and forest mortality. *New Phytol.* **2018**, *218*, 15–28. [CrossRef]

19.   Liu, Y.Y.; Wang, A.Y.; An, Y.N.; Lian, P.Y.; Wu, D.D.; Zhu, J.J.; Meinzer, F.C.; Hao, G.Y. Hydraulics play an important role in causing low growth rate and dieback of aging *Pinus sylvestris* var. mongolica trees in plantations of Northeast China. *Plant Cell Environ.* **2018**, *41*, 1500–1511. [CrossRef] [PubMed]

20.   Kaiser, W.M. Effects of water deficit on photosynthetic capacity. *Physiol. Plant.* **1987**, *71*, 142–149. [CrossRef]

21.   Chaves, M.M.; Maroco, J.P.; Pereira, J.S. Understanding plant responses to drought—From genes to the whole plant. *Funct. Plant Biol.* **2003**, *30*, 239–264. [CrossRef] [PubMed]

22.   Kramer, P.J. *Water Relation of Plants*; Academic Press: New York, NY, USA, 1983.

23.   Peramaki, M.; Nikinmaa, E.; Sevanto, S.; Ilvesniemi, H.; Siivola, E.; Hari, P.; Vesala, T. Tree stem diameter variations and transpiration in Scots pine: An analysis using a dynamic sap flow model. *Tree Physiol.* **2001**, *21*, 889–897. [CrossRef] [PubMed]

24.   Daudet, F.A.; Améglio, T.; Cochard, H.; Archilla, O.; Lacointe, A. Experimental analysis of the role of water and carbon in tree stem diameter variations. *J. Exp. Bot.* **2005**, *56*, 135–144. [CrossRef]

25.   Čermák, J.; Kučera, J.; Bauerle, W.L.; Phillips, N.; Hinckley, T.M. Tree water storage and its diurnal dynamics related to sap flow and changes in stem volume in old-growth Douglas-fir trees. *Tree Physiol.* **2007**, *27*, 181–198. [CrossRef] [PubMed]

26. Peramaki, M.; Vesala, T.; Nikinmaa, E. Modeling the dynamics of pressure propagation and diameter variation in tree sapwood. *Tree Physiol.* **2005**, *25*, 1091–1099. [CrossRef] [PubMed]

27. Hinckley, T.; Lassoie, J. Radial growth in conifers and deciduous trees: A comparison. *Mitt.-Vienna Forstl. Bundesversuchsanstalt* **1981**, *142*, 17–56.

28. Whitehead, D.; Jarvis, P.G. Coniferous Forests and Plantations. In *Water Deficits and Plant Growth*; Kozlowski, T.T., Ed.; Academic Press: New York, NY, USA, 1981; Volume VI, pp. 49–152.

29. Herzog, K.M.; Thum, R.; Häsler, R. Diurnal changes in the radius of a subalpine Norway spruce stem: Their relation to the sap flow and their use to estimate transpiration. *Trees* **1995**, *10*, 94–101. [CrossRef]

30. Zweifel, R.; Zimmermann, L.; Newbery, D.M. Modeling tree water deficit from microclimate: An approach to quantifying drought stress. *Tree Physiol.* **2005**, *25*, 147–156. [CrossRef] [PubMed]

31. Ehrenberger, W.; Rüger, S.; Fitzke, R.; Vollenweider, P.; Günthardt-Goerg, M.S.; Kuster, T.; Zimmermann, U.; Arend, M. Concomitant dendrometer and leaf patch pressure probe measurements reveal the effect of microclimate and soil moisture on diurnal stem water and leaf turgor variations in young oak trees. *Funct. Plant. Biol.* **2012**, *39*, 297–305. [CrossRef] [PubMed]

32. Sánchez-Costa, E.; Poyatos, R.; Sabaté, S. Contrasting growth and water use strategies in four co-occurring Mediterranean tree species revealed by concurrent measurements of sap flow and stem diameter variations. *Agric. For. Meteorol.* **2015**, *207*, 24–37. [CrossRef]

33. Sevanto, S.; Hölttä, T.; Markkanen, T.; Perämäki, M.; Nikinmaa, E.; Vesala, T. Relationships between diurnal xylem diameter variation and environmental factors in Scots pine. *Boreal Environ. Res.* **2005**, *10*, 447–458.

34. Steppe, K.; Sterck, F.; Deslauriers, A. Diel growth dynamics in tree stems: Linking anatomy and ecophysiology. *Trends Plant Sci.* **2015**, *20*, 335–343. [CrossRef]

35. Zweifel, R.; Drew, D.M.; Schweingruber, F.; Downes, G.M. Xylem as the main origin of stem radius changes in eucalyptus. *Funct. Plant Biol.* **2014**, *41*, 520–534. [CrossRef]

36. Zweifel, R.; Item, H.; Häsler, R. Stem radius changes and their relation to stored water in stems of young Norway spruce trees. *Trees Struct. Funct.* **2000**, *15*, 50–57. [CrossRef]

37. Deslauriers, A.; Morin, H.; Urbinati, C.; Carrer, M. Daily weather response of balsam fir (*Abies balsamea* (L.) Mill.) stem radius increment from dendrometer analysis in the boreal forests of Quebec (Canada). *Trees* **2003**, *17*, 477–484. [CrossRef]

38. Turcotte, A.; Morin, H.; Krause, C.; Deslauriers, A.; Thibeault-Martel, M. The timing of spring rehydration and its relation with the onset of wood formation in black spruce. *Agric. For. Meteorol.* **2009**, *149*, 1403–1409. [CrossRef]

39. Köcher, P.; Horna, V.; Leuschner, C. Environmental control of daily stem growth patterns in five temperate broad-leaved tree species. *Tree Physiol.* **2012**, *32*, 1021–1032. [CrossRef]

40. Zweifel, R.; Item, H.; Häsler, R. Link between diurnal stem radius changes and tree water relations. *Tree Physiol.* **2001**, *21*, 869–877. [CrossRef]

41. Drew, D.M.; Downes, G.M. The use precision dendrometers in research on daily stem size and wood property variation: A review. *Dendrochronologia* **2009**, *27*, 159–172. [CrossRef]

42. Downes, G.; Beadle, C.; Worledge, D. Daily stem growth patterns in irrigated *Eucalyptus globulus* and *E. nitens* in relation to climate. *Trees* **1999**, *14*, 102–111. [CrossRef]

43. Deslauriers, A.; Rossi, S.; Anfondillo, T. Dendrometer and intra-annual tree growth: What kind of information can be inferred? *Dendrochronologia* **2007**, *25*, 113–124. [CrossRef]

44. Leštianska, A.; Merganičová, K.; Merganič, J.; Střelcová, K. Intra-annual patterns of weather and daily radial growth changes of Norway spruce and their relationship in the Western Carpathian mountain region over a period of 2008–2012. *J. For. Sci.* **2015**, *61*, 315–324. [CrossRef]

45. Oberhuber, W.; Gruber, A.; Kofler, W.; Swidrak, I. Radial stem growth in response to microclimate and soil moisture in a drought-prone mixed coniferous forest at an inner Alpine site. *Eur. J. For. Res.* **2014**, *133*, 467–479. [CrossRef]

46. Lukáčik, I. Arborétum Borová hora—História, súčasnosť a perspektívy. In *Dendroflora of Central Europe—Utilization of Knowledge in Research, Education and Practice*; Lukáčik, I., Sarvašová, I., Eds.; Vydavateľstvo TU vo Zvolene: Zvolen, Slovakia, 2015; pp. 9–19.

47. Kölling, C. Klimahüllen für 27 Baumarten. *AFZ-DerWald* **2007**, *62*, 1242–1245.

48. Oberhuber, W.; Hammerle, A.; Kofler, W. Tree water status and growth of saplings and mature Norway spruce (*Picea abies*) at a dry distribution limit. *Front. Plant Sci.* **2015**, *6*, 703. [CrossRef] [PubMed]

49. van der Maaten, E.; van der Maaten-Theunissen, M.; Smiljanić, M.; Rossi, S.; Simard, S.; Wilmking, M.; Deslauriers, A.; Fonti, P.; von Arx, G.; Bouriaud, O. dendrometeR: Analyzing the pulse of tree in R. *Dendrochronologia* **2016**, *40*, 12–16. [CrossRef]

50. Torrence, C.; Compo, G.P. A practical guide to wavelet analysis. *Bull. Am. Meteorol. Soc.* **1998**, *79*, 61–78. [CrossRef]

51. Rösch, A.; Schmidbauer, H. WaveletComp1.1: A Guided Tour through the Rpackage. 2018. Available online: http://www.Hsstat.Com/Projects/WaveletComp/WaveletComp_guided_tour.Pdf (accessed on 12 September 2020).

52. Eilmann, B.; Rigling, A. Tree-growth analyses to estimate tree species' drought tolerance. *Tree Physiol.* **2012**, *32*, 178–187. [CrossRef]

53. Vieira, J.; Rossi, S.; Campelo, F.; Freitas, H.; Nabais, C. Seasonal and daily cycles of stem radial variation of pinus pinaster in a drought-prone environment. *Agric. For. Meteorol.* **2013**, *180*, 173–181. [CrossRef]

54. Ibáñez, I.; Clark, J.S.; Dietze, M.C.; Feeley, K.; Hersh, M.; LaDeau, S.; McBride, A.; Welch, N.E.; Wolosin, M.S. Predicting biodiversity change: Outside the climate envelope, beyond the species-area curve. *Ecology* **2006**, *87*, 1896–1906. [CrossRef]

55. Walentowski, H.; Falk, W.; Mette, T.; Kunz, J.; Bräuning, A.; Melnardus, C.; Zang, C.; Sucliffe, L.; Leuschner, C. Assessing future suitability of tree species under climate change by multiple methods: A case study in southern Germany. *Ann. For. Res* **2017**, *60*, 101–126. [CrossRef]

56. Begum, S.; Nakaba, S.; Yamagishi, Y.; Oribe, Y.; Funada, R. Regulation of cambial activity in relation to environmental conditions: Understanding the role of temperature in wood formation of trees. *Physiol. Plant.* **2013**, *147*, 46–54. [CrossRef] [PubMed]

57. Körner, C.; Basler, D. Phenology under global warming. *Science* **2010**, *327*, 1461–1462. [CrossRef]

58. Kozlowski, T.T. *Water Deficits and Plant Growth. Vol III. Plant Responce and Control of Water Balance*; Academic Press: New York, NY, USA, 1976; 368p.

59. Zweifel, R.; Zimmermann, L.; Zeugin, F.; Newberry, D.M. Intra-annual radial growth and water relations of trees: Implications towards a growth mechanism. *J. Exp. Bot.* **2006**, *57*, 1445–1459. [CrossRef] [PubMed]

60. Devine, W.D.; Harrington, C.A. Factors affecting diurnal stem contraction in young Douglas-fir. *Agric. For. Meteorol.* **2011**, *151*, 414–419. [CrossRef]

61. Chaves, M.M.; Pereira, J.S.; Maroco, J.; Rodrigues, M.L.; Ricardo, C.P.P.; Osorio, M.L.; Carvalho, I.; Faria, T.; Pinheiro, C. How plants cope with water stress in the field? Photosynthesis and growth. *Ann. Bot.* **2002**, *89*, 907–916. [CrossRef] [PubMed]

62. Cherubini, P.; Gartner, B.L.; Tognetti, R.; Bräker, O.U.; Schoch, W.; Innes, J.L. Identification, measurement and interpretation of tree rings in woody species from Mediterranean climates. *Biol. Rev.* **2003**, *78*, 119–148. [CrossRef] [PubMed]

63. McDowell, N.; Pockman, W.T.; Allen, C.D.; Breshears, D.D.; Cobb, N.; Kolb, T.; Plaut, J.; Sperry, J.; West, A.; Williams, D.G.; et al. Mechanisms of plant survival and mortality during drought: Why do some plants survive while others succumb to drought? *New Phytol.* **2008**, *178*, 719–739. [CrossRef]

64. Gričar, J.; Čufar, K. Seasonal dynamics of phloem and xylem formation in silver fir and Norway spruce as affected by drought. *Russ. J. Plant Physiol.* **2008**, *55*, 538–543. [CrossRef]

65. Swidrak, I.; Gruber, A.; Oberhuber, W. Xylem and phloem phenology in co-occurring conifers exposed to drought. *Trees* **2014**, *28*, 1161–1171. [CrossRef]

66. Köcher, P.; Horna, V.; Leuschner, C. Stem water storage in five coexisting temperate broad-leaved tree species: Significance, temporal dynamics and dependence on tree functional traits. *Tree Physiol.* **2013**, *33*, 817–832. [CrossRef]

67. Anfodillo, T.; Rento, S.; Carraro, V.; Furlanetto, L.; Urbinati, C.; Carrer, M. Tree water relations and climatic variations at the alpine timberline: Seasonal changes of sap flux and xylem water potential in *Larix decidua* Miller, *Picea abies* (L.) Karst and *Pinus cembra* L. *Ann. For. Sci.* **1998**, *55*, 159–172. [CrossRef]

68. Saulnier, M.; Corona, C.; Stoffel, M.; Guibal, F.; Edouard, J.-L. Climate-growth relationships in a Larix decidua Mill. Network in the French Alps. *Sci. Total Environ.* **2019**, *664*, 554–566. [CrossRef]

69. Migliavacca, M.; Cremonese, E.; Colombo, R.; Busetto, L.; Galvagno, M.; Ganis, L.; Meroni, M.; Pari, E.; Rossini, M.; Siniscalco, C.; et al. European larch phenology in the Alps: Can we grasp the role of ecological factors by combining field observations and inverse modelling? *Int. J. Biometeorol.* **2008**, *52*, 587–605. [CrossRef] [PubMed]

70. Battipaglia, G.; Saurer, M.; Cherubini, P.; Siegwolf, R.T.W.; Cotrufo, M.F. Tree rings indicate different drought resistance of a native (*Abies alba* Mill.) and a non-native (*Picea abies* (L.) Karst.) species co-occurring at a dry site in Southern Italy. *For. Ecol. Manag.* **2009**, *257*, 820–828. [CrossRef]

71. Breshears, D.D.; Adams, H.D.; Eamus, D.; McDowell, N.; Law, D.J.; Will, R.E.; Williams, A.P.; Zou, C.B. The critical amplifying role of increasing atmospheric moisture demand on tree mortality and associated regional die-off. *Front. Plant Sci.* **2013**, *4*, 266. [CrossRef] [PubMed]

72. Oberhuber, W.; Gruber, A. Climatic influences on intra-annual stem radial increment of *Pinus sylvestris* (L.) exposed to drought. *Trees* **2010**, *24*, 887–898. [CrossRef]

73. Steppe, K.; De Pauw, D.J.W.; Lemeur, R.; Vanrolleghem, P.A. A mathematical model linking tree sap flow dynamics to daily stem diameter fluctuations and radial stem growth. *Tree Physiol.* **2006**, *26*, 257–273. [CrossRef]

74. Lindner, M.; Garcia-Gonzalo, J.; Kolstrom, M.; Green, T.; Reguera, R.; Maroschek, M.; Seidl, R.; Lexer, M.J.; Netherer, S.; Schopf, A.; et al. *Impacts of Climate Change on European Forests and Options for Adaptation*; Report to the European Commission Directorate-General for Agriculture and Rural Development; European Forestry Institute: Joensuu, Finland, 2008; 173p.

75. Bošeľa, M.; Lukáč, M.; Castagneri, D.; Sedmák, R.; Biber, P.; Carrer, P.; Konôpka, B.; Nola, P.; Thomas, A.; Nagel, T.A.; et al. Contrasting effects of environmental change on the radial growth of co-occurring beech and fir trees across Europe. *Sci. Total Environ.* **2018**, *615*, 1460–1469. [CrossRef]

76. Moreno, F.; Conejero, W.; Martín-Palomo, M.J.; Girón, I.F.; Torrecillas, A. Maximum daily trunk shrinkage reference values for irrigation scheduling in olive trees. *Agric. Water Manag.* **2006**, *84*, 290–294. [CrossRef]

77. Carpenter, S.R.; Brock, W.A. Rising variance: A leading indicator of ecological transition. *Ecol. Lett.* **2006**, *9*, 311–318. [CrossRef]

Publisher's Note: MDPI stays neutral with regard to jurisdictional claims in published maps and institutional affiliations.

Article

Autumn Phenological Response of European Beech to Summer Drought and Heat

Veronika Lukasová [1],*, Jaroslav Vido [2], Jana Škvareninová [3], Svetlana Bičárová [1], Helena Hlavatá [4], Peter Borsányi [5] and Jaroslav Škvarenina [2],*

[1] Earth Science Institute, Slovak Academy of Sciences, Stará Lesná, 059 60 Tatranská Lomnica, Slovakia; bicarova@ta3.sk

[2] Faculty of Forestry, Technical University in Zvolen, T. G. Masaryka 24, 960 53 Zvolen, Slovakia; vido@tuzvo.sk

[3] Faculty of Ecology and Environmental Sciences, Technical University in Zvolen, T. G. Masaryka 24, 960 53 Zvolen, Slovakia; skvareninova@tuzvo.sk

[4] Slovak Hydrometeorological Institute, Ďumbierska 26, 040 01 Košice, Slovakia; Helena.Hlavata@shmu.sk

[5] Slovak Hydrometeorological Institute, Zelená 5, 974 04 Banská Bystrica, Slovakia; Peter.Borsanyi@shmu.sk

* Correspondence: vlukasova@ta3.sk (V.L.); jaroslav.skvarenina@tuzvo.sk (J.Š.); Tel.: +42-190-211-7277 (V.L.)

Received: 6 August 2020; Accepted: 16 September 2020; Published: 18 September 2020

Abstract: The changes in precipitation and temperature regimes brought on by the current climate change have influenced ecosystems globally. The consequences of climate change on plant phenology have been widely investigated during the last few years. However, the underlying causes of the timing of autumn phenology have not been fully clarified yet. Here, we focused on the onset (10%) of leaf colouring—LCO—(Biologische Bundesanstalt, Bundessortenamt und Chemische Industrie (BBCH) 92) of European beech (*Fagus sylvatica*, L.) as an important native tree species growing throughout Europe. Studied beech stands are located along the natural distribution range of the European beech in Western Carpathians (Slovakia) at different altitudes from lowlands (300 m a.s.l.) to uplands (1050 m a.s.l.) and climatic regions from warm to cold. To define limiting climate conditions for LCO, we established several bioclimatic indices as indicators of meteorological drought: climatic water balance (CWB), standardized precipitation index (SPI), standardized precipitation-evapotranspiration index (SPEI), dry period index (DPI), and heat waves (HW). In addition, meteorological variables such as monthly mean temperatures and precipitation totals were taken into account. Throughout the 23-year period (1996–2018) of ground-based phenological observations of temperate beech forests, the timing of LCO was significantly delayed ($p \leq 0.05$) in the middle to high altitudes, while in the lowest altitude, it remained unchanged. Over the last decade, 2009–2018, LCO in middle altitudes started at comparable to low altitudes and, at several years, even later. This resulted mainly from the significant negative effect of drought prior to this phenological phase ($p \leq 0.01$) expressed through a 1-month SPI in September (SPI_{IX}) at the stand at the low-altitude and warm-climatic region. Our results indicate that the meteorological drought conditioned by lower total precipitation and higher evapotranspirative demands in the warmer climate advance leaf senescence. However, at present time, growth in rising temperature and precipitation is acceptable for most beech stands at middle to high altitudes. Beech utilizes these conditions and postpones the LCO by 0.3–0.5 and 0.6–1.2 day per year at high and middle altitudes, respectively. Although we show the commencing negative effect of drought at mid-altitudes with lower (below 700 mm) total annual precipitation, the trend of LCO in favourable warm climates is still significantly delayed. The ongoing warming trend of summer months suggests further intensification of drought as has started to occur in middle altitudes, spreading from the continual increase of evapotranspiration over the next decades.

Keywords: leaf colouring; phenological shifts; climate change; *Fagus sylvatica* L.; altitude; trends

1. Introduction

Under climate change, there is an alteration in the temperature and precipitation regimes. The mean global air temperature has increased over the last decades, and the occurrence of temperature extremes has increased [1]. Furthermore, continued enhancement of frequency and severity of drought events, particularly in the Northern hemisphere, is expected [2–4]. Models of future climate for Central Europe predict increased frequency and severity of heat stress and drought [5,6]. For most of the temperate continental zone, an increase of annual mean temperature of 3–4 °C might be expected until the end of the century [7,8]. The increase in temperature is accompanied by enhanced evapotranspirative demand imposed on the forest [9] and agricultural ecosystems [10], and these, if undersupplied with water, will dry [11]. Heat waves and droughts have significant impacts on their own, but when they occur simultaneously, their negative effects could be greatly intensified. Although drought stress highly depends on meteorological conditions, it relates to the power of tolerance and resistance of plants.

This study focuses on the European beech (*Fagus sylvatica*, L., ssp. *sylvatica*) response to such changing climate evaluated in natural conditions. European beech is one of the most important native tree species in Europe with the spatial distribution covering most of the European continent [12]. Due to high genetic diversity and phenotypic plasticity, *F. sylvatica* is adapted to different environments and altitudinal zones within its natural range [13]. This makes it a good reference species for large-scale studies of plastic and adaptive responses in its fitness-related traits to climate change over its full distribution range [14]. With regard to the altitudinal distribution of European beech, it is assumed that beech forests growing at low altitudes will suffer from drought [8,15–17]. Forests exposed to drought are mostly endangered by biotic factors, such as documented declination of beech vitality followed by the biotic damage and mortality in Hungarian and Slovenian forests [15,18]. Furthermore, under limited water availability, beech may suffer from xylem embolism, restricted nutrient uptake capacity, and reduced growth. Similarly, negative impact on nutrient uptake and growth of beech could have waterlogged soils due to higher than average precipitation throughout the growing season [19]. In beech forests, the silvicultural practices today must be aware of the potential risks, which a changing climate may impose on sustainable forest management. Compared to other tree species, the competitive capacity of beech might be reduced under expected future climate conditions [19].

Since trees as long-lived plants adapt to the climatic seasonality through phenology, it is considered a good indicator of changing climate conditions [20–25]. Under climate actions, plants could shift the onset of phenological phases and thus extend or shorten their growing seasons as the photosynthetically active period modulates ecosystem carbon balance [26,27]. In temperate forests, the temperature [28–31] was determined to be the main driving factor for the onset of leafing, although the effect of photoperiod [32] and soil water content [33] as prominent factors are sometimes evoked. However, the clear determination of the response of leaf senescence to the associated environmental controls is more difficult. We considered potential indices in order to predict autumn phenological response to climate conditions and extreme events, those based on water movement in the plant–atmosphere system (evapotranspiration), those based on water availability (precipitation), and those based on temperature requirements and forcing (temperature).

We used three decades of meteorological observations (1989–2018) along with phenological observations (1996–2018) of beech stands growing at different altitudes and climate conditions in Central Europe (Slovakia) to analyse the consequences of drought and heat on beech autumn leaf colouration. We hypothesize the following:

(1) The climate change causes significant changes in temperature and precipitation regimes and thus affects the timing of the onset of leaf colouring in European beech stands.

(2) Temperature-, evapotranspiration- and precipitation-based bioclimatological indices and variables of summer months can explain the shifts of the onset of leaf colouring.

The results of the analysis provide an opportunity to consider under which conditions beech can utilize the changing climate to extend the vegetation period and where it is limited and less productive.

2. Materials and Methods

2.1. Field Observations and Stand Descriptions

The studied beech stands are located in the Western Carpathians (Slovakia), in the latitude range between 48° and 49° N and in the longitude range between 17° and 22° E (Figure 1), and belong to the phenological network of the Slovak Hydrometeorological Institute (SHMI). SHMI is a specialized organization providing hydrological and meteorological services at the national and international levels. The phenological database of SHMI administrates a set of continual phenological observation data in natural conditions of the main ecosystems from 218 phenological stations [34]. This species-specific study focuses on seven beech-dominated (*Fagus sylvatica*, L.) stands located at different altitudes ranging between 300 m and 1050 m a.s.l. in three main climatic areas: warm, moderately warm, and cold. The characteristic of the climate was obtained from the Atlas of the Country of the Slovak Republic [35]. Temperature and precipitation conditions at each stand are characterized by temperature and precipitation normal in Table 1. Stands in this study are marked by the abbreviation of location and the altitude (Figure 1 and Table 1). The soil type present at all stands is deep cambisol with a slightly acidic to acidic reaction [36]. According to the altitude, we differentiated stands as low (below 500 m), middle (500–1000 m), and high altitude (above 1000 m). Here, we analysed a 23-year long (1996–2018) sequence of phenological data to evaluate the factors driving the onset of leaf colouring (LCO)) in different altitudes.

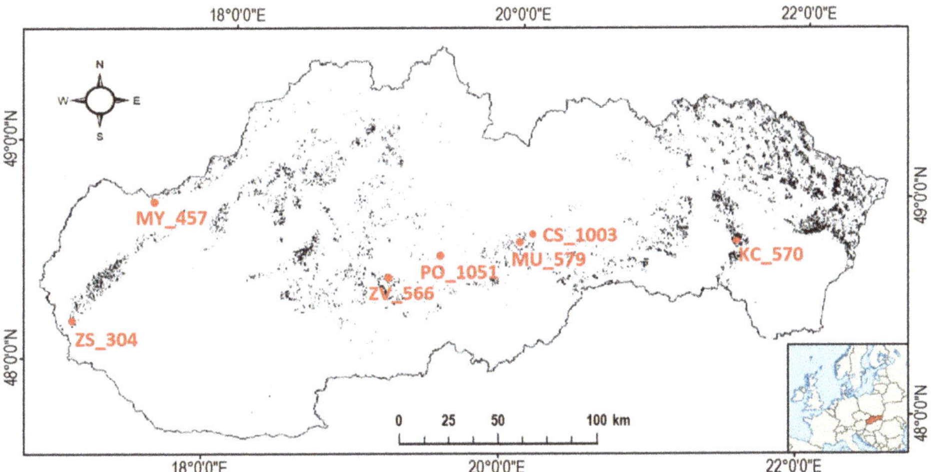

Figure 1. Seven phenological stations of Slovak Hydrometeorological Institute (SHMI) with beech forests: numerical values in each of the seven locations mark the altitude. Black colour on the map marks forests where European beech dominates (>60%) in the forest tree species composition.

The evaluated phenological phase LCO represents the beginning of autumn phenological phases. According to the methodology of Slovak Hydrometeorological Institute (SHMI) [37], LCO occurs when 10% of leaves change their colour from green to yellow, red, or brown. According to the international phenological scale BBCH (Biologische Bundesanstalt, Bundessortenamt und Chemische Industrie) [38], this phenological phase corresponds to the code BBCH 92. The ground phenological measurements consist of visual observations of phenological phases performed by the same observer on the same 10 individual trees from the main canopy. The main canopy is formed by mature tree crowns of the upper layer. The onset days of the phenological phases of the individuals are noticed and averaged over this group of 10 trees. Only beech stands with trees of ages above 50 years creating the main canopy were included in the analysis. The position of the trees is inside the forest, at least 50 m from the forest edge. Phenological observations are carried out regularly during the onset of spring vegetative and generative phenological phases occur every 2–3 days; in the autumn, they occur at least once a week. The SHMI manages the observations performed by volunteers.

Table 1. The basic characteristics of beech stands were obtained from the website of the National Forest Center (© NLC Zvolen, 2020) [39]. The stand identifier (ID) consist of the abbreviations of the stand location and the altitude. The characteristics of soils obtained from the soil portal (© VÚPOP Bratislava): retention capacity (Ret.) and permeability (Perm.) of soil have 3 main degrees—low (L), medium (M), and high (H)—and 2 intermediates—low to medium (L–M) and medium to high (M–H) [35].

Beech Stand Location /Characteristics	Železná Studienka	Myjava	Zvolen	Kecerovce	Muráň	Červená Skala	Po'ana
Altitude (m a.s.l.)	304	457	566	570	579	1003	1051
Stand ID	ZS_304	MY_457	ZV_566	KC_570	MU_579	CS_1003	PO_1051
Soil type	C [1]	C	C	C	C	C	C
Soil Ret./Perm.	L–M/M–H	H/M	H/M	H/M	H/M	H/M	M/M
Age	150	65	80	75	115	75	155
Tree height (m)	26	19	25	22	28	15	24
Trunk diameter (cm)	36	22	23	25	32	18	45
Climatic area	W [2]	M [3]	M	M	M	C [4]	C
TNy [5] (°C)	10.1	9	8	8.1	7.9	5.2	5.7
TNs [6] (°C)	18	16.3	15.8	15.9	15.9	12.7	13
PNy [7] (mm)	675	729	745	703	689	857	863
PNs [8] (mm)	176	176	190	239	198	241	263

[1] Cambisol, [2] warm, [3] moderately warm, [4] cold, [5] yearly temperature normal, [6] temperature normal of summer months (July–September), [7] yearly precipitation normal, and [8] precipitation normal of summer months (July–September).

We derived four meteorological drought indices which may act as weather stressors inducing the shifts in the beginning of leaf colouring (LCO) in beech-dominated stands: climatic water balance (CWB), standardized precipitation index (SPI), standardized precipitation-evapotranspiration index (SPEI) and dry period index (DPI), and the heat stress factor—heatwaves (HW). The meteorological data in daily and monthly time-steps entering into the calculations of those indices were acquired from the network of automatic meteorological stations of SHMI. The crucial criteria to linking data from the phenological and the meteorological stations were a maximum distance less than 20 km, similar altitude, and the same climatic area. The indices were modified according to their suitability to correlate with phenological events since they are usually calculated for a period of a year. The time scales are distinguished with indices bearing the number of analysed months.

2.2. Standardized Precipitation Index (SPI)

Standardized precipitation index (SPI) is a precipitation-based index developed by McKee et al. [40] to quantify a precipitation deficit for several time scales. It was designed to be an indicator of drought that recognizes the importance of time scales in the analysis of water availability and water use [41]. In this study, SPI is used as an indicator of meteorological drought, which is able to recognize both short-term and long-term precipitation deficits. In this study, we used SPI calculated for 1 and 3 months prior to leaf senescence to determine conditions inducing drought stress in plants leading to earlier beginning of leaf colouration. The time scales of SPI were set to cover the last months before leaf senescence start to determine conditions inducing drought stress in plants leading to an earlier beginning of leaf colouration. Since LCO generally begins at the beginning of October, we calculated the SPI of September and SPI of three summer months—July to September—and distinguished it with as index bearing the number of analysed months (SPI_{IX} and SPI_{VII-IX}). To compute the SPI, precipitation data series were fitted to a gamma probability distribution function as follows:

$$G(x) = \int_0^x g(t)dt \tag{1}$$

where

$$g(x) = \frac{1}{\beta^\alpha \Gamma(\alpha)} x^{\alpha-1} e^{-x/\beta} \tag{2}$$

$$\Gamma(\alpha) = \sum_0^\infty x^{\alpha-1} e^{-x} dx \tag{3}$$

where x is precipitation in mm at a certain time scale, α represents the shape parameter, β is a scale parameter, and $\Gamma(\alpha)$ defines the gamma function. Since the gamma function is not defined for $x = 0$, the cumulative probability is calculated as follow:

$$H(x) = q + (1-q)G(x) \tag{4}$$

where q is the probability of a zero estimated by m/n and m is the number of zeros in precipitation time series n.

Finally, the cumulative probability $H(x)$ is transformed to the standard normal value (Z) or SPI using an approximation by Abramowitz and Stegun [42]:

$$Z = SPI = -(t - \frac{c_0 + c_1 t + c_2 t^2}{1 + d_1 t + d_2 t^2 + d_3 t^3}) \text{ for } 0 < H(x) \leq 0.5 \tag{5}$$

and

$$Z = SPI = +(t - \frac{c_0 + c_1 t + c_2 t^2}{1 + d_1 t + d_2 t^2 + d_3 t^3}) \text{ for } 0.5 < H(x) < 1.0 \tag{6}$$

where

$$t = \sqrt{\ln(\frac{1}{(H(x))^2})} \text{ for } 0 < H(x) \leq 0.5 \tag{7}$$

$$t = \sqrt{\ln(\frac{1}{(1-H(x))^2})} \text{ for } 0.5 < H(x) < 1.0 \tag{8}$$

$c_0 = 2.515517$, $c_1 = 0.802853$, $c_2 = 0.010328$, $d_1 = 1.432788$, $d_2 = 0.189269$, and $d_3 = 0.001308$.

Meteorological drought was characterized according to the scale defined for SPI and SPEI (Table 2) [40].

Table 2. Classification of meteorological drought expressed by standardized precipitation index (SPI) and standardized precipitation-evapotranspiration index (SPEI) [43].

SPI/SPEI	Drought Category
≥ 2	Extremely wet
1.5 to 1.99	Severely wet
1.0 to 1.49	Moderately wet
-0.99 to 0.99	Near normal
-1.49 to -1.0	Moderately dry
-1.99 to -1.5	Severely dry
≤ -2	Extremely dry

2.3. Standardized Precipitation-Evapotranspiration Index (SPEI)

Temperature rise has a strong role in affecting the severity of droughts [44,45]. In this study, the importance of the effect of temperature on drought conditions through the potential evapotranspiration (PET) was taken into account by the use of the standardized precipitation-evapotranspiration index (SPEI). A crucial advantage of the SPEI over other widely used drought indices that consider the effect of PET on drought severity is that its multi-scalar characteristics enable the identification of different drought types and impacts in the context of global warming [46]. In the original version of the SPEI, the Thornthwaite equation [47] was applied to obtain the PET in SPEIbase v1.0. With a value for PET_i, a simple measure of the water surplus or deficit (D_i) for the analysed month i was calculated using a precipitation (P_i) as follows:

$$D_i = P_i - PET_i \tag{9}$$

The calculated Di values are combined at different time scales, following the same procedure as that for the SPI. Standardization of SPEI using the probability density function of a three parameter Log-logistic distributed variable is described in Vicente-Serrano et al. [46]. The time scales for SPEI calculation were set to correspond with the SPI–SPEI of September ($SPEI_{IX}$) and the SPEI of July to September $SPEI_{VII-IX}$.

2.4. Climatic Water Balance (CWB)

CWB is an indicator of drought referring to the balance between income of water from precipitation and the outflow of water by potential evapotranspiration [16,48]. The potential evapotranspiration can be generally defined as the amount of water that could evaporate and transpire from a vegetated landscape without restrictions other than the atmospheric demand [49]. The negative CWB indicates the lack of disposed water in the environment. The climatic water balance was calculated according to the modified method of Thornthwaite and Mather [50], where the CWB_m is the difference between the precipitation total (P_m) and potential evapotranspiration (PET_m) of month m:

$$CWB_m = P_m - PET_m \tag{10}$$

The potential evapotranspiration of the month PET_m in mm was calculated as follows:

$$PET_m = 0.535 \times f\left(\frac{10.T_m}{I}\right)^a \tag{11}$$

where T_m (°C) is the monthly mean temperature of analysed months and f is the correction factor depending on the month length and latitude:

$$f = k \times s_0 \tag{12}$$

where k is a coefficient of number of days and s_0 is the maximum duration of sunshine during the day (h).

I is the temperature index as a sum of values of the monthly temperature indices:

$$I = \sum\nolimits_{j=1}^{12} i_j; \text{ where} : i_j = \left(\frac{T_m}{5}\right)^{1,514} \tag{13}$$

a is an exponent calculated from the temperature index I:

$$a = \left(0.0675 \times I^3 - 7.71 \times I^2 + 1792 \times I + 47239\right) \times 10^{-5} \tag{14}$$

In this study, we calculated *CWB* for September (CWB_{IX}) and the incidence of meteorological drought was indicated, when the potential evapotranspiration transcended the actual precipitation. We created a classification scale differentiated into seven categories similar to that for SPI and SPEI. The thresholds for the drought levels were determined in consideration of the average monthly precipitation totals and average potential evapotranspiration totals (Table 3).

Table 3. Classification of meteorological drought designed for climatic water balance (CWB) of a single month.

CWB (mm)	Category
≥90	Extremely wet
60 to <90	Severely wet
30 to <60	Moderately wet
>−30 to <30	Near normal
>−60 to −30	Moderately dry
>−90 to −60	Severely dry
≤−90	Extremely dry

2.5. Dry Period Index (DPI)

We introduced DPI as an indicator of drought capable of identifying a small amount of rainfall as well as its irregular distribution over the season, and it complements the SPI_{VII-IX} and $SPEI_{VII-IX}$ data. It is understood that, the more dry days in a row and the less precipitation over the period, the bigger the stress introduced on a tree is. The dry period characterizes a period of the year with precipitation less than 0.3 mm per day [51,52], provided that the precipitation is intercepted by aboveground parts of trees and evaporated as an unproductive evaporation [52,53]. In this study, we evaluated dry periods (DP) as periods of 5 or more consecutive days with precipitation less than 0.3 mm evaluated over three summer months: July–September. Duration of each DP was noticed and summarized over the studied period: DP = n (days ≥ 5, p ≤ 0.3 mm). Dry period index (DPI) considers the total number of days in DP and the precipitation total over the studied season as follow:

$$DPI = \frac{\sum DP_i}{\sum P_m} \tag{15}$$

where DP_i is the length of dry period i and P_m is the precipitation total (mm) of a month m. A classification scale was created to divide DPI values into four stages of drought (Table 4).

Table 4. Classification of meteorological drought expressed by dry period index (DPI).

DPI	Drought Category
≤10	Near normal
>10 to 20	Moderately dry
>20 to 30	Severely dry
>30	Extremely dry

2.6. Heatwaves (HW)

A heatwave (HW) is a temperature-based indicator of meteorological drought inducing drying of the environment. In particular, heat waves occurring in summer result in an increased evapotranspirative demand from soil and plants, which, if not met by adequate water resources in the soil, eventually results in the development of concurrent drought stress [54]. The heatwave involves prolonged periods of extremely high temperatures for a particular region [11]. Regarding to the variety of HW definitions, here, we prefer the definition introduced by the Expert Team on Climate Change Detection and Indices (ETCCDI) adjusted according to Russo et al. [55]. Heatwave was characterized as 3 or more consecutive days that are above the 90th percentile temperature of the normal period between 1981–2010. The advantage of using relative-based thresholds such as high percentile values is allowance of the measurement of heatwaves across various locations over long periods when temperatures are substantially higher than normal. We calculated the 90th percentile from the daily average and maximum temperatures of the normal period 1981–2010 for all individual beech stands and noted all HWs longer than 3 days. The individual heatwave days were then aggregated into the total number of heatwave days over the hottest months: June–August. We used HW_{AVG} when calculating HW from average daily temperatures and HW_{MAX} when HW was calculated from the maximum daily temperatures. According to the duration of HW, we created the classification scale differentiated into 5 stages of HW strengths (Table 5). It is understood that, the more days in heatwave, the more stress could be induced to the tree. Since the occurrence of HWs depends on atmospheric circulation patterns [55], we evaluated the spatial difference of their incidence.

Table 5. Classification of heatwaves (HW) according to the total number (n) of HW days in the season.

n (HW Days)	Categories of Heatwaves
0 to 2	No HW
3 to 10	Weak HW
11 to 20	Moderate HW
21 to 30	Severe HW
>30	Extreme HW

2.7. Statistical Analysis

The linear regression between autumn leaf colouring and bioclimatic indices and variables and the spring leaf onset were analysed. The shifts in leaf phenology per 1 degree increase in temperature (day $°C^{-1}$) were determined using the regression coefficient of the corresponding linear regressions to compare the phenological sensitivities between altitudes (stands). The significant correlation was determined at $p \leq 0.05$ and was highly significant at $p \leq 0.01$.

The age of the studied beech trees is over 50 years. To examine the evolution of temperature and precipitation over the last 30 years, these meteorological variables were compared to the normal from the earlier growing stages of studied beech stands (1961–1990). The differences between actual and normal summer temperatures and precipitations were tested with the paired sample *t*-test. The H_0 hypothesis supposed stable climatic conditions, while H_A expected significant differences at the p level ≤ 0.05.

The evolution of beech phenology over the period between 1996–2018 along an altitudinal gradient and trends of bioclimatic indices and variables over 1989–2018 was evaluated by the nonparametric Mann–Kendal (M-K) trend test with the significance level set for $p \leq 0.05$ and was highly significant for $p \leq 0.01$. The shifts in leaf phenology per year increase were calculated using the Sen's slopes of the M-K trend and compared between stands (altitudes).

3. Results

3.1. Precipitation and Indices Indicating the Drought

Compared to the normal period of 1961–1990, the precipitation totals in the summer periods (P_{VII-IX}) over the studied recent 30 years indicate the increase of precipitation, particularly over the last 25 years (Figure 2). Using the paired sample T-test, significant positive differences ($p \leq 0.05$) from normal (PN_{VII-IX} in 1961–1990) were identified at ZV_304, MY_457, KC_570, MU_579, and CS_1003 (Table 6). At the stands ZV_566 and PO_1051, most summers had precipitation below normal. The linear temporal trends of the precipitation differences from normal were not significant because of the great interannual variability of precipitation. However, the regression coefficients obtained solely positive values, indicating the increasing precipitation at all stands (Figure 2) compared to 1961–1990. Although the precipitations itself rose (Figure 2 and Table A1), the M-K trend tests revealed nonsignificant changes of meteorological drought indices and variables over time (Table A1).

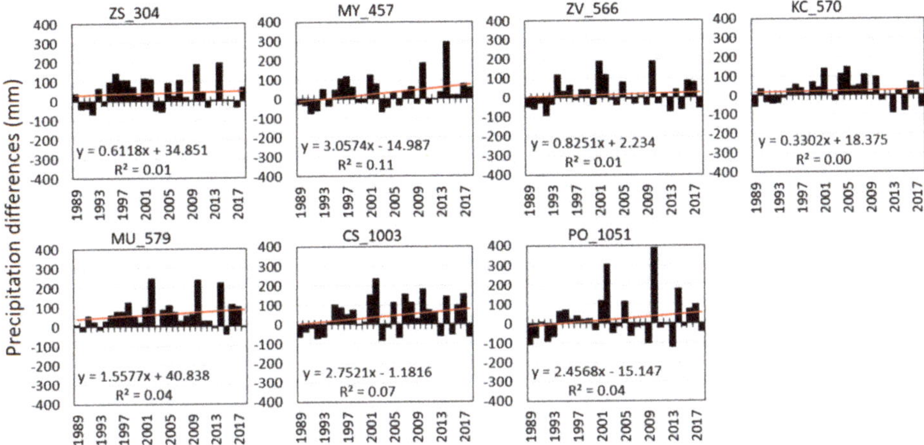

Figure 2. Difference of precipitation during summer seasons in 1989–2018 (P_{VII-IX}) compared to the normal of 1961–1990 (PN_{VII-IX}): the black bars show the precipitation difference in mm. The red line trends indicate the linear relations including their numerical representation.

Table 6. Differences of the summer precipitation total (P_{VII-IX}) from the precipitation normal of summer months (PN_{VII-IX} of 1961–1990): significant differences with $p \leq 0.1$ (**) and $p \leq 0.5$ (*) are highlighted with bold, $t_{(\alpha;f-1)}$ is the critical value, and t is the test statistics.

Stand	$t_{(\alpha;f-1)}$	t	p
ZS_304	2.04	**3.38**	**0.00 **
MY_457	2.04	**2.25**	**0.04 *
ZV_566	2.04	1.14	0.28
KC_570	2.04	**2.08**	**0.05 *
MU_579	2.04	**5.11**	**0.00 **
CS_1003	2.04	**2.62**	**0.02 *
PO_1051	2.04	1.14	0.28

When we analysed the indices of September (Figure 3a–c), the incidence of severely to extremely dry Septembers were according to $SPEI_{IX}$ only in 2011 and according to SPI_{IX} in 2006 and 2011. However, the CWB_{IX} indicated Septembers in eight years to be severely to extremely dry (1989, 1991, 1999, 2003, 2004, 2006, 2009, and 2011). Furthermore, it implied the moderate to severely dry Septembers in many more years compared to the SPI_{IX} or $SPEI_{IX}$ (Figure 3a–c), suggesting the potential higher rainfall deficit in environment. On the contrary, severely to extremely wet Septembers according to all indices occurred in four years (2001, 2007, 2010, and 2014).

When we evaluated the meteorological drought of summer seasons (July to September) in 1989–2018, the SPI_{VII-IX} identified seasons 2011 (only MU_570) and 2013 to be severely to extremely dry while $SPEI_{VII-IX}$ classified only the season in 1992 (Figure 3d,e). By introducing the dry period index (DPI_{VII-IX}), we were able to identify droughts that appeared over longer periods. DPI_{VII-IX} hinted that the beech stands at low- to mid-altitudes are threatened by the combination of dry periods and low precipitation, resulting into the severe to extreme drought. At the stands at low to mid-altitudes, ZS_304, MY_457, and ZV_566, this drought occurred commonly throughout the entire studied period 1989–2018. However, in the last decade, severe to extreme drought started to occur more frequently at the rest of mid-altitude stands: KC_570 and MU_579 (Figure 3f). The longest dry period took 24 days, and it occurred twice in ZS_304 (2000 and 2013) and once each at MY_457 (2013) and PO_1051

(1990). Regarding the higher precipitations at high altitudes, such a long period without precipitation at PO_1051 induced only moderate dryness (Figure 3f).

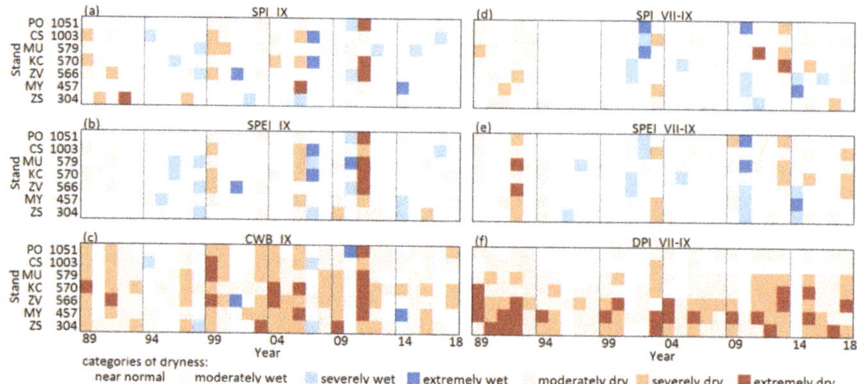

Figure 3. Meteorological drought of one month (September) and three months (July to September) prior to onset of leaf colouring indicated using the indices: standardized precipitation index (SPI$_{IX}$) (**a**), standardized precipitation evapotranspiration index (SPEI$_{IX}$) (**b**), climatic water balance (CWB$_{IX}$) (**c**), SPI$_{VII-IX}$ (**d**), SPEI$_{VII-IX}$, (**e**) and dry periods index DPI$_{VII-IX}$ (**f**) for each stand over period between 1989–2018. In each mosaic, the stands are ordered with increasing altitude from the bottom to the top.

3.2. Temperature and Heat Waves

Compared to the temperature normal of 1961–1990 (TN$_{VII-IX}$), the average July to September temperatures (T$_{VII-IX}$) of 1989–2018 exceeded normal in almost all analysed years (Figure 4). The temperatures at the beginning of the studied period suggested were balanced with two single warm years: 1992 and 1994. However, the amplified warming over next two decades culminated at the end of this period. In some years, we noticed the incidence of temperatures higher by more than 3 °C compared to the TN$_{VII-IX}$ of 1961–1990 (Figure 4). The paired sample t-test of the differences of T$_{VII-IX}$ from normal TN$_{VII-IX}$ (1961–1990) revealed significant differences at all altitudes (Table 7).

Table 7. Differences of the average summer temperature (T$_{VII-IX}$) from the temperature normal of summer months (TN$_{VII-IX}$ of 1961–1990): significant differences with $p \leq 0.1$ (**) are highlighted with bold, $t_{(\alpha;f-1)}$ is the critical value, and t is the test statistics.

Stand	$t_{(\alpha;f-1)}$	t	p
ZS_304	2.04	5.37	**0.00 ****
MY_457	2.04	5.97	**0.00 ****
ZV_566	2.04	5.31	**0.00 ****
KC_570	2.04	6.27	**0.00 ****
MU_579	2.04	7.30	**0.00 ****
CS_1003	2.04	5.89	**0.00 ****
PO_1051	2.04	7.00	**0.00 ****

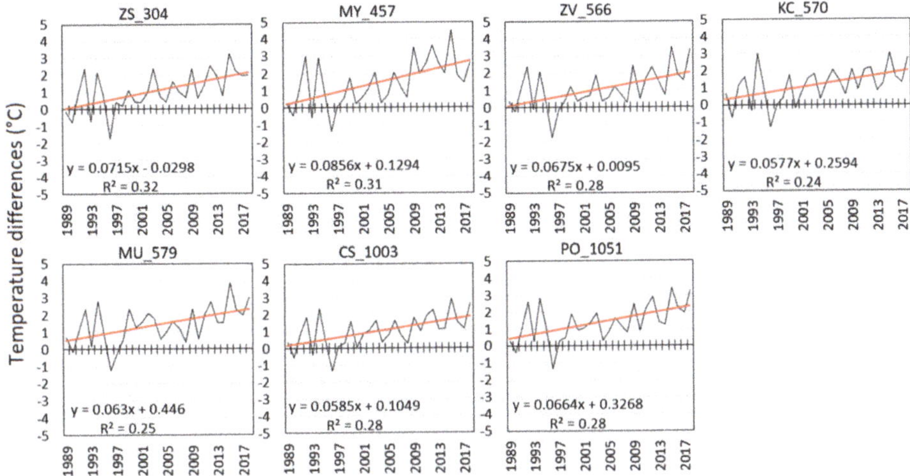

Figure 4. Differences of temperature during summer seasons in 1989–2018 ($T_{VII–IX}$) compared to the normal of 1961–1990 ($TN_{VII–IX}$): the black line shows the temperature difference in °C. The red line trends indicate the linear relations including their numerical representation.

The increasing duration of heatwaves (Figure 5) accompanied the indicated warming during the studied period. Comparing the HWs based on average daily temperatures to those based on daily maxima, we have not found any significant differences between them, (t = −0.31 < $t_{(\alpha;f-1)}$) = 1.655, and both show the increased incidence of HWs in the last decade. In the first (1989–1998) and the second (1999–2008) decades of the studied period, there were only one severe HW per decade: in 1994 at KC_570 and in 2006 at MY_457 (Figure 5). In 1994, the longest single HW of the studied period occurred, when the above-normal maximal daily temperature was 20 days in a row in the most eastern stand, KC_570. On the contrary, over the last decade, the severe to extreme HWs occurred in 2010 at MY_457 and in 2012, 2013, 2015, and 2017 in most stands. In the comparison with the previous two decades, the five-times increase of HWs in the last decade showed an evident intensification of HWs over the studied period. The effect of HWs differed spatially. While in 2015 the strongest HW presented itself in all stands, in the other years (2010, 2012, 2013, and 2017) it affected only some areas. In 2010 and 2013, HW mostly affected the western stands. In 2012 and 2017, HW took action in most western and central stands. The eastern stand KC_570, however, was not affected over these years.

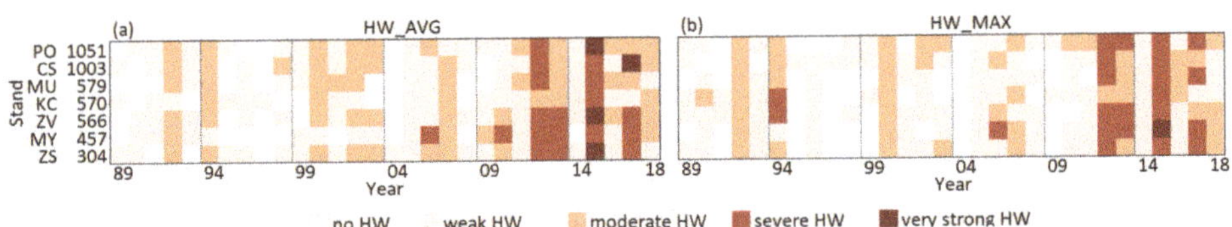

Figure 5. Heatwaves (HW) from June to August over the studied period 1989–2018. On the figure (**a**) are shown temporal changes of HW_{AVG} when HW was calculated from the average daily temperatures and on the figure (**b**) are shown temporal changes of HW_{MAX} when HW was calculated from the maximum daily temperatures.

The M-K trend test revealed a significant increase of most of the temperature-based variables over the summer seasons (Table 8). The trend of HW_{AVG} rose significantly ($p \leq 0.05$) at all stands, while HW_{MAX} showed significant ($p \leq 0.05$) increasing trend only at low (ZS_304 and MY_457) and high (CS_1003 and PO_1051) altitudes, and the middle altitudes between 500 and 1000 m a.s.l. did not. The trend tests of the average temperatures revealed highly significant growth of T_{VI-IX} and T_{VII-IX} with $p \leq 0.01$ and of $T_{VIII-IX}$ with $p \leq 0.05$. The average temperature of September (T_{IX}) shows differences between altitudes. The significant increasing trend ($p \leq 0.05$) occurred only at mid-altitudes (ZV_566, KC_570, and MU_579) while, at altitudes below 500 m a.s.l. (ZS_304 and MY_457) and around 1000 m a.s.l. (CS_1003 and PO_1051), did not. (Table 8).

Table 8. Mann–Kendal (M-K) trend test of the heatwaves (HW_{AVG} and HW_{MAX}) and average temperatures (T_{VI-IX}, T_{VII-IX}, $T_{VIII-IX}$, and T_{IX}) prior to onset of leaf colouring during 1989–2018. Significant trends with $p \leq 0.1$ (**) and $p \leq 0.5$ (*) are highlighted with bold.

	ZS_304 K tau	*p*	MY_457 K tau	*p*	ZV_566 K tau	*p*	KC_570 K tau	*p*	MU_579 K tau	*p*	CS_1003 K tau	*p*	PO_1051 K tau	*p*
HW_{AVG}	0.33	**0.01** **	0.40	**0.00** **	0.41	**0.00** **	0.32	**0.02** *	0.29	**0.03** *	0.32	**0.02** *	0.39	**0.00** **
HW_{MAX}	0.30	**0.03** *	0.38	**0.00** **	0.23	0.09	0.13	0.35	0.26	0.06	0.26	**0.05** *	0.39	**0.00** **
T_{VI-IX}	0.48	**0.00** **	0.48	**0.00** **	0.48	**0.00** **	0.41	**0.00** **	0.41	**0.00** **	0.46	**0.00** **	0.44	**0.00** **
T_{VII-IX}	0.39	**0.00** **	0.41	**0.00** **	0.39	**0.00** **	0.36	**0.01** **	0.32	**0.01** **	0.40	**0.00** **	0.41	**0.00** **
$T_{VIII-IX}$	0.27	**0.04** *	0.29	**0.02** *	0.25	**0.05** *	0.36	**0.01** **	0.28	**0.03** *	0.31	**0.02** *	0.31	**0.02** *
T_{IX}	0.14	0.28	0.19	0.14	0.27	**0.04** *	0.25	**0.05** *	0.25	**0.05** *	0.21	0.11	0.22	0.09

3.3. Phenological Response

The onset of leaf colouring (LCO) differed between altitudes (Figure 6a). Beech in stands at altitudes above 1000 m a.s.l. exhibited a beginning of leaf colouring first, on average, on day of year (DOY) 251 and 255 for CS_1003 and PO_1051, respectively. Over the studied period, the LCO was delayed in two aspects: from the highest to the lowest altitudes and from the beginning to the end of studied period (Figure 6a). From Figure 6a, it is evident that, in the last decade, LCO at the middle altitudes started at comparable to low altitudes and, in the last several years, even earlier.

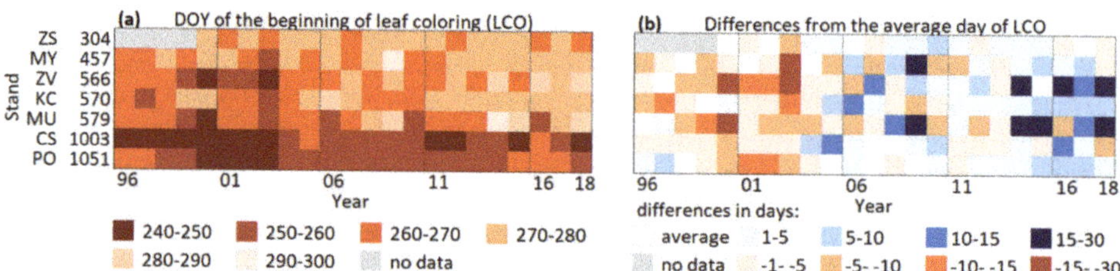

Figure 6. The onset of leaf colouring (LCO) at beech stands during the period 1996–2018 (**a**) and the differences from the average DOY (Day of Year) of LCO (**b**): the prevailing blue colour after 2006 in (**b**) highlights later LCO at beech stands, especially at middle altitudes.

Differences from the average DOY (day of year) of LCO varied in the range of ±30 days. The most pronounced shift from below-average to above-average DOY of LCO was revealed at stands at middle altitudes: ZV_566 and MU_579 (Figure 6b). Here, the significant trend that delayed LCO by 1.2 and 0.8 day per year was indicated by S-slope for ZV_566 and MU_579, respectively (Table 9). Moreover, significant shifts to later LCOs were detected at stands MY_457, KC_570, and PO_1051 with S-slopes of 0.4, 0.6, and 0.5 day per year, respectively (Table 9). This indicates a positive reaction of beech at these stands to climate change. The unchanged LCO with nonsignificant trend was observed at the lowest altitudes: in stand ZS_304 (Table 9). Here, the interannual standard deviation from the average onset day over the 19-year period (2000–2018) was only 3 days.

Table 9. M-K trend test of the onset of leaf colouring (LCO): significant trends with $p \leq 0.1$ (**) and $p \leq 0.5$ (*) are highlighted in bold.

Stand	K-tau	p	S-Slope
ZS_304	0.12	0.51	0.1
MY_457	0.34	**0.03 ***	0.4
ZV_566	0.52	**0.00 ***	1.2
KC_570	0.49	**0.00 ***	0.6
MU_579	0.31	**0.04 ***	0.8
CS_1003	0.21	0.18	0.3
PO_1051	0.39	**0.01 ***	0.5

At stand ZS_304, the significant correlation ($p \leq 0.01$) between LCO and SPI_{IX} referring to the drought conditions one month prior to colouring and the near significant correlations ($p \leq 0.10$) with most of the meteorological drought indices (Table 10) indicate the advance of LCO in the environment with less water supply. Similarly, the near significant correlation ($p \leq 0.10$) was revealed with the temperature of September (T_{IX}), when warming by 1 °C resulted in the 1-day earlier onset of LCO (Table 10). Following that, we assume that beech growing at ZS_304 reaches its climatic limit for growing season termination, since the increasing temperature and incidence of meteorological drought act negatively to LCO and evoke earlier beginning. This stand is located in the warm and moderate moist climatic region at the lower limit of the beech vertical distribution range (Table 1). Furthermore, the soil with lower retention capacity and higher permeability limit the water availability here. We compared the environmental conditions in ZS_304 with the MU_579, where the precipitation-based indices correlate with LCO with $p \leq 0.05$ for SPI_{IX} and $p \leq 0.10$ for $SPEI_{IX}$ (Table 10). Although the annual precipitation normal of 1981–2010 between these two stands differed only by 14 mm, the temperature normal was lower by the 2.2 °C in favour of the MU_579 in the moderate warm region (Table 1). Here, we suppose that the synergic effect of the lower temperatures, conditioning lower evapotranspirative demands together with the high retention and medium permeability of soil at the beech stands with the same precipitation totals, have resulted in delays of LCO compared to ZS_304. However, limitation by the significant effect of water supplement may cause a slower (0.8 day per year) trend in MU_579 compared to ZV_566 (1.2 day per year), where the precipitation normal is exceeded by 56 mm in the comparable temperature and soil conditions.

The range of temperature effects on leaf senescence was wide and depended on regional climate, since it increased from −1 day per °C in ZS_304 to up to +4 days per °C for beech stands at the middle altitudes, with no limits for disposed water (KC_570). The ongoing warming in moderate warm regions will cause additional enhanced requirements in the environment for disposed water and could result in a shift to earlier LCOs when there is a lack of disposed water at the end of summer.

Despite the amplified warming of summer months at all beech stands, significant correlations with the onset of leaf colouring was noticed at only one stand. The Pearson correlation coefficient in the stand KC_570 indicated increasing temperatures, and the strength of heatwaves indicated the later onset of LCO.

Table 10. Pearson correlation between onset of leaf colouring (LCO) and the bioclimatic indicators (BioClim) of meteorological drought: standardized precipitation indices (SPI), climate water balances (CWB), dry period index (DPI), precipitation totals (P) and heatwaves (HW), and average temperatures (T) over summer season. Significant correlations with $p \leq 0.1$ (**) and $p \leq 0.5$ (*) are highlighted with bold.

Stand/BioClim	ZS_304	MY_457	ZV_566	KC_570	MU_579	CS_1003	PO_1051
SPI_{IX}	0.60 **	−0.12	0.15	−0.21	**0.42 ***	0.02	0.21
$SPI_{VII–IX}$	0.31	−0.04	0.16	−0.18	−0.01	0.23	−0.07
$SPEI_{IX}$	0.25	−0.15	0.08	−0.30	0.36	−0.02	0.12
$SPEI_{VII–IX}$	0.37	−0.18	0.05	−0.25	−0.09	0.22	−0.06
CWB_{IX}	0.18	−0.11	0.08	−0.20	0.25	0.03	0.16
$CWB_{VII–IX}$	0.41	−0.13	−0.07	−0.23	−0.05	0.22	−0.10
$DPI_{VII–IX}$	−0.43	0.10	0.10	0.35	−0.04	−0.21	0.02
$P_{VII–IX}$	0.41	−0.06	0.18	−0.15	−0.01	0.23	−0.09
HW_{AVG}	0.01	0.25	0.22	**0.59 ****	−0.02	−0.18	0.16
HW_{MAX}	−0.09	0.20	−0.01	**0.56 ****	0.08	−0.24	0.18
$T_{VI–IX}$	−0.23	0.30	0.30	**0.51 ***	0.15	−0.05	0.11
$T_{VII–IX}$	−0.23	0.37	0.30	**0.53 ***	0.23	−0.02	0.08
$T_{VIII–IX}$	−0.40	0.30	0.23	**0.48 ***	0.21	−0.16	0.08
T_{IX}	−0.16	0.30	0.31	**0.50 ***	0.13	0.14	0.19

4. Discussion

Considering the plant phenology as a complex phenomenon, when plant species react specifically to a combination of factors acting at the same time [56,57], in this study, we used a number of bioclimatic variables based on temperature, precipitation, and evapotranspiration to reveal the climatic limits for mature beech stands. As expected, the warming climate caused lengthening of the growing seasons of European beech which is a tolerant tree species. However, as the results indicate, the lengthening only presents itself in optimal climate conditions.

Our results revealed a wide range of temperature effects on the onset of leaf colouring, which are in coincidence with previously published results [58–60]. At the lowest altitudes with water supplement and retention limitations, LCO advanced by −1 day per °C. On the contrary, at stands at the middle altitudes without precipitation limitation and high soil water retention, LCO was delayed by up to +4 days per °C of warming. The susceptibility of the LCO of European beech to increasing temperatures in environments optimally supplied with water allows beech to utilize this warming to extend the growing season. Such an extension appears at the studied stands at middle to high altitudes, where warming is accompanied by a sufficient amount of total annual precipitation (above 700 mm). These findings agree with the theory of Fu et al. [58], who, taking advantage of temperature manipulative experiments on beech seedlings, concluded that, in the absence of water and nutrient limitation, temperature is a dominant factor controlling the leaf senescence process in European beech. The temporal trends reflecting the impact of climate change revealed that stands at middle altitudes postponed the onset of leaf colouring by 0.4–1.2 day per year. At high altitudes, the trend of LCO delayed by 0.3–0.5 day per year.

Warming of the studied period precede the increasing number of heatwaves, especially in the last decade, 2009–2018 (Figure 5). This is in agreement with numerous studies reporting a number of severe heatwaves that recently occurred over various global regions [55,61–63]. Only several of the most severe heatwaves identified across Europe since 1950 [55,61,64,65] hit the region of Western Carpathians, where the studied beech stands are located. The HWs in 1994, 2006, 2010, 2015, and 2017 were linked to those observed in other European countries. Furthermore, most of the study stands were affected by severe to extreme HWs in 2012 and 2013, while the extreme HWs in 2003, 2007, and 2014 did not affect the studied region. Although the longest single HW occurred in 1994 at KC_570 when maximum daily above-normal temperatures was 20 days in a row, the strongest

HW of all was that in 2015 with more than 30 days in HWs over the summer season (Figure 5). In 2015, the large temperature extremes of the HW induced earlier senescence in numerous tree species in Central Europe, especially Slovakia and the Czech Republic [66,67]. A potential negative effect of HW is the decrease of net photosynthesis [68] and the alteration of long-term continental carbon balances [69]. In this study, we expected that this reduction in photosynthesis and production might relate to the potential advances in onset of autumn colouring as a reaction to the heatwave-induced leaf damage. In regard to the observed onset day of leaf colouring, the regression analyses did not reveal any significant negative effect of HWs based on the average as well as maximal daily temperatures. However, the opposite reaction was revealed at KC_570, where the correlation coefficients indicate a very significant positive effect of increasing temperature as well as heatwaves on the postponed LCO. Such a positive effect of HWs on the prolonging of the growing season at this stand was supported by sufficient precipitation over the summer season (Table 1), which is comparable to the precipitation at high altitudes (CS_1003 and PO_1051). This stand is located in eastern Slovakia, where the effect of continental climate causes higher precipitation totals over summer months compared to more western stands [70]. The trend of HW_{AVG} and HW_{MAX} increased significantly at all stands, except for middle altitudes between 500 and 1000 m a.s.l. where HW_{MAX} increased but was not significant.

European beech is known for tolerating neither severe droughts nor water-logged soils due to higher than average seasonal precipitation. Compared to the other dominant tree species with higher tolerance to soil water deficit, e.g., oak, the competitive capacity of beech might be reduced under the expected future climate conditions [19] if a strong increase in the frequency of moderate soil water stress occurs [60]. Despite the projected declining trend of summer precipitations [7], the summer precipitation P_{VII-IX} at the studied stands positively deviates from the normal PN_{VII-IX} (1961–1990) (Figure 2). This increase, however, will probably be insufficient regarding the rising evapotranspiration requirements in a warming environment, since the evapotranspiration is a leading output component of the water balance controlling the amount of water running off from the territory of beech stands [9].

The climatic water balance (CWB_{IX}) as a difference between precipitation and evapotranspiration indicates considerable predominance of severely dry Septembers, although according to the classification scale SPI_{IX} and $SPEI_{IX}$, the severe to extreme droughts related to given stand conditions appeared in only a few years. It is worth noting that the stand ZS_304 is in the xerothermic locality at the lower limit of the natural occurrence of beech in the Western Carpathians. Here, the significant correlation of LCO with SPI_{IX} and $SPEI_{IX}$ and the near significant correlation with the other precipitation- and evapotranspiration-based indices and variables as well as with the temperature variable $T_{VIII-IX}$ (Table 10) are clear evidence of the negative effect of low (although increasing; Figure 2) precipitation totals and increasing temperature. Not only stands at low altitudes but also those at middle altitudes with precipitation normal (1981–2010) below 700 mm suffer from drought. As has been presented in previous studies [8,15], the deterioration of climate conditions for European beech occurs usually at low altitudes with xeric climatic limits. However, regarding temperatures which are more favourable for beech at the middle altitudes, the effect of precipitation below 700 mm is less pronounced, since the evapotranspiration demands are lower there. Therefore, the effect of warming environment leads to a delaying trend of LCO at stand MU_579, despite the significant negative effect of meteorological drought (Table 10). However, the warming trend of summer months (Table 8 and Figure 4) suggests a continued increase in temperature, and it is probable that the beech stands that profit from the warming today will suffer from a deteriorating effect in the next decades. The ongoing warming in moderately warm regions will cause additional enhanced water requirements in environment and could result in a shift to earlier LCOs when there is a lack of water at the end of summer.

The dry period index (DPI_{VII-IX}) hinted that these lowest stands (ZS_304, MY_457, and ZV_566) are mostly threatened by severe to extreme drought periods. However, over the last decade, these drought periods occurred at the other stands at the middle altitude (KC_570 and MU_579; Figure 3f). Although European beech is known to have mechanisms to control water deficit, this species does not tolerate severe drought [4,71]. Under limited water availability, beech may suffer from xylem embolism, which is more likely to hit provenances from the Northern Europe rather than those from Southern Europe [72]. Some studies indicated the physiological response of beech to drought starting after five consecutive weeks of continuous meteorological drought [73]. However, over the studied 30-year period, in the natural environmental conditions of our studied stands, such a long period without precipitation did not occur. The longest dry periods with precipitation less than 0.3 mm did not exceeded 24 days in a row. The occurrence of extremely dry periods at low to middle altitudes appears to be a common phenomenon over the last 30 years, and the length of the longest period as well as DPI_{VII-IX} have not significantly increased over time.

In addition to the above indices, it would be beneficial to utilize the data from soil water potential (SWP) measurements in an evaluation of the phenological response to drought [74]. We compared our results with the SWP data measured during 2012–2014 [73] close to beech stand ZV_566, published by Vido et al. [73]. In 2013, the relationship between extreme dry periods (DPI_{VII-IX}, Figure 3), early LCO (Figure 6), and low SWP (Figure 3 in [73]) could be indicated. On the contrary, in 2014, when the DPI_{VII-IX} values were close to normal (Figure 3) and the SWP did not indicate drought (Figure 3 in [73]), the considerable delay of LCO was observed (Figure 6). The difference of LCOs between 2013 and 2014 is 22 days. Following that, we advise to use SWP measurement data in the future research by considering climate change effects at the end of growing season in beech forests.

5. Conclusions

Despite the important role of autumn senescence in fulfilling the ecosystem functions, current knowledge of the drivers of leaf senescence is still limited. There is a large spread in predictions of how the length of the growing season will vary under future climate conditions. In this study, we analysed the mechanisms by which the climate controls the onset of autumn phenology of European beech. This helps in predictions of the effects of changing climate on its future distribution in Western Carpathians to find a balance between forest production requirements and environmental demands of beech in forest management decisions. The concurrent lengthening of canopy duration (later LCO) in beech stands that we revealed in this study appears to be inclined towards the forest production requirements. Furthermore, our results suggest that climate change alter the period with carbon assimilation. Particularly at middle altitudes, we documented the differences in the onset of leaf colouration between the start and end of the studied period 1989–2018 up to 30 days.

We found a significant increasing trend of summer temperatures up to 4 °C throughout the period 1989–2010 and, simultaneously, significant positive differences of summer precipitation from the normal period of 1961–1990. However, it is probable that these significantly higher summer precipitations will be insufficient for beech in the upcoming decades. We assume that the ongoing warming of summer months will enhance evapotranspiration demands and will dry the environment. We expect that the negative effect of meteorological drought on the beech autumn phenological phases, which was found to already occur at low altitudes and has started at the middle altitudes, will further advance. This will cause an earlier onset of leaf senescence and shortening of the growing season. Foresters should be prepared for the production loses that the deteriorating effects of climate change could bring over next decades. Since the trees long-living organisms, it is appropriate to consider a gradual change of tree species compositions in the areas threatened by meteorological drought towards more drought-tolerant tree species such as oak.

Following the results, we conclude that European beech autumn phenology at mid- to high altitudes currently profit from the climate warming, but it is susceptible to drought at the low- to mid-altitudes. The results of this species-specific study will contribute to the adjustment of adaptation and mitigation policies in forestry and nature conservation and to validation of evolutionary and ecological hypotheses related to climate change effects.

Author Contributions: Conceptualization, V.L. and J.Š. (Jaroslav Škvarenina); methodology, V.L. and J.Š. (Jana Škvareninová); software, J.V.; formal analysis, S.B.; investigation, V.L., J.V., and S.B.; resources, J.Š. (Jaroslav Škvarenina), J.Š. (Jana Škvareninová); data curation, H.H. and P.B.; writing—original draft preparation, V.L.;

writing—review and editing, J.V., S.B., J.Š. (Jaroslav Škvarenina), and J.Š. (Jana Škvareninová); visualization, V.L.; supervision, J.Š. (Jaroslav Škvarenina) project administration, V.L.; funding acquisition, J.Š. (Jaroslav Škvarenina), J.Š. (Jana Škvareninová), and S.B. All authors have read and agreed to the published version of the manuscript.

Funding: This research was funded by research grants of The Ministry of Education, Science, Research, and Sport of the Slovak Republic: VEGA 1/0370/18, VEGA 1/0111/18, VEGA 1/0500/19, and VEGA 2/0015/18 and by grants from the Slovak Research and Development Agency no. APVV-18-0347 and APVV-15-0425.

Acknowledgments: We thank Peter Zelina for comments on early versions of the manuscript and English editing. Additional support was provided from the Stefan Schwarz fund for postdoctoral researchers awarded by the Slovak Academy of Sciences to Veronika Lukasová and from the contract HZ GIEFS provided by Svetlana Bičarová.

Conflicts of Interest: The authors declare no conflict of interest.

Abbreviations

DOY	Day of Year
CWB	Climatic Water Balance
SPI	Standardized Precipitation Index
SPEI	Standardized Precipitation-Evapotranspiration Index
DPI	Dry Periods Index
P	Precipitation
PN	Precipitation Normal
HW	Heatwave
T	Temperature
TN	Temperature Normal
LCO	Onset of Leaf Colouring
M-K	Mann–Kendal
S-slope	Sen's slope
SWP	Soil Water Potential

Appendix A

Table A1. Temporal trend of the meteorological drought indices and precipitation.

	ZS_304		MY_457		ZV_566		KC_570		MU_579		CS_1003		PO_1051	
	Ktau	p	Ktau	p	Ktau	p	Ktau	p	Ktau	p	Ktau	p	Ktau	p
SPI_{IX}	0.09	0.47	0.04	0.75	0.10	0.46	−0.07	0.60	0.14	0.28	0.03	0.86	0.00	1.00
$SPI_{VII–IX}$	0.06	0.67	0.22	0.09	0.16	0.21	0.10	0.43	0.15	0.26	0.21	0.11	0.12	0.36
$SPEI_{IX}$	−0.06	0.65	−0.01	0.97	0.01	0.97	−0.11	0.39	0.08	0.52	−0.02	0.87	−0.04	0.78
$SPEI_{VII–IX}$	−0.09	0.49	0.08	0.55	0.05	0.71	−0.04	0.76	−0.04	0.78	0.15	0.25	0.00	1.00
CWB_{IX}	−0.01	0.94	0.02	0.92	0.06	0.65	−0.13	0.34	0.11	0.42	0.03	0.86	−0.02	0.89
$CWB_{VII–IX}$	−0.05	0.72	0.09	0.48	−0.01	0.94	−0.03	0.86	0.03	0.86	0.17	0.20	0.03	0.83
$DPI_{VII–IX}$	−0.01	0.94	−0.12	0.36	−0.15	0.27	0.02	0.89	0.07	0.60	0.21	0.10	0.08	0.57
$P_{VII–IX}$	0.02	0.92	0.22	0.09	0.16	0.21	0.10	0.44	0.13	0.32	0.21	0.11	0.12	0.36

References

1. IPCC. *Climate Change 2013: The Physical Science Basis. Contribution of Working Group I to the Fifth Assessment Report of the Intergovernmental Panel on Climate Change*; Stocker, T.F., Qin, D., Plattner, G.-K., Tignor, M., Allen, S.K., Boschung, J., Nauels, A., Xia, Y., Bex, V., Midgley, P.M., Eds.; Cambridge University Press: Cambridge, UK; New York, NY, USA, 2013; p. 1535.
2. Saxe, H.; Cannell, M.G.R.; Johnsen, Ø.; Ryan, M.G.; Vourlitis, G. Tree and forest functioning in response to global warming. *N. Phytol.* **2001**, *149*, 369–400. [CrossRef]
3. Salinger, S. Increasing climate variability and change: Reducing the vulnerability. *Clim. Change* **2005**, *70*, 1–3. [CrossRef]
4. Bréda, N.; Huc, R.; Granier, A.; Drenier, E. Temperate forest trees and stands under severe drought: A review of ecophysiological responses, adaptation processes and long-term consequences. *Ann. For. Sci.* **2006**, *63*, 625–644. [CrossRef]

5. Teskey, R.; Wertin, T.; Bauweraerts, I.; Ameye, M.; McGuire, M.A.; Steppe, K. Responses of tree species to heat waves and extreme heat events. *Plant Cell Environ.* **2015**, *38*, 1699–1712. [CrossRef] [PubMed]

6. Škvarenina, J.; Tomlain, J.; Hrvoľ, J.; Škvareninová, J. Occurence of dry and wet periods in altitudinal vegetation stages of West Carpathians in Slovakia: Time-Series Analysis 1951–2005. In *Bioclimatology and Natural Hazards*; Strelcová, K., Matyas, C., Kleidon, A., Lapin, M., Matejka, F., Blazenec, M., Škvarenina, J., Holecy, J., Eds.; Springer: Amsterdam, The Netherlands, 2009; pp. 97–106.

7. Lindne, M.; Maroschek, M.; Netherer, S.; Kremer, A.; Barbati, A.; Garcia-Gonzalo, J.; Seidl, R.; Delzon, S.; Corona, P.; Kolstro, M.; et al. Climate change impacts, adaptive capacity, and vulnerability of European forest ecosystems. *For. Ecol. Manag.* **2010**, *259*, 698–709. [CrossRef]

8. Hlásny, T.; Mátyás, C.; Seidl, R.; Kulla, L.; Merganičová, K.; Trombik, J.; Dobor, L.; Barcza, Z.; Konôpka, B. Climate change increases the drought risk in Central European forests: What are the options for adaptation? *For. J.* **2014**, *60*, 5–18. [CrossRef]

9. Střelcová, K.; Mind'áš, J.; Škvarenina, J. Influence of tree transpiration on mass water balance of mixed mountain forests of the West Carpathians. *Biologia* **2006**, *19*, S305–S310. [CrossRef]

10. Streda, T.; Stredova, H.; Chuchma, F.; Kucera, J.; Roznovsky, J. Smart method of agricultural drought regionalization: A winter wheat case study. *Contrib. Geophys. Geod.* **2019**, *49*, 25–36. [CrossRef]

11. Perkins, S.E. A review on the scientific understanding of heatwaves—Their measurement, driving mechanisms, and changes at the global scale. *Atmos. Res.* **2015**, *164–165*, 242–267. [CrossRef]

12. Gomory, D.; Kukla, J.; Schieber, B. Taxanómia, fylogenéza a rozšírenie buka v Európa a na Slovensku (Taxonomy, phylogeny and distribution of beech in Europe and in Slovakia). In *Buk a Bukové Ekosystémy Slovenska (Beech and Beech Ecosystems of Slovakia)*; Barna, M., Kuflan, J., Bublinec, E., Eds.; Veda: Bratislava, Slovakia, 2011; pp. 37–62.

13. Kramer, K. Phenotypic plasticity of the phenology of seven European tree species in relation to climatic warming. *Plant Cell Environ.* **1995**, *18*, 93–104. [CrossRef]

14. Robson, T.M.; Garzón, M.B.; BeechCOSTe52 Database Consortium. Phenotypic trait variation measured on European genetic trials of Fagus sylvatica L. *Sci. Data* **2018**, *5*, 180149. [CrossRef] [PubMed]

15. Mátyás, C.; Berki, I.; Czúcz, B.; Gálos, B.; Móricz, N.; Rastovits, E. Future of Beech in Southeast Europe from the Perspective of Evolutionary Ecology. *Acta Silv. Lignaria Hung.* **2010**, *6*, 91–110.

16. Škvarenina, J.; Križová, E.; Tomlain, J. Impact of the climate change on the water balance of altitudinal vegetation stages in Slovakia. *Ekologia* **2004**, *23*, 13–29.

17. Lukasová, V.; Vasiľová, I.; Bucha, T.; Snopková, Z.; Škvarenina, J. Effect of biometeorological variables on the onset of phenophases derived from MODIS data and visual observations. *Cent. Eur. For. J.* **2014**, *60*, 39–51. [CrossRef]

18. Lakatos, F.; Molnár, M. Mass mortality of beech in South–West Hungary. *Acta Silv. Lignaria Hung.* **2009**, *5*, 75–82.

19. Geßler, A.; Keitel, C.; Kreuzwieser, J.; Matyssek, R.; Seiler, W.; Rennenberg, H. Potential risks for European beech (Fagus sylvatica L.) in a changing climate. *Trees* **2007**, *21*, 1–11. [CrossRef]

20. Chmielewski, F.M.; Rotzer, T. Response of tree phenology to climate change across Europe. *Agric. For. Meteorol.* **2001**, *108*, 101–112. [CrossRef]

21. Ahas, R.; Aasa, A.; Menzel, A.; Fedotova, V.G.; Scheifinger, H. Changes in European spring phenology. *Int. J. Climatol.* **2002**, *22*, 1727–1738. [CrossRef]

22. Menzel, A. Phenology: Its importance to the global change community. *Clim. Chang.* **2002**, *54*, 379–385. [CrossRef]

23. Schwartz, M.D.; Ahas, R.; Aasa, A. Onset of spring starting earlier across the northern hemisphere. *Glob. Chang. Biol.* **2006**, *12*, 343–351. [CrossRef]

24. Babálová, D.; Škvareninová, J.; Fazekaš, J.; Vyskot, I. The dynamics of the phenological development of four woody species in south-west and central Slovakia. *Sustainability* **2018**, *10*, 1497. [CrossRef]

25. Lukasová, V.; Bucha, T.; Škvareninová, J.; Škvarenina, J. Validation and Application of European Beech Phenological Metrics Derived from MODIS Data along an Altitudinal Gradient. *Forests* **2019**, *10*, 60. [CrossRef]

26. Myneni, R.B.; Keeling, C.; Tucker, C.J.; Asrar, G.; Nemani, R.R. Increased plant growth in the northern high latitudes from 1981 to 1991. *Nature* **1997**, *386*, 698. [CrossRef]

27. Richardson, A.D.; Black, T.A.; Ciais, P.; Delbart, N.; Friedl, M.A.; Gobron, N.; Varlagin, A. Influence of spring and autumn phenological transitions on forest ecosystem productivity. *Philos. Trans. R. Soc. Lond. Ser. B Biol. Sci.* **2010**, *365*, 3227–3246. [CrossRef]

28. Menzel, A. Trends in phenological phasesin Europe between 1951 and 1996. *Int. J. Biometeorol.* **2000**, *46*, 76–81. [CrossRef]

29. Chmielewski, F.M.; Rötzer, T. Phenological trends in Europe in relation to climatic changes. *Agrometeorol. Schr.* **2000**, *7*, 1–15.

30. Čufar, K.; de Luis, M.; Saz, M.A.; Črepinšek, Z.; Kajfež-Bogataj, L. Temporal shifts in leaf phenology of beech (Fagus sylvatica) depend on elevation. *Trees* **2012**, *26*, 1091–1100. [CrossRef]

31. Schieber, B.; Kubov, M.; Janík, R. Effects of climate warming on vegetative phenology of the common beech Fagus sylvatica in a submontane forest of the Western Carpathians: Two-decade analysis. *Pol. J. Ecol.* **2017**, *65*, 339–351. [CrossRef]

32. Vitasse, Y.; Basler, D. What role for photoperiod in the bud burst phenology of European beech. *Eur. J. For. Res.* **2013**, *132*, 1–8. [CrossRef]

33. Peñuelas, J.; Filella, I.; Zhang, X.; Llorens, L.; Ogaya, R.; Lloret, F.; Comas, P.; Estiarte, M.; Terradas, J. Compex spatiotemporal phenological shifts as a response to rainfall changes. *N. Phytol.* **2004**, *161*, 837–846. [CrossRef]

34. Slovak Hydrometeorological Institute. Available online: http://www.shmu.sk/en/?page=1793 (accessed on 10 July 2020).

35. Enviroportál. Available online: https://geo.enviroportal.sk/atlassr/ (accessed on 9 September 2020).

36. Pôdny Portál. Available online: http://www.podnemapy.sk/portal/prave_menu/podna_mapa/podna_mapa.aspx (accessed on 6 September 2020).

37. Kolektív. *Návod na Fenologické Pozorovanie Lesných Rastlín (Instructions for Phenological Observation of Forest Plants)*; SHMU: Bratislava, Slovakia, 1984; p. 23.

38. Meier, U. *Growth Stages of Mono- and Dicotyledonous Plants, BBCH Monograph*; Blackwell Wissenschafts-Verlag: Berlin, Germany; Wien, Austria, 1997; p. 662.

39. LGIS. Available online: http://gis.nlcsk.org/lgis/ (accessed on 3 September 2020).

40. McKee, T.B.; Doesken, N.J.; Kleist, J. Drought monitoring with multiple time scales. In Proceedings of the 9th Conference on Applied Climatology, Dallas, TX, USA, 15–20 January 1995; pp. 233–236.

41. Guttman, N.B. Comparing the Palmer drought index and the Standardized precipitation index. *J. Am. Water Resour. Assoc.* **1998**, *34*, 113–121. [CrossRef]

42. Abramowitz, M.; Stegun, I.A. *Handbook of Mathematical Functions with Formulas, Graphs and Mathematical Tables*; Doves Publications Inc.: New York, NY, USA, 1965; p. 1046.

43. Lloyd-Hughes, B.; Saunders, M.A. A drought climatology for Europe. *Int. J. Climatol.* **2002**, *22*, 1571–1592. [CrossRef]

44. Rebetez, M.; Mayer, H.; Dupont, O.; Schindler, D.; Gartner, K.; Kropp, J.P.; Menzel, A. Heat and drought 2003 in Europe: A climate synthesis. *Ann. For. Sci.* **2006**, *63*, 569–577. [CrossRef]

45. Dubrovsky, M.; Svoboda, M.D.; Trnka, M.; Hayes, M.J.; Wilhite, D.A.; Žalud, Z.; Hlavinka, P. Application of relative drought indices in assessing climate-change impacts on drought conditions in Czechia. *Theor. Appl. Climatol.* **2008**, *96*, 155–171. [CrossRef]

46. Vicente-Serrano, S.M.; Beguería, S.; López-Moreno, J.I. A Multiscalar Drought Index Sensitive to Global Warming: The Standardized Precipitation Evapotranspiration Index. *J. Clim.* **2010**, *23*, 1696–1718. [CrossRef]

47. Thornthwaite, C.W. An approach towards a rational classification of climate. *Geogr. Rev.* **1948**, *38*, 55–94. [CrossRef]

48. Thornthwaite, C.W.; Mather, J.R. Instructions and tables for computing potential evapotranspiration and the water balance. *Publ. Climatol.* **1957**, *10*, 132.

49. Penman, H.L. Natural evaporation from open water, bare soil and grass. *Proc. R. Soc. Lond. Ser. A Math. Phys. Sci.* **1948**, *193*, 120–145.

50. Novák, V. *Vyparovanie Vody v Prírode a Metódy Jeho Určovania (Water Evaporation in Nature and Methods of its Determination)*; Veda: Bratislava, Slovakia, 1995; p. 260.

51. Voss, R.; May, W.; Roeckner, E. Enhanced resolution modeling study on antropogenic climate change: Changes in extremes of the hydrological cycle. *Int. J. Climatol.* **2002**, *22*, 755–777. [CrossRef]

52. Brezianska, K.; Vitková, L. Analýza bezzrážkových období a ich vplyv na zásobu vody v pôde na Záhorskej nížine (Analyse of periods without precipitation and their influence on soil water storage at Záhorská lowland). *Acta Hydrol. Slovaca* **2015**, *16*, 260–266.

53. Šútor, J.; Šurda, P.; Štekauerová, V. Vplyv bezzrážkových období na dynamiku zásob vody v zóne aerácie pôdy (Influence of precipitation-free periods on the dynamics of water reserves in the soil aeration zone). *Acta Hydrol. Slovaca* **2011**, *12*, 22–28.

54. Rennenberg, H.; Loreto, F.; Poole, A.; Brilli, F.; Fares, S.; Beniwal, S.R.; Gessler, A. Physiological responses of forest trees to heat and drought. *Plant Biol.* **2006**, *8*, 556–571. [CrossRef] [PubMed]

55. Russo, S.; Sillmann, J.; Fisher, E.M. Top ten European heatwaves since 1950 and their occurrence in the coming decades. *Environ. Res. Lett.* **2015**, *10*, 124003. [CrossRef]

56. Valencia-Barrera, R.M.; Comtois, P.; Fernández-González, D. Bioclimatic indices as a tool in pollen forecasting. *Int. J. Biometeorol.* **2002**, *46*, 171–175. [CrossRef]

57. Puchałka, R.; Koprowski, M.; Gričar, J.; Przybylak, R. Does tree-ring formation follow leaf phenology in Pedunculate oak (Quercus robur L.)? *Eur. J. For. Res.* **2017**, *136*, 259–268. [CrossRef]

58. Fu, Y.H.; Piao, S.; Delpierre, N.; Hao, F.; Hanninen, H.; Liu, Y.; Sun, W.; Janssens, I.A. Larger temperature response of autumn leaf senescence than spring leaf-out phenology. *Glob. Chang. Biol.* **2017**, *24*, 2159–2168. [CrossRef]

59. Vitasse, Y.; Porté, A.J.; Kremer, A.; Michalet, R.; Delzon, S. Responses of canopy duration to temperature changes in four temperate tree species: Relative contributions of spring and autumn leaf phenology. *Oecologia* **2009**, *161*, 187–198. [CrossRef]

60. Dolschak, K.; Gartner, K.; Berger, T.W. The impact of rising temperatures on water balance and phenology of European beech (Fagus sylvatica L.) stands. *Model. Earth Syst. Environ.* **2019**, *5*, 1347–1363. [CrossRef]

61. Barriopedro, D.; Fischer, E.M.; Luterbacher, J.; Trigo, R.M.; García-Herrera, R. The hot summer of 2010: Redrawing the temperature record map of Europe. *Science* **2011**, *332*, 220–224. [CrossRef]

62. Miralles, D.; Teuling, A.; van Heerwaarden, C.; de Arellano, J.V.-G. Mega-heatwave temperatures due to combined soil desiccation and atmospheric heat accumulation. *Nat. Geosci.* **2014**, *7*, 345–349. [CrossRef]

63. Roznovsky, J.; Litschmann, T.; Stredova, H.; Streda, T.; Salas, P.; Horka, M. Microclimate Evaluation of the Hradec Kralove City using HUMIDEX. *Contrib. Geophys. Geod.* **2017**, *47*, 231–246. [CrossRef]

64. Fisher, E.M. Contribution of land-atmosphere coupling to recent European summer heat waves. *Geophys. Res. Lett.* **2007**, *34*, L06707. [CrossRef]

65. Sánchez-Benítez, A.; García-Herrera, R.; Barriopedro, D.; Sousa, P.; Trigo, R. June 2017: The Earliest European Summer Mega-heatwave of Reanalysis Period. *Geophys. Res. Lett.* **2018**, *45*, 1955–1962. [CrossRef]

66. Škvareninová, J.; Babálová, D.; Valach, J.; Snopková, Z. Impact of temperature and wetness of summer months on autumn vegetative phenological phases of selected species in Fageto-Quercetum in the years 2011–2015. *Acta Univ. Agric. Silvic. Mendel. Brun.* **2017**, *65*, 939–946. [CrossRef]

67. Škvareninová, J.; Hlavatá, H.; Jančo, M.; Škvarenina, J. Impact of climatological drought on the leaves yellowing phenophase selected tree species. *Acta Hydrol. Slovaca* **2018**, *19*, 220–226.

68. Pšidová, E.; Živčák, M.; Stojnić, S.; Orlović, S.; Gomory, D.; Kučerová, J.; Ditmarová, Ľ.; Střelcová, K.; Brestič, M.; Kalaji, H.M. Altitude of origin influences the responses of PSII photochemistry to heat waves in European beech (*Fagus sylvatica* L.). *Environ. Exp. Bot.* **2018**, *152*, 97–106. [CrossRef]

69. Ciais, P.; Reichstein, M.; Viovy, N.; Granier, A.; Ogée, J.; Allard, V.; Aubinet, M.; Buchmann, N.; Bernhofer, C.; Carrara, A.; et al. Europe-wide reduction in primary productivity caused by the heat and drought in 2003. *Nature* **2005**, *437*, 529–533. [CrossRef]

70. ForestPortal. Available online: http://www.forestportal.sk/lesne-hospodarstvo/informacie-o-lesoch/zakladne-informacie-o-lesoch/Pages/klima.aspx. (accessed on 4 September 2020).

71. Fotelli, M.N.; Rennenberg, H.; Holst, T.; Mayer, H.; Gessler, A. Carbon isotope composition of various tissues of beech (Fagus sylvatica) regeneration is indicative of recent environmental conditions within the forest understorey. *N. Phytol.* **2003**, *159*, 229–244. [CrossRef]

72. Stojnić, S.; Suchocka, M.; Benito-Garzón, M.; Torres-Ruiz, J.M.; Cochard, H.; Bolte, A.; Cocozza, C.; Cvjetković, B.; de Luis, M.; Martinez-Vilalta, J.; et al. Variation in xylem vulnerability to embolism in European beech from geographically marginal populations. *Tree Physiol.* **2018**, *38*, 173–185. [CrossRef]

73. Vido, J.; Střelcová, K.; Nalevanková, P.; Leštianska, A.; Kandrík, R.; Pástorová, A.; Škvarenina, J.; Tadesse, T. Identifying the relationships of climate and physiological responses of a beech forest using the Standardised Precipitation Index: A case study for Slovakia. *J. Hydrol. Hydromech.* **2016**, *64*, 246–251. [CrossRef]

74. Vilhar, U. Comparison of drought stress indices in beech forests: A modelling study. *iForests* **2016**, *9*, 635–642. [CrossRef]

 *water*

Article

Impact of Water Deficit on Seasonal and Diurnal Dynamics of European Beech Transpiration and Time-Lag Effect between Stand Transpiration and Environmental Drivers

Paulína Nalevanková [1,2,*], Zuzana Sitková [3], Jíři Kučera [4] and Katarína Střelcová [1]

[1]   Department of Natural Environment, Faculty of Forestry, Technical University in Zvolen, T.G. Masaryka 24, 960 01 Zvolen, Slovakia

[2]   Oikos NGO, Environmental Laboratory, Na Karasíny 247/21, 971 01 Prievidza, Slovakia

[3]   National Forest Centre, Forest Research Institute, T.G. Masaryka 22, 960 01 Zvolen, Slovakia

[4]   Environmental Measuring Systems, Ltd., Kociánka 85/39, 612 00 Brno, Czech Republic

*   Correspondence: xnalevankova@tuzvo.sk; Tel.: +421-45-5206490

Received: 17 September 2020; Accepted: 3 December 2020; Published: 8 December 2020

Abstract: In-situ measurements of tree sap flow enable the analysis of derived forest transpiration and also the water state of the entire ecosystem. The process of water transport (by sap flow) and transpiration through vegetation organisms are strongly influenced by the synergistic effect of numerous external factors, some of which are predicted to alter due to climate change. The study was carried out by in-situ monitoring sap flow and related environmental factors in the years 2014 and 2015 on a research plot in Bienska dolina (Slovakia). We evaluated the relationship between derived transpiration of the adult beech (*Fagus sylvatica* L.) forest stand, environmental conditions, and soil water deficit. Seasonal beech transpiration (from May to September) achieved 59% of potential evapotranspiration (PET) in 2014 and 46% in 2015. Our study confirmed that soil water deficit leads to a radical limitation of transpiration and fundamentally affects the relationship between transpiration and environmental drivers. The ratio of transpiration (E) against PET was significantly affected by a deficit of soil water and in dry September 2015 decreased to the value of 0.2. The maximum monthly value (0.8) of E/PET was recorded in August and September 2014. It was demonstrated that a time lag exists between the course of transpiration and environmental factors on a diurnal basis. An application of the time lags within the analysis increased the strength of the association between transpiration and the variables. However, the length of these time lags changed in conditions of soil drought (on average by 25 min). Transpiration is driven by energy income and connected evaporative demand, provided a sufficient amount of extractable soil water. A multiple regression model constructed from measured global radiation (R_S), air temperature (AT), and air humidity (RH) explained 69% of the variability in beech stand transpiration (entire season), whereas (R_S) was the primary driving force. The same factors that were shifted in time explained 73% of the transpiration variability. Cross-correlation analysis of data measured in time without water deficit demonstrated a tighter dependency of transpiration (E) on environmental drivers shifted in time (-60 min R_S, $+40$ min RH and $+20$ min vapour pressure deficit against E). Due to an occurrence and duration of soil water stress, the dependence of transpiration on the environmental variables became weaker, and at the same time, the time lags were prolonged. Hence, the course of transpiration lagged behind the course of global radiation by 60 ($R^2 = 0.76$) and 80 ($R^2 = 0.69$) minutes in conditions without and with water deficit, respectively.

Keywords: forest stand transpiration; European beech; environmental drivers; drought; time lag

---

1. Introduction

Plant transpiration is a major component of terrestrial ecosystem evapotranspiration and represents a significant water loss term of the water balance [1]. In terrestrial conditions, 39% of precipitation returns to the atmosphere through transpiration, which accounts for 61% of evapotranspiration, on average. Besides, forest ecosystem transpiration can contribute 50–70% of terrestrial evapotranspiration [2]. Therefore, most water evaporating from ecosystems transits through plant organisms and its concrete volume is regulated by plants [3]. At the same time, evapotranspiration can account for over 50% of the total water loss in most terrestrial ecosystems and belongs to the primary water-loss components of the water cycle [4,5].

Evapotranspiration can be estimated relatively precisely, but it requires a high investment in equipment, well-trained research personnel and know-how [6]. The methods of direct measurement of evapotranspiration are currently rather limited, particularly in woody crops [6,7]. However, transpiration can be estimated utilising sap flow monitoring. The physiological process of transpiration and water flux through xylem are well known. In recent years, several methods for the precise measurement of tree trunk sap flow have been developed and successfully applied in research [8,9]. Modern systems are relatively affordable, undemanding to use, and some provide instant information on sap flow without demanding recalculations [10]. Hence, at a larger time scale (i.e., days and longer), the amount of water flow through the trunk is approximately equal to canopy transpiration [11]. Therefore, the monitoring of sap flow is a highly useful tool to observe and investigate the impact of environmental conditions, including extreme events such as drought, on forest water balance and its stress load [12,13]. Sap flow measurements have become a standard tool for ecosystem research [14].

A detailed study of transpiration, its limitations, and the dynamics in connection to changing environmental conditions is important for understanding the impact of vegetation on hydrology and can help improve ecosystem water balance modelling and to predict plant responses to climate change [1,9,15,16]. Transpiration is influenced by the synergic effect of environmental factors, vegetation characteristics (leaf area index (LAI), tree age and vitality, stocking density, root system) and management (thinning, pruning) [17]. The impact of environmental factors is significant and can be divided into atmospheric evaporative demands (potential evapotranspiration and vapour pressure deficit) [16,18] and soil moisture, the lack of which leads to limited plant transpiration [19–21]. However, some woody species can uptake water even from groundwater or bedrock fractures [22,23]. Environmental factors are predicted to change in connection to human-induced global warming, which according to the IPCC report, reached approximately 1 °C (likely between 0.8 °C and 1.2 °C) above pre-industrial levels in 2017, increasing at 0.2 °C per decade (high confidence) [24]. Various climate scenarios predict a substantial change in precipitation distribution and in average temperatures, which can cause more prolonged and severe summer drought [25,26]. A warmer atmosphere means increased evaporation demands and is accompanied by a lack of water supply, leading to an intensification of water-related plant stress [27,28]. Water deficit results in drought stress, substantially limiting the physiological processes of plants and phytomass formation and may result in a long-term decrease in forest productivity [29,30] and an increase in drought-induced mortality [31–33].

As aforementioned, the main weather parameters that affect transpiration, together with crop characteristics and water supply, are radiation, air temperature, humidity, and wind speed. A common effect of these parameters can be expressed via the evaporation power of the atmosphere (potential evapotranspiration or reference crop evapotranspiration) [34]. Over the years, several research papers have been published focused on modelling the water use (transpiration or sap flow) of various tree species [3,35–38]. The simulation of tree water use is carried out using several environmental variables applying various approaches [38]. Some studies point out that within the relationship between sap flow (or transpiration) and environmental variables exists a time lag of varying range and direction [3,39–41]. This time lag is the main characteristic of the hysteresis phenomenon, which was detected and described in many studies within different species and ecosystems [40,42,43].

Overall, the diurnal course of sap flow can lag behind the course of the environmental driver or can occur in advance. The time lag between measured sap flow and environmental factors can be caused by the priority use of trunk water storage for transpiration, particularly in the morning [40,44]. At the same time, hysteresis is considered to be a conservation mechanism to prevent dehydration during high evaporative demand [45]. However, prior studies demonstrated that time lags were affected by drought, solar radiation, and high evaporation demands [42,46]. The study of these time lags can advance our understanding and, therefore, improve the response of vegetation to changing climate factors, and the inclusion of the detected lags can enhance water flux modelling [40,41].

Our study focused on assessing the seasonal and diurnal dynamics of sap flow-derived transpiration of mature beech stand in relation to changing environmental drivers with an emphasis on the occurrence of drought. European beech (*Fagus sylvatica* L.) belongs to the essential broadleaved species in Europe, from both an economic and ecological point of view [30,47]. Beech increases the stability of forest stands, which is reflected in its pan-European application in the silvicultural concepts of reconstructing non-natural monocultural conifer stands (mostly spruce) to stable and resilient mixed forests more resistant to predicted climate change [48–50]. On the other hand, beech decline in Europe was reported in connection with a drought-related decrease in tree vitality, which causes subsequent increased endanger due to biotic factors, such as insects and wood decay fungi [51–53].

In Slovakia, European beech, with a share of almost 34% (with an increasing trend), is the most commonly occurring tree species and has the largest share of wood production within deciduous tree species [50]. Beech forests, due to their broad areas of spread, significantly contribute to the hydrological cycle by water transpiration and interception, with transpiration having the largest share regarding beech stand total evapotranspiration [54]. In the context of climate change, it is predicted that beech ecosystems will be threatened all over Europe [55] and will be exposed to severe stress from drought because shallow-rooting beech is considered a drought-sensitive species [47,56,57]. Drought episodes and increased temperature during the growing season negatively affect beech stands' water balance, growth, competitiveness, and resilience. The effect is increased during the prolonged duration of the dry season and has been repeated for several years in a row [47,57–59]. The limitation in beech growth as a consequence of drought and resulting reduced tree competitiveness was documented already in the 20th century [60]. Similarly, recent research results confirm a 17% reduction of annual gross primary beech production due to the 2003 drought [61]. A study from 2019 concluded that the European beech is, in particular, sensitive to a lack of precipitation and is less sensitive to heat stress [61].

The main objectives of this study were, therefore, to (i) describe the seasonal and diurnal dynamics of beech transpiration in relation to the course of environmental factors affecting transpiration, focusing on the drought effect, (ii) detect the time lags between the transpiration and meteorological drivers, and (iii) assess the impact of water deficit on time lags.

2. Materials and Methods

2.1. Experimental Stand

The study was conducted in the area of Bienska dolina, situated in the central part of the Slovak Republic at an elevation of 450 m a.s.l. with coordinates 48°36′43″ N and 19°03′59″ E (Figure 1). The locality is classified as a 3rd oak-beech altitudinal forest zone (classification according to Zlatník [62]) with Haplic Cambisol (Humic, Eutric, Endoskeletic, and Siltic) formed on volcanic parent material (andesite and andesitic tuffs) [63]. The depth of soil reaches a maximal 66 cm. The textural class of the fine-earth fraction is detected as silt loam in the topsoil or loam in the subsoil. Upper horizons are characterised by the presence of only a small amount of coarse rock fragments, whereas coarse gravel and stones are plentiful in the subsoil (C horizon, up to 80%). The abundance of roots in the upper 30–35 cm of soil is relatively high and decreases in deeper soil layers.

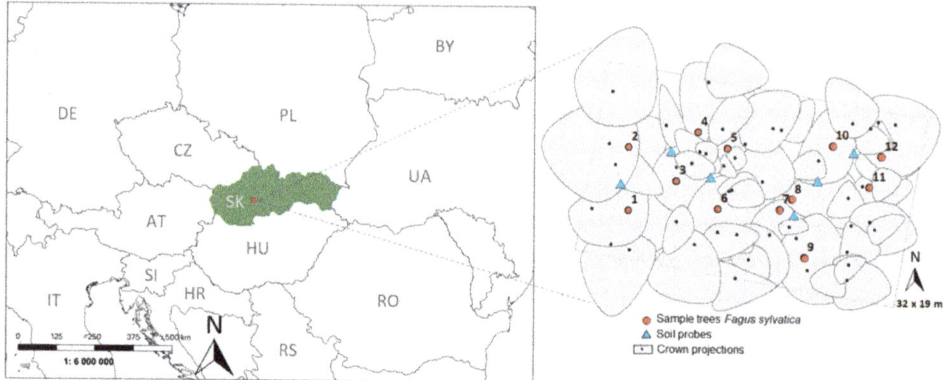

Figure 1. Location of the experimental plot Bienska dolina in central Slovakia (SK), Central Europe (**left**) and design of the experimental plot showing crown projections of all beech trees, position of selected sample trees and soil probes (**right**).

An experimental plot with an area of 608 m² was established in a 68-year-old European beech forest, which is dominated by *Fagus sylvatica* L. with a minor admixture of *Quercus petraea* (Matt.) Liebl. (sessile oak, 10%) and *Larix decidua* (Mill.) (larch, 5%). However, these admixture species occur outside the delineated plot. The slope is east facing and, on average reaches a maximum of 30%. The relative stocking density of the stand is 0.85 with a maximum leaf area index of 6.1 in July (LAI-2200 Plant Canopy Analyzer, LI-COR Biosciences, Lincoln, NE, USA). The total stand volume of beech is 282 m³ ha⁻¹. The understory vegetation is poor and is characterised by the rare occurrence of beech seedlings and typical herbs species, such as *Asarum europaeum* and *Dentaria bulbifera*.

Within the area of the experimental plot (608 m²), 56 trees were measured using Field-Map technology, which combines real-time GIS software with electronic equipment for mapping and dendrometric measurements (IFER Ltd., Jílové u Prahy, Czech Republic). Using this technology, the main dendrometric characteristics (tree height, diameter at breast height (DBH) and crown projection) of trees were focused upon and documented (Figure 1, right). The measured DBH of beech trees at the plot varied from 5.7 to 42.3 cm, and heights were from 8.4 to 29.3 m. The distribution of trees within the DBH classes is presented in Figure 2. From these trees, 12 representative sample trees were chosen based on DBH, height and their location to apply sap flow measurement systems. Sample trees were characterised by a mean tree height of 26.3 m ± 1.3 m (from 24.7 to 29.1 m) and an average diameter (DBH) of 32.4 cm ± 4.8 cm (from 27.1 to 42.3 cm).

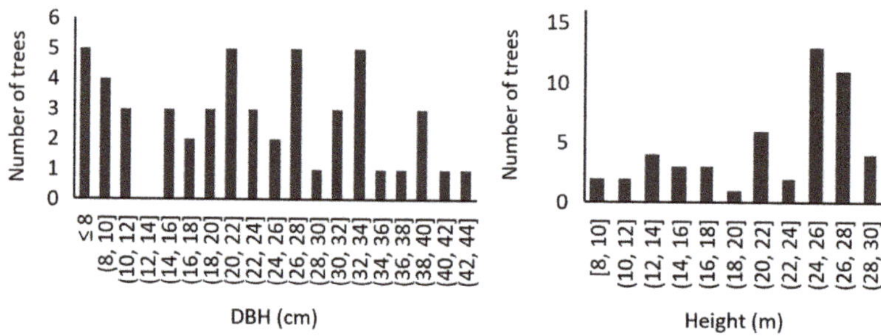

Figure 2. Number of trees in DBH and height classes (interval of the class is 3 cm and 3 m, respectively) within the experimental plot area (608 m²).

The climate of Bienska dolina is classified as slightly warm and moderately humid [64]. Figure 3 shows the 1961–1990 reference period's monthly averages (normals) of mean air temperature (*AT*) and precipitation totals (*P*) for the area, calculated based on data from a nearby station belonging to a network of professional meteorological stations of the Slovak Hydrometeorological Institute. The long-term mean annual air temperature and annual sum of precipitation are 7.3 °C and 690 mm, respectively. The normal values during the vegetation season (May–September) are 14.8 °C for mean temperature and 250 mm for precipitation. The rainiest month according to the reference period is June, with an average precipitation of 90 mm, and the warmest month is July, with an average temperature of 17 °C.

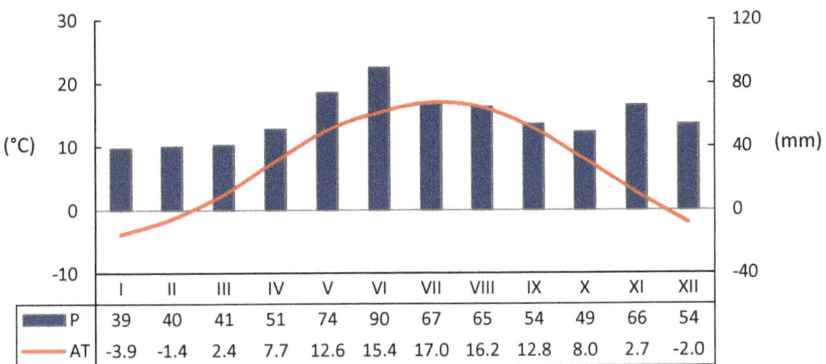

Figure 3. Monthly averages (normals) of mean air temperature (*AT*, °C) and precipitation totals (*P*, mm) for the Bienska dolina area, calculated based on reference period 1961–1990 data (January–December; in Roman numerals I–XII).

2.2. Measured and Derived Environmental Variables

Meteorological variables were during the years 2014 and 2015 measured within an open grass area situated ca. 250 m from the experimental stand by an automatic meteorological station. The station was equipped with an air temperature (*AT*, °C) and relative humidity (*RH*, %) sensor and a global radiation (R_S, W m^{-2}) sensor (EMS33 and EMS11; Environmental Measuring System (EMS Brno) Ltd., Brno, Czech Republic), which were situated at a height of 2 m above ground (low cut grass). Wind speed (*u*, m s^{-1}) was monitored using a 034B Wind Sensor (Met One Instruments Inc., Grants Pass, OR, USA) at a height of 2 m, and precipitation (*P*, mm) was measured using a rain gauge type 370 with a collecting area of 320 cm^2 placed at a height of 1 m above ground (Met One Instruments Inc.). The interval of measurements was 5 min, and data were stored every 20 min in the data logger edgeBox V12 (EMS Brno Ltd.) powered by a 12 V solar-charged battery.

Potential evapotranspiration (*PET*, mm h^{-1}) as a variable representing theoretical atmospheric evaporative demands unaffected by soil water deficit was calculated according to the Penman equation [65] (Equation (1)).

$$PET = \frac{\Delta}{\Delta + \gamma}R_n + \frac{\gamma}{\Delta + \gamma}\frac{6.43\,(1 + 0.536\,u)\,VPD}{\lambda} \tag{1}$$

where Δ is the slope of the saturation vapour pressure curve (kPa K^{-1}), *Rn* is the net radiation (W m^{-2}) estimated as 77% of the global incoming solar radiation [34] and *u* is the wind speed measured at 2 m height. The used psychometric constant (γ) was 66 Pa K^{-1} and the latent heat of vaporization (λ) was 2.45 MJ kg^{-1} [34].

VPD is the vapour pressure deficit (kPa) calculated by Equation (2).

$$VPD = e_s - e_a \qquad (2)$$

where e_s is the saturated vapour pressure at a given air temperature and e_a is the vapour pressure of the free-flowing air.

Soil water potential (*SWP*, MPa) was continuously measured using measuring sets containing three calibrated gypsum blocs (Delmhorst Inc., Towaco, NJ, USA), and data were stored at 60-min intervals in a data logger (MicroLog SP3, EMS Brno Ltd., Brno, Czech Republic). *SWP* values varied from 0 up to –1.5 MPa (the lowest measurable limit of the equipment). Measurements were provided in six different soil profiles distributed across the research plot and were conducted at soil depths of 15, 30, and 50 cm with respect to the depth of soil, which is maximal at 66 cm. In this paper, we used average values of the three depths of the research plot.

Daily relative extractable soil water (*REW*, dimensionless) at the stand scale was calculated using the forest water balance model BILJOU© (https://appgeodb.nancy.inra.fr/biljou/), a detailed description of which is given in Granier et al. [66]. The model required daily meteorological data (measured wind speed, precipitation, air temperature, relative humidity, global radiation), maximum leaf area index (LAI), and day of the year (DOY) of the onset of phenophase budburst (DOY 105) and leaf fall (DOY 298), maximum extractable soil water (98 mm in the entire root zone), vertical root distribution, and soil properties (determined by soil analysis). *REW* is widely used to quantify drought intensity and varies between 1 (when the soil is at field capacity) and 0 (permanent wilting point). When *REW* drops below a critical threshold of 0.4 (*REW* falls below 40% of maximum extractable water), a soil water deficit is assumed to occur, and transpiration is gradually reduced due to stomatal closure [31,66]. The model calculates *REW* according to Equation (3).

$$REW = \frac{EW}{EW_M} \qquad (3)$$

where *EW* is the actual extractable soil water in the rooting zone (*EW* = available soil water–minimum soil water (i.e., lower limit of water availability)), and EW_M is the maximum extractable soil water (EW_M = soil water content at field capacity–minimum soil water (i.e., lower limit of water availability)).

2.3. Sap Flow and Scaling up to Forest Stand Level

On the selected 12 sample trees, at a stem height of ca. 2 m, EMS51A Sap Flow systems connected to a 16-channel datalogger RailBox V16 manufactured by EMS Brno Ltd. (Brno, Czech Republic) were installed. The system uses a tissue heat balance method (THB, [8,67]) based on volume (three-dimensional) heating of the stem segment [10] to measure the values of volumetric sap flow directly in kg of water per a specific period and per one centimetre of stem circumference. The sap flow in Bienska dolina was measured at 5-min intervals and logged as 20-min averages in the datalogger. To obtain sap flow per the entire tree, we recalculated the raw values of sap flow (kg h^{-1} cm^{-1}) according to the corresponding circumference of the individual sample trees (kg h^{-1} tree^{-1}). In this paper, we processed the data of sap flow measurements performed during the growing season (from May to end of September) for the years 2014 and 2015.

The system used is designed for large trees with a stem diameter larger than 12 cm, whereas its underlying theory is clear with the absence of uncertain empirical parameters and without the need for field calibration as stated by the manufacturer (http://www.emsbrno.cz/p.axd/en/Sap.Flow.large.trees.html) and Tatarinov et al. [10]. The system consists of a controlling unit (MicroSet8X), sap flow sensor and a set of stainless steel electrodes. The measuring area was protected against direct sunlight and water using reflective insulation. The system (via controlling unit) maintains a pre-set (constant) temperature difference (used value 1 K) between a defined spatial sector of sapwood and reference probes (with an accuracy better than 1%) by electronic control using the variable electric heating power of the controller as the primary signal to quantify sap flux via the heat capacity of water [9]. The sapwood section was internally heated with alternating current (average power consumption 0.3–0.4 W by temperature difference 1 K) by three stainless steel plate electrodes 25-mm wide and 1-mm thick (available in three lengths: 60, 70, and 80 mm). The three electrodes in series plus one reference electrode 10 cm below were inserted at distances exactly 2 cm into the sapwood, passing through the xylem tissues. The central electrode was placed in a radial direction relative to the tree trunk (Figure 4). Into the geometrical centre of these electrodes, the thermosensor needles were inserted to measure temperature differences between the upper and lower electrodes [8,9]. The volume heating method provides a relatively stable temperature field under different values of affected factors (such as wood heat conductivity, sap flow radial profile and temperature gradients), despite this approach requiring relatively higher power consumption [10]. This method eliminates errors in measuring the dynamics of sap flow due to the principle of maintaining a constant temperature for the heated part (no time required to reach steady-state) [10,67].

The general equation that describes the heat balance of xylem takes the following form:

$$P = Q_r \, dT \, c_w + dT \, z \tag{4}$$

where P is the power of heat input (W), Q_r is the amount of water passing through the heated volume (kg s^{-1}), dT is the temperature difference within the measuring point (K), c_w is the specific heat of water (J kg^{-1} K^{-1}), and z represents the coefficient heat losses from the measuring point (W K^{-1}).

Water passing through the measuring point (in terms of volume or mass) is calculated from the power input and temperature rise of water passing through the heated space. The sap flow (kg s^{-1} cm^{-1}) calculation (Equation (5)) is derived from Equation (4).

$$Q = \frac{P}{c_w \, d \, dT} - \frac{z}{c_w} \tag{5}$$

where d is the effective width of the measuring point (5.5 cm), and z/c_w expresses heat losses from the sensor, which is set when Q is equal to zero (P, c_w and dT are listed in the description of Equation (4)) [67,68].

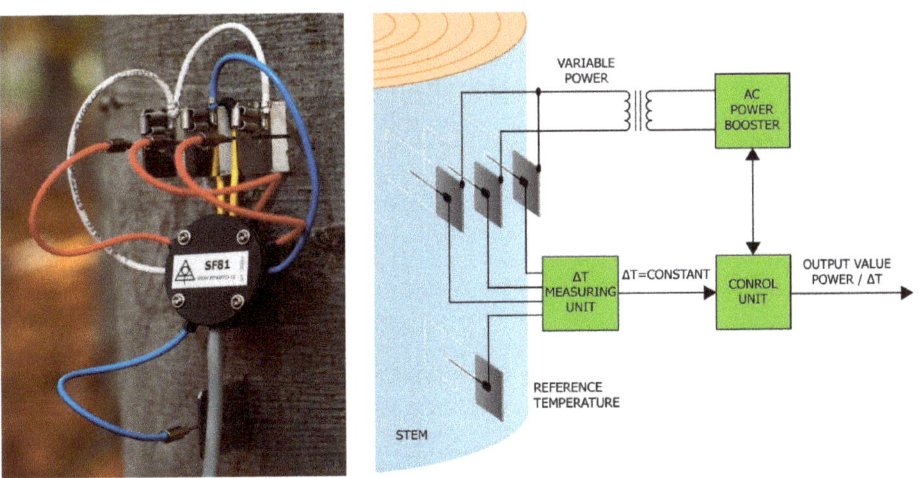

Figure 4. Installed Sap Flow System produced by EMS Brno Ltd. (Brno, Czech Republic) and the flows chart of measuring process. Sap flow values are calculated directly from the P/dT ratio.

Based on these measurements, we performed the upscaling from individual tree sap flow to the forest stand on the base of sap flow distribution at the diameter at breast height (DBH; diameter of the tree at a height at 1.3 m) classes [8]. The first step was to obtain the values of sap flow for mean trees (Q_{mean}, kg h^{-1}) of m DBH classes by means of regression relating the sap flow of the measured sample tree (Q_{sample}, kg h^{-1}) to the chosen biometric parameter—DBH. These regression analyses were performed with data obtained during a period of constant and sufficient soil water separately for the years 2014 and 2015. Derived Q_{mean} values were multiplied by numbers of trees in DBH classes (n_i) and then summarised within the stand area unit of 1 ha (Equation (6)). The result represents the sap flow values of the entire forest stand (Q_{stand}).

$$Q_{stand} = \sum_{i=1}^{i=m} (Q_{mean})_i \, n_i \tag{6}$$

By dividing gained stand sap flow (Q_{stand}) by the sum of sap flow directly measured (Q_{sample}), we determined the non-dimensional coefficient S (-) (Equation (7)). Finally, we multiplied the measured sap flow data (20-min) by the coefficient S and obtained stand-level values, which we considered to be equal to stand transpiration (E, mm h^{-1} (kg m^{-2} h^{-1})).

$$S = \frac{\sum Q_{stand}}{\sum Q_{sample}} \tag{7}$$

2.4. Data Processing and Statistical Evaluation

Base data processing was performed via Mini32 software produced by EMS Brno Ltd. (Brno, Czech Republic) compatible with all used equipment. Statistical analyses were carried out using the statistical software Statistica® 12 (Statsoft, Tulsa, OK, USA) and Statgraphics centurion 18 (Statpoint Technologies, Inc. The Plains, VA, USA) and for all analyses, $p < 0.001$ was considered significant unless otherwise stated.

The principal component analysis (PCA) was used to analyse the environmental conditions that drive transpiration. The analysis was performed on data of the following variables: global radiation, air temperature, air humidity, precipitation, potential evapotranspiration, vapour pressure deficit and soil water potential. PCA is a classic method used to reduce the dimensionality of the data in order to understand better the underlying factors affecting those variables. To determine the number of principal components to retain, we used eigenvalues. An eigenvalue which is greater than or equal to 1 indicated that principal components account for more variance than one of the original variables.

A backward stepwise linear regression procedure was used to determine the influence of the chosen variables on stand transpiration and to select a subset containing only significant predictors. The importance of the predictor can be evaluated by comparing the determination coefficient (R^2) of the model fit before and after removing a predictor.

The time lags between the environmental variables and transpiration were estimated by performing a time-series cross-correlation analysis. The synchronised time series of environmental variables and transpiration were shifted in time (20-min step interval) until the highest R^2 was reached, and the corresponding time lag was found. The correlation analysis was performed to compare how can time lag change the strength of the association between transpiration and the variables and, at the same time, to detect the change within lags in the response of transpiration to atmospheric drivers due to different soil moisture conditions. Therefore, we analysed the relation of transpiration to variables measured in the corresponding time (sap flow and driver measured at the same time–time lag 0) and also variables shifted in time (in a step of 20 min, time lag +/−20 min, +/−40 min, etc.). In this study, a negative (−) time lag means that the variable is shifted on the time axis to the left—the course (or peak) of variable occurs earlier in the day than the peak of transpiration. A positive (+) time lag indicates that the variable is shifted on the time axis to the right (against transpiration)—the course (or peak) of the variable occurs later in the day than the peak of transpiration.

To evaluate the impact of water deficit on the relationship between transpiration and the main affecting variables, the data were divided into two categories based on the indication of water deficit by *REW* values (REW > 0.4, REW < 0.4). In the period when the value of *REW* dropped below the threshold of 0.4, it was assumed that water stress occurred [66].

3. Results and Discussion

3.1. Seasonal Variation of Environmental Conditions

The growing seasons (May–September) of the years 2014 and 2015 were characterised by mean air temperatures that exceeded the monthly long-term means, except for August 2014, when the mean temperature was slightly (0.2 °C) below normal (Figure 5). The warmest months compared to the normal occurred in 2015, whereas in July, August, and September, the deviations from normal were at a level of +3.5 °C, +4.3 °C, and +2.1 °C. These high temperatures were accompanied by a lack of precipitation, which was documented by the negative deviations from the normal in monthly precipitation totals. The precipitation totals were 62, 26, and 13 mm below normal, in June, August, and September, respectively. Vice versa, in 2014, the monthly precipitation was below the long-term value only in June. The highest positive deviation, 112 mm, from the normal rainfall was detected in July 2014, which reached 168% of normal. The seasonal (May–September) total of precipitation was 534 mm in 2014 and 279 mm in 2015, whereas the precipitation normal for the locality was 350 mm.

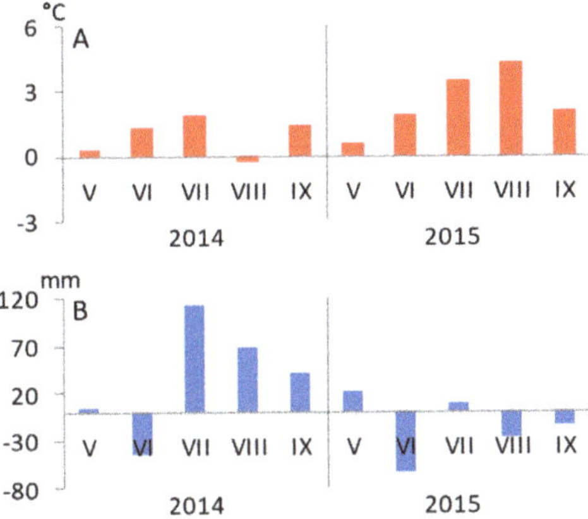

Figure 5. Deviations in monthly (May–September; in Roman numerals V–IX) air temperature (part **A**; °C) and precipitation totals (part **B**; mm) from long-term mean (normal) of 1961–1990.

Figure 6 shows the seasonal courses of the atmospheric variables as global radiation (*R_S*), air temperature (*AT*), vapour pressure deficit (*VPD*) and potential evapotranspiration (*PET*). Precipitation distribution within the growing seasons is depicted together with the course of average soil water potential (Figure 6, bottom). Significantly different conditions were observed in two years of interest, which resulted mainly from unequal precipitation occurrences. In comparison with the previous years, the 2015 growing season was characterised by a lack of precipitation (−71 mm compared to normal) and higher values of air temperatures, global radiation and, hence, also of vapour pressure deficit and potential evapotranspiration, particularly during July, August, and September. That resulted in the occurrence of dry episodes, as was evidenced by reduced values of *SWP* (the hourly values of soil water potential almost reached the value of −1.5 MPa, which is a limit of the equipment) and relative extractable water (*REW*). In the forest, the soil water deficit (or water stress) was assumed to occur when *REW* fell below the threshold of 0.4 [66], inducing stomatal regulation in forest trees [56]. However, [69] reported a slightly lower threshold value for a coniferous forest. Based on the REW threshold of 0.4, we detected 20 days in 2014 and 98 days in 2015 with a water deficit. The first day, when the water deficit was found, was DOY 166 (day of the year, 15 June) and DOY 197 (28 June) in 2014 and 2015, respectively. Drought episodes at Bienska dolina plot were rarely interrupted by summer storms, particularly in 2015.

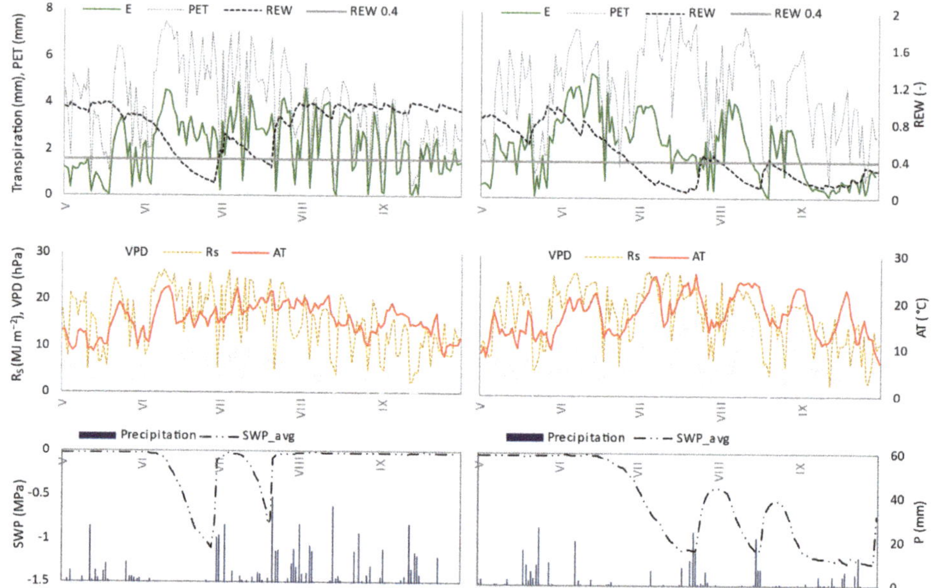

Figure 6. Seasonal dynamics of transpiration and potential evapotranspiration (*PET*) during vegetation season of 2014 (**left**) and 2015 (**right**) and the course of relative extractable water (*REW*) with a marked threshold of 0.4 (**top**); daily values of global radiation (R_S) vapour pressure deficit (*VPD*) and average air temperature (*AT*) (**middle**); daily values of average soil water potential and precipitation totals (*P*) (**bottom**).

The daily maximum of *VPD* was 12.8 hPa and 18.8 hPa in 2014 (11 June) and 2015 (7 June), respectively. The values of *PET* varied in 2014 from 0.5 mm to 7.5 mm per day and in 2015 between 0.6 mm and 8.2 mm per day (with maximal value on 11 and 7 June, respectively). The total amounts of *PET* per entire season (May–September) were 609 and 716 mm, respectively.

3.2. Seasonal Dynamics of Transpiration

The temporal courses of the daily stand transpiration (*E*) determined based on sap flow measurements of adult beech trees are shown in Figure 6. Transpiration and the meteorological factors have a synchronised dynamic within the growing season, particularly when soil moisture is not a limiting factor. In conditions with a sufficient water supply, the course of transpiration had almost identical dynamics as the course of *PET* and also *VPD* (*PET* includes the effect of *VPD*, R_S, and wind speed), which are the expression of atmospheric evaporative demand. Therefore, with the increasing *PET*, an increase in *E* was observed. The high evaporative demand related to the high value of global radiation and a corresponding increase in *AT*, and on the other hand, the decrease in air humidity. However, a temporary reduction in transpiration was detected during rainfall events due to an increase in the water content of the air and the associated changes in atmospheric conditions. Similarly, [70], in their study from Spain, confirmed that beech is strongly affected by water deficit and high temperatures, but these limitations are weakened when cloudiness occurs.

Within the study period, rapid decrease cases in transpiration were observed, particularly in 2015, which were clearly associated with soil water limitations (low values of *SWP* and *REW* below 0.4) and were a physiological response of beech stand to water stress. The decrease in *E* was also associated with high evaporative demand, and the reduction in transpiration was more pronounced as the drought period was prolonged or with the culmination of water deficit. A substantial limitation of beech forest transpiration linked to drought conditions has also been described in many publications [12,19,71,72]. Under drought conditions and/or high atmospheric evaporative demands, plants regulate the transpiration of water by stomatal closure. Stomatal regulation of *E* is a vital mechanism that allows plants to regulate and optimise CO_2 assimilation versus water loss by evaporation [73]. The total transpiration of the entire measured season in 2014 (May–September) was 359 mm, whereas in 2015, the cumulative values of transpiration reached only 183 mm. Thus, potential seasonal evapotranspiration was 609 and 716 mm in 2014 and 2015, respectively. Hence, the ratio *E/PET* in 2014 and 2015 was 0.6 and 0.5, respectively. Granier et al. [71] specified beech stand transpiration derived from sap flow measurement as 76 and 72% of total evapotranspiration (eddy covariance technique) for the season from May to October. On average, the transpiration of temperate deciduous forest accounts for 67% of evapotranspiration [2]. In our case, seasonal beech transpiration (from May to September) achieved 59% of *PET* in 2014, whereas in 2015, it was only 46% (Table 1). Potential evapotranspiration, according to Penman [65], represents the theoretical value of water, which can be evaporated from the ecosystem provided little or negligible resistance to water flux and no soil water limitation. In times of drought, the plants try to protect against excessive water loss, the stomatal resistance is increasing. The ratio *E/PET* in Bienska dolina was markable as decreased in July, August, and most in September of 2015. The ratio of transpiration against *PET* was affected by a deficit of soil water documented by the decreased values of *SWP* and *REW* dropped below 0.4. This effect of *REW* to *E/PET* ratio was described also by [17] in a sessile oak stand (*Quercus petraea*).

Table 1. Monthly and seasonal (May–September) sums of transpiration (*E*) and potential evapotranspiration (*PET*) in years 2014 and 2015; the ratio *E/PET*.

	2014			2015		
	E (mm)	**PET (mm)**	**E/PET**	**E (mm)**	**PET (mm)**	**E/PET**
May	46	119	0.4	58	123	0.5
June	86	165	0.5	106	156	0.7
July	90	149	0.6	72	187	0.4
August	81	104	0.8	73	153	0.5
September	55	72	0.8	19	96	0.2
Season	359	609	0.59	328	716	0.46

The highest achieved value of the ratio *E/PET* was 0.8 and occurred in August and September 2014, the months rich for precipitation and with average evaporation demands. m, beech transpiration represented 80% of potential evapotranspiration in the high air evaporative demands and conforming soil moisture conditions. In contrast, the smallest ratio, 0.2, was documented in September 2015, when, due to intensive and prolonged drought duration, the transpiration was only 19 mm compared to the value of potential evapotranspiration of 96 mm.

Within the studied seasons, also observed were relatively higher values of transpiration during the time of reduced values of *REW* (mostly in 2015). However, significantly reduced *E/PET* ratio indicated the regulation of transpiration. This phenomenon usually occurred after precipitation events and at the beginning of the period with the decrease in soil water availability. Assuming correctly determined *REW* values, we can therefore suppose that it was related to the use of water from tree reservoirs in drought conditions, similarly as was described in [12]. In some cases, however, we cannot strictly rule out the usage of water from deeper soil layers or groundwater fractures.

For a detailed display of the diurnal course of transpiration, we chose one sunny/clear day and one day with the occurrence of clouds (rain) in the period with a sufficient supply of soil water (DOY 260 and DOY 216) and also in the period in which there was a deficit of soil water (*REW* decreased under the value of 0.4, DOY 242 and DOY 254). Figure 7 shows the 20-min variations in transpiration and meteorological factors within days. On clear days, the courses of transpiration and global radiation and vapour pressure deficit were synchronous but lagged in time. The diurnal course of transpiration on clear days had one peak, which occurred after the peak of global radiation. On clear days, the course of transpiration followed the R_S course. With sunrise in the morning and the increase in R_S, transpiration began to increase and peaked around noon: within sufficient soil water conditions at 13:00 and in conditions of water deficit at 14:00. In comparison, the maximum of R_S occurred at 12:00 and 11:20, respectively. Hence, the course of transpiration lagged behind R_S by 60 min when water was not the limiting factor and 160 min in conditions of soil water deficiency. The maximums of *AT* and *VPD* were reached at 14:20; thus, the peak occurred 80 min later in day than the peak of transpiration. On the other hand, during drought conditions, it was at 13:40, indicating that the maximal values of *AT* and *VPD* occurred 20 min sooner than maximum *E*. The transpiration rate was significantly limited during soil water deficit compared to days with sufficient soil water supply.

During cloudy days, the typical bell-shaped curve of transpiration was not observed. High air humidity and low *VPD* values caused significant attenuation of transpiration or it falling to negligible values.

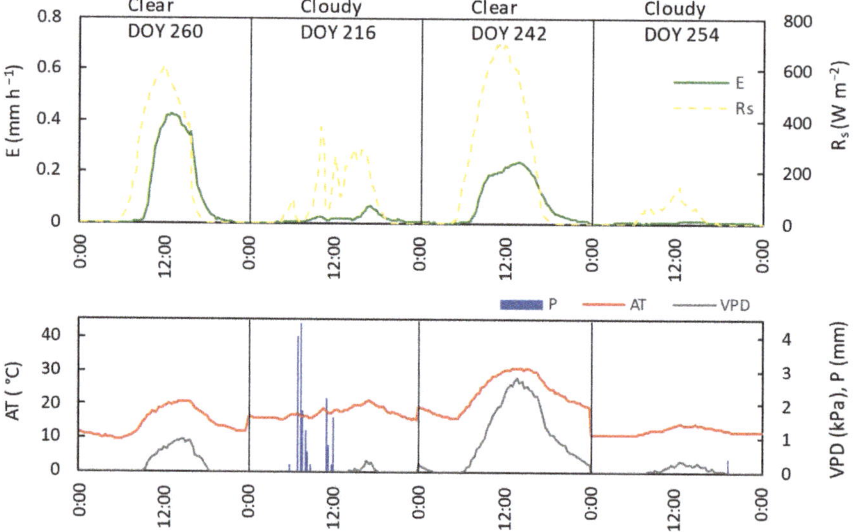

Figure 7. Diurnal dynamics of transpiration (*E*, 20-min data) and global radiation (R_S), air temperature (*AT*), vapour pressure deficit (*VPD*) and precipitation (*P*) during sunny/clear days and cloudy days in period without soil water deficit (**left**, DOY 260 and 216) and with soil water deficit (**right**, DOY 242 and 254). The data are provided in Central European Time (UTC+1).

3.3. Relation of Transpiration to Environmental Conditions

Principal component analysis was used to extract a smaller number of common factors, which can represent a large percentage of the variability in the original variables. Therefore, to express the covariances amongst the variables in terms of a small number of meaningful factors. The first three PCA components, only whose eigenvalues were greater than or equal to 1, could together explain 83%

of the variance in the data set (Table 2). Up to 53% of the variability could be explained by the first component. The second and third components explained 16% and 14%, respectively. According to Table 3, the first component was positively related mostly to *VPD*, R_S, *PET*, *AT* and negatively to *RH*. We can conclude, therefore, that the high factor loadings that occurred in the first component represent atmospheric conditions and evaporative demands. The second component explained 16% and was high and positively related to soil water potential and represents the demands for soil water moisture. The third component was positively related to precipitation. Global radiation, vapour pressure deficit, air temperature, and relative humidity were the main atmospheric factors and were included in the subsequent analyses.

Table 2. Eigenvalues and explained variance by three components on the environmental data (PCA).

Component No.	Eigenvalue	Percent of Variance	Cumulative Percentage
1	3.7	53	53
2	1.1	16	69
3	1.0	14	83

Table 3. Table of component weights (PCA).

Variables	Component 1	Component 2	Component 3
R_S	0.45	−0.23	−0.03
AT	0.42	0.09	0.17
RH	−0.46	−0.16	0.00
P	−0.07	−0.19	0.97
VPD	0.48	0.18	0.06
PET	0.42	−0.36	−0.06
SWP	0.05	0.85	0.15

A backward stepwise regression was performed to detect the interactive control of stand transpiration by environmental drivers. The input variables were R_S, *AT*, *RH*, and *VPD*. The results of the analysis are shown in Table 4, which provides the regression equations (models) and determination coefficients (R^2). All three variables together explained 69% of the variability in stand transpiration (upper part of table). R^2 significantly decreased when R_S was excluded from the parameters; hence, R_S was the primary driver. The same factors, but shifted in time, explained 73% of the transpiration variability (lower part of the table). The time lags between environmental variables and transpiration were estimated by performing a time-series cross-correlation analysis. In this context, ref [40] described that three variables, namely net radiation, air temperature and vapour pressure deficit together explained 97% of the variations in hourly plant water use in a humid headwater catchment. They also confirmed the major impact of radiation based on the decreased R^2 of the model when radiation was not considered in regression analysis. The second most important factor was *VPD*. The study from a water-limited desert ecosystem [43] found that R_S, *AT* and *VPD* together explained 84% and 77% of the variance in stand transpiration in 2014 and 2015, respectively. Both studies have consistently stated that water use (sap flux or stand transpiration) was primarily controlled by the availability of energy.

Table 4. Results of fitting a multiple linear regression model to describe the relationship between stand transpiration and global radiation (R_S), air temperature (AT) and relative humidity (RH) as independent variables (variables without time shift: top, variables shifted in time: bottom).

Variable	R^2	Model $\quad$ *p < 0.001*
R_S, AT, RH	0.690	$E = 0.0536679 + 0.000295379\ R_S + 0.00499328\ AT - 0.00127058\ RH$
R_S, AT	0.670	$E = 0.0834326 + 0.000341555\ R_S + 0.00679228\ AT$
R_S, RH	0.666	$E = 0.171546 + 0.000330795\ R_S - 0.00181746\ RH$
AT, RH	0.532	$E = 0.163561 + 0.00883183\ AT - 0.00279198\ RH$
Shifted variable	R^2	**Model** $\quad$ *p < 0.001*
$R_{S\text{-}60}, AT_{+40}, RH_{+20}$	0.729	$E = 0.00707624 + 0.00037642\ R_{S\text{-}60} + 0.00342536\ AT_{+40} - 0.000541175\ RH_{+20}$
$R_{S\text{-}60}, AT_{+40}$	0.726	$E = -0.0479013 + 0.000403582\ R_{S\text{-}60} + 0.00389637\ AT_{+40}$
$R_{S\text{-}60}, RH_{+20}$	0.718	$E = 0.07679 + 0.000410532\ R_{S\text{-}60} - 0.000797576\ RH_{\text{-}40}$
AT_{+40}, RH_{+20}	0.537	$E = 0.169809 + 0.00875072\ AT_{+40} - 0.00285242\ RH_{+20}$

As shown in Table 5 and Figure 8, a strong dependence was demonstrated between transpiration and meteorological factors. The correlation between E and VPD in conditions without soil water deficit was linear ($R^2 = 0.72$) and in conditions with water deficit, the relationship was best described by a polynomial function ($R^2 = 0.55$). The correlation between E and R_S was linear in both cases. In contrast, the correlation between E and AT and RH was polynomial in both cases, but within the relationship between E versus RH, the correlation was negative. The main drivers of transpiration process are mostly considered R_S and VPD (e.g., [17,40,43,74]). The linearity between transpiration and environmental variables was also observed in other studies [40,43,75,76]. For example, [43] observed linear correlation of stand transpiration and R_S, but the relationship between E and VPD was well fitted by a two-degree polynomial function, as well as the relationship between E and AT.

Table 5. Regression analysis of stand transpiration and environmental variables shifted in time for 2 categories of soil water conditions: without water deficit versus water deficit.

Variable	Without Water Deficit, REW > 0.4			Water Deficit, REW < 0.4		
	Time Lag	R^2	Equation	Time Lag	R^2	Equation
R_s	−60 min	0.76	$y = 0.0035 + 0.0005x$ *	−80 min	0.69	$y = 0.0099 + 0.0004x$ *
AT	-	0.64	$y = 0.0955 - 0.0237x + 0.0014x^2$ *	+40 min	0.42	$y = -0.0408 + 0.0005x + 0.0003x^2$ *
RH	+40 min	0.54	$y = 0.6651 - 0.0094x + 2.8194 \times 10^{-5}x^2$ *	+60 min	0.49	$y = 0.5425 - 0.0105x + 5 \times 10^{-5}x^2$ *
VPD	+20 min	0.72	$y = -0.0001 + 0.0002x$ *	+40 min	0.55	$y = -0.0207 + 0.1723x - 0.0262x^2$ *

$^* p = 0.0000.$

The strongest dependence of transpiration on the environmental variables was primarily found in conditions of sufficient water supply (without water deficit, REW > 0.4). The highest strength of linear correlation was between E and R_S shifted in time by −60 min, where 76% of the variability in transpiration can be explained by changes in global radiation ($R^2 = 0.76$). Hence, transpiration was most affected by R_S, which occurred 60 min 'before'. In the same manner, but in the opposite direction (positive), the tightness of the relationship between transpiration and VPD was higher when VPD was shifted in time by +20 min ($R^2 = 0.72$) (the VPD 20 min 'after' transpiration). Similarly, the tightest relationship between E and RH was confirmed by RH shifted in time of +40 min. The time lag was not confirmed within air temperature, although it is possible that the time shift was less than 20 min. Global radiation is the main driver that determines other variables, and their course tends to lag behind. That may be the reason why only time lag of R_S is negative.

The effect of drought stress was also reflected by a notable decrease in the dependence of transpiration from the tested variables. R^2 decreased to 0.69 (from 0.76), 0.42 (from 0.64), 0.49 (from 0.54) and to 0.55 (from 0.72) for E versus R_S, AT, RH and VPD, respectively (Table 5). These observations are consistent with the conclusions of [74]. Besides that, based on linear or polynomial regression, as a consequence of drought stress, the time shifts were prolonged by an average of 25 min (Figure 9). Hence, the course of transpiration lagged behind the course of global radiation by 60 min and by 80 min in conditions without water deficit and with water deficit, respectively. This phenomenon may be related to the plant's water-saving strategy. At the same time, water from the trunk's reserves is used more during the soil water deficit [12].

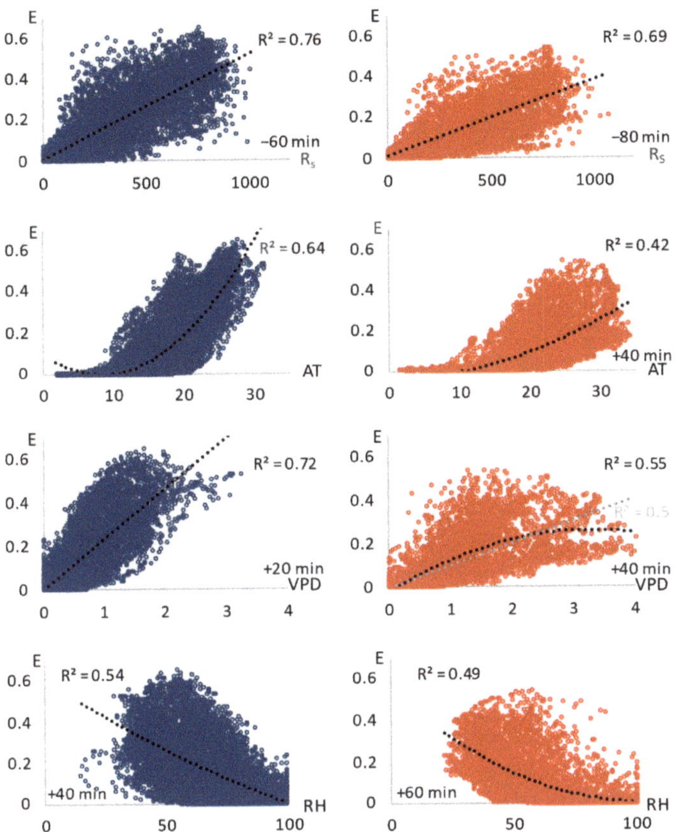

Figure 8. Relationships between 20-min values of transpiration (E) computed from sap flow measurements (in 2014, 2015) and environmental drivers: global radiation (R_S), air temperature (AT) vapour pressure deficit (VPD) and relative humidity (RH) shifted in time (+ or − min); data divided into two categories according to the state of *REW*: left/blue when REW > 0.4 (without water deficit) and right/red when REW < 0.4 (with water deficit).

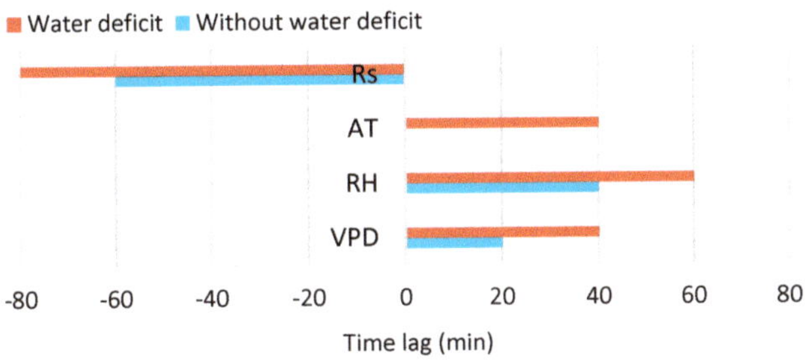

Figure 9. Changes in time lags for conditions with water deficit (REW < 0.4) and without water deficit (REW > 0.4).

Several authors have observed the time shift in the daily course of transpiration compared to the course of various environmental factors. Xu and Yu [43] studied hysteresis loops in shrub species and described the time legs in 2014 between transpiration and R_S, AT and VPD at an average of 1.5, 4.5, and 4.5 h, respectively. This study from desert ecosystem pointed that in 2015 maximum transpiration occurred after maximum R_S for 1.5 h, but before AT and VPD (for 2 h), which approximately corresponded with our results. The authors found seasonal patterns and as a possible cause marked stem water storage. Carrasco et al. [77] and Xu and Yu [43] also reported the water storage in the tree trunk having a role in time shift. Similarly, [40] analysed the relationships between transpiration of Scots pine and meteorological variables and also observed time lags and their seasonality. They found that sap flow lagged behind daily changes in meteorological variables indicated by phase shifts. Some studies claimed that the time shifts could be connected to the water conservation or self-protection mechanism of plants through stomatal control. Therefore, when the transpiration/sap flow reached the maxima earlier than environmental drivers (mainly VPD), further increases in the latter did not lead to more water loss by transpiration due to stomatal closure in response to meteorological changes [75,76]. In our study, the results pointed to the occurrence of the transpiration maxima before the peak of VPD, AT and RH. On the other hand, the situation was different for the courses of R_S. At the same time, the impact of the soil water deficit to the time lags was confirmed. O'Grady et al. [42] also described larger lags in dry than in wet seasons in Australian savannas (eucalyptus). A similar effect was identified by [78].

The relationship between E and VPD in the conditions of sufficient soil moisture was best described by a linear equation (coincided with [43]), and in the period of drought stress, this relationship changed and was described by a polynomial function. During water non-limited conditions, the stand transpiration increased linearly with the increase in VPD, because stomatal control of transpiration was minimal, and in drought conditions, the transpiration increased rapidly only until an unspecified threshold in VPD was gained. Then, transpiration was already inhibited. This finding may indicate that high values of VPD drive the stomatal regulation of transpiration. Similarly, [71] stated that in conditions of well-watered soil, beech shows a strong link between transpiration and VPD, but in conditions with limited soil water, canopy conductance decreases and is weakly related to atmospheric water demands. MacKay et al. [16] marked VPD as the main control of water loss in the forest, whereas soil moisture had an effect when the content of soil water decreased below a site-specific threshold during the growing season. Many authors have described the threshold control of VPD on transpiration, although the threshold value tends to be slightly different (1.2–2 kPa) [43,77,79,80]. However, [81] reported that hourly sap flow in northern hardwood forest declined in dry soil when VPD exceeded 1 kPa.

Transpiration is driven by energy income and connected evaporative demand, provided enough extractable soil water. Global radiation is the main driver that determines the course of other variables, and their course tends to lag behind. When assessing the causes of the delayed course of transpiration, it is also necessary to take into account that sap flow measurements take place on the tree trunk, whereas evaporation occurs in the tree canopy (the evaporating surface involves leaves). As was confirmed, there exist time shifts in sap flow measured in different parts of the tree (stem base versus crown part) caused by the use of water stored in the trunk for transpiration. The stored water is depleted in the morning and refilled during the night (nocturnal sap flow), depending on the weather. The amount of stored water depends on the species and dimensions of trees and varies in time. Čermák et al. [82] stated that on a daily basis (but not for the entire +growing season), the volume of water withdrawn from storage was equivalent to the water refilled to storage. The storage water on clear days was ca. 23% of the daily sap flow in old-growth Douglas-fir tree. For comparison, [83] for Douglas-fir stated water stored in xylem was 20 to 25% of total daily water use in 60-m trees, whereas in 15-m trees it was only 7%. For Oregon white oak stored water accounted for 10–23% of total daily water use in 25-m tall trees, whereas stored water comprised 9–3% in 10-m trees. The research of [77] indicated that tree trunk water storage contributes from 6 to 28% of the daily water budget of large trees depending on the species and [44] found the strong relationship between stored water use and *DBH*, sapwood area and leaf area. Wang et al. [40] determined nocturnal sap flow as 17% of the total sap flow on average in Scots pine in a humid low-energy headwater catchment and stated that trunk water storage would contribute to hysteresis and time lags between sap flow and meteorological forcing.

We can conclude that both the water storage and its use for morning transpiration and also atmospheric drivers can cause time lags. However, it is important to recognise their existence to incorporate them into water flux modelling and when investigating plant responses to changing environmental drivers.

4. Conclusions

This paper has evaluated the seasonal and diurnal dynamics of transpiration as a component of potential evapotranspiration and examined the environmental drivers' control of beech transpiration and the impact of water deficit on this relation. The experiment was based on the in-situ measurement of sap flow and accompanying measurement of environmental variables and soil water potential. It was found that seasonal beech transpiration (from May to September) achieved 59% of potential evapotranspiration in 2014, whereas in 2015, it was only 46%. During the studied growing seasons 2014 and 2015, soil water deficit led to the radical limitation of transpiration and affected the relationship between transpiration and environmental drivers. The ratio of transpiration (*E*) against potential evapotranspiration (*PET*) was significantly affected by the deficit of soil water and in dry September 2015 decreased to the value of 0.2. The maximum monthly value (0.8) of *E/PET* was recorded in August and September 2014. These months were characterised by above-normal precipitation totals, *REW* values above 0.4 and *SWP* values close to 0 MPa.

A time lag was demonstrated between the course of transpiration and environmental factors on a diurnal basis. Performing a time series cross-correlation, time lags for environmental variables were observed. An application of the time lags within the analysis increased the strength of the association between transpiration and the variables. We determined that the variation in beech transpiration was tightly associated with alterations in global radiation (*R*$_S$), air temperature (*AT*) and air humidity (*RH*). A multiple regression model constructed from these three environmental variables explained 69% of the variability in the beech stand transpiration. When we used time-shifted variables in the model (based on the cross-correlation), the model explained 73% of the transpiration variability.

Author Contributions: Conceptualization, P.N.; methodology, P.N.; software, P.N. and J.K.; formal analysis, P.N.; investigation, P.N.; data curation, P.N., Z.S. and J.K.; writing—original draft preparation, P.N.; writing—review and editing, P.N., Z.S. and K.S.; visualization, P.N.; sap flow system scheme, J.K. (EMS Brno Ltd.); supervision, K.S.; funding acquisition, K.S. All authors have read and agreed to the published version of the manuscript.

Funding: This research was funded by VEGA research projects funded by the Science Grant Agency of the Ministry of Education, Science, Research and Sport of the Slovak Republic No. 1/0370/18 and by Slovak Research and Development Agency under the contract No. APVV-16-0325, APVV-18-0390, APVV-19-0183, and project LignoSilva: Centre of Excellence of Forest-based Industry, ITMS: 313011S735 supported by the Research & Development Operational Programme funded by the ERDF.

Conflicts of Interest: The authors declare no conflict of interest.

References

1. Jasechko, S.; Sharp, Z.D.; Gibson, J.J.; Birks, S.J.; Yi, Y.; Fawcett, P.J. Terrestrial water fluxes dominated by transpiration. *Nature* **2013**, *496*, 347–350. [CrossRef] [PubMed]
2. Schlesinger, W.H.; Jasechko, S. Transpiration in the global water cycle. *Agric. For. Meteorol.* **2014**, *189–190*, 115–117. [CrossRef]
3. Kučera, J.; Brito, P.; Jiménez, M.S.; Urban, J. Direct Penman–Monteith parameterization for estimating stomatal conductance and modeling sap flow. *Trees* **2016**. [CrossRef]
4. Lu, J.; Sun, G.; McNulty, S.G.; Amatya, D.M. Modeling actual evapotranspiration from forested watersheds across the southeastern United States. *J. Am. Water Resour. Assoc.* **2003**, *39*, 886–896. [CrossRef]

5. Zheng, H.; Yu, G.; Wang, Q.; Zhu, X.; He, H.; Wang, Y.; Zhang, J.; Li, Y.; Zhao, L.; Zhao, F.; et al. Spatial variation in annual actual evapotranspiration of terrestrial ecosystems in China: Results from eddy covariance measurements. *J. Geogr. Sci.* **2016**, *26*, 1391–1411. [CrossRef]

6. Ferreira, M.I. Stress coefficients for soil water balance combined with water stress indicators for irrigation scheduling of woody crops. *Horticulturae* **2017**, *3*, 38. [CrossRef]

7. Ferreira, M.I.; Paço, T.A.; Silvestre, J.; Silva, R.M. Evapotranspiration estimates and water stress indicators for irrigation scheduling in woody plants. In *Agricultural Water Management Research Trends*; Nova Science Publishers, Inc.: New York, NY, USA, 2008; pp. 129–170.

8. Čermák, J.; Kučera, J.; Nadezhdina, N. Sap flow measurements with some thermodynamic methods, flow integration within trees and scaling up from sample trees to entire forest stands. *Trees* **2004**, *18*, 529–546. [CrossRef]

9. Köstner, B.; Falge, E.; Alsheimer, M. Sap Flow Measurements. *Struct. Role Submerg. Macrophytes Lakes* **2017**, 99–112. [CrossRef]

10. Tatarinov, F.A.; Kučera, J.; Cienciala, E. The analysis of physical background of tree sap flow measurement based on thermal methods. *Meas. Sci. Technol.* **2005**, *16*, 1157–1169. [CrossRef]

11. Kaufmann, M.R.; Kelliher, F.M. Measuring transpiration rates. In *Techniques and Approaches in Forest Tree Ecophysiology*; Lassoie, J.P., Hinckley, T.M., Eds.; CRC Press: Boca Raton, FL, USA, 1991; pp. 117–140.

12. Betsch, P.; Bonal, D.; Breda, N.; Montpied, P.; Peiffer, M.; Tuzet, A.; Granier, A. Drought effects on water relations in beech: The contribution of exchangeable water reservoirs. *Agric. For. Meteorol.* **2011**, *151*, 531–543. [CrossRef]

13. De Swaef, T.; De Schepper, V.; Vandegehuchte, M.W.; Steppe, K. Stem diameter variations as a versatile research tool in ecophysiology. *Tree Physiol.* **2015**, *35*, 1047–1061. [CrossRef] [PubMed]

14. Tenhunen, J.D.; Valentini, R.; Köstner, B.; Zimmermann, R.; Granier, A. Variation in forest gas exchange at landscape to continental scales. *Ann. For. Sci.* **1998**, *55*, 1–11. [CrossRef]

15. Davi, H.; Dufrêne, E.; Granier, A.; Le Dantec, V.; Barbaroux, C.; François, C.; Bréda, N. Modelling carbon and water cycles in a beech forest. Part II: Validation of the main processes from organ to stand scale. *Ecol. Modell.* **2005**, *185*, 387–405. [CrossRef]

16. MacKay, S.L.; Arain, M.A.; Khomik, M.; Brodeur, J.J.; Schumacher, J.; Hartmann, H.; Peichl, M. The impact of induced drought on transpiration and growth in a temperate pine plantation forest. *Hydrol. Process.* **2012**, *26*, 1779–1791. [CrossRef]

17. Breda, N.; Granier, A. Intra and interannual variations of transpiration, leaf area index and radial growth of sessile oak stand (Quercus petraea). *Ann. Sci. For.* **1996**, *53*, 521–536. [CrossRef]

18. Eamus, D.; Boulain, N.; Cleverly, J.; Breshears, D.D. Global change-type drought-induced tree mortality: Vapor pressure deficit is more important than temperature per se in causing decline in tree health. *Ecol. Evol.* **2013**, *3*, 2711–2729. [CrossRef] [PubMed]

19. Kirchen, G.; Calvaruso, C.; Granier, A.; Redon, P.O.; Van der Heijden, G.; Bréda, N.; Turpault, M.P. Local soil type variability controls the water budget and stand productivity in a beech forest. *For. Ecol. Manag.* **2017**, *390*, 89–103. [CrossRef]

20. Jiao, L.; Lu, N.; Fang, W.; Li, Z.; Wang, J.; Jin, Z. Determining the independent impact of soil water on forest transpiration: A case study of a black locust plantation in the Loess Plateau, China. *J. Hydrol.* **2019**, *572*, 671–681. [CrossRef]

21. Lüttschwager, D.; Jochheim, H. Drought primarily reduces canopy transpiration of exposed beech trees and decreases the share of water uptake from deeper soil layers. *Forests* **2020**, *11*, 537. [CrossRef]

22. Schwinning, S. The ecohydrology of roots in rocks. *Ecohydrology* **2010**, *3*, 238–245. [CrossRef]

23. Eliades, M.; Bruggeman, A.; Lubczynski, M.W.; Christou, A.; Camera, C.; Djuma, H. The water balance components of Mediterranean pine trees on a steep mountain slope during two hydrologically contrasting years. *J. Hydrol.* **2018**, *562*, 712–724. [CrossRef]

24. IPCC 2018. Global warming of 1.5 °C. An IPCC Special Report on the impacts of global warming of 1.5 °C above pre-industrial levels and related global greenhouse gas emission pathways, in the context of strengthening the global response to the threat of climate change. In *Summary for Policymakers*; Masson-Delmotte, V., Zhai, P., Pörtner, H.O., Roberts, D., Skea, J., Shukla, P.R., Pirani, A., Moufouma-Okia, W., Péan, C., Pidcock, R., et al., Eds.; World Meteorological Organization: Geneva, Switzerland, 2019; p. 32.

25. Rowell, D.P.; Jones, R.G. Causes and uncertainty of future summer drying over Europe. *Clim. Dyn.* **2006**, *27*, 281–299. [CrossRef]

26. IPCC 2014. *Climate Change 2014: Impacts, Adaptation, and Vulnerability. Part. B: Regional Aspects. Europe. Contribution of Working Group II to the Fifth Assessment Report of the Intergovernmental Panel on Climate Change*; Cambridge University Press: Cambridge, UK, 2014.

27. Centritto, M.; Tognetti, R.; Leitgeb, E.; Střelcová, K.; Cohen, S. Chapter 3 Above Ground Processes: Anticipating Climate Change Influences. In *Forest Management and the Water Cycle: An Ecosystem-Based Approach*; Bredemeier, M., Cohen, S., Godbold, D.L., Lode, E., Pichler, V., Schleppi, P., Eds.; Springer: Dordrecht, The Netherlands, 2011; Volume 212, pp. 263–289, ISBN 978-90-481-9833-7.

28. Williams, A.P.; Allen, C.D.; Macalady, A.K.; Griffin, D.; Woodhouse, C.A.; Meko, D.M.; Swetnam, T.W.; Rauscher, S.A.; Seager, R.; Grissino-Mayer, H.D.; et al. Temperature as a potent driver of regional forest drought stress and tree mortality. *Nat. Clim. Chang.* **2013**, *3*, 292–297. [CrossRef]

29. Bertini, G.; Amoriello, T.; Fabbio, G.; Piovosi, M. Forest growth and climate change: Evidences from the ICP-Forests intensive monitoring in Italy. *IForest* **2011**, *4*, 262–267. [CrossRef]

30. Chen, K.; Dorado-Linan, I.; Akhmetzyanov, L.; Gea-Izquierdo, G.; Zlatanov, T.; Menzell, A. Influence of climate drivers and the North Atlantic Oscillation on beech growth at marginal sites across the Mediterranean. *Clim. Res.* **2015**, *66*, 229–242. [CrossRef]

31. Bréda, N.; Huc, R.; Granier, A.; Dreyer, E. Temperate forest tree and stands under severe drought: A review of ecophysiological responses, adaptation processes and long-term consequences. *Ann. Sci. For.* **2006**, *63*, 625–644. [CrossRef]

32. McDowell, N.G.; Beerling, D.J.; Breshears, D.D.; Fisher, R.A.; Raffa, K.F.; Stitt, M. The interdependence of mechanisms underlying climate-driven vegetation mortality. *Trends Ecol. Evol.* **2011**, *26*, 523–532. [CrossRef]

33. Cailleret, M.; Jansen, S.; Robert, E.M.R.; Desoto, L.; Aakala, T.; Antos, J.A.; Beikircher, B.; Bigler, C.; Bugmann, H.; Caccianiga, M.; et al. A synthesis of radial growth patterns preceding tree mortality. *Glob. Chang. Biol.* **2017**, *23*, 1675–1690. [CrossRef]

34. Allen, R.G.; Pereira, L.S.; Raes, D.; Smith, M. *Crop. Evapotranspiration—Guidelines for Computing Crop Water Requirements—FAO Irrigation and Drainage Paper 56*; Food and Agriculture Organization of the United Nations: Rome, Italy, 1998.

35. Buckley, T.N.; Mott, K.A.; Farquhar, G.D. A hydromechanical and biochemical model of stomatal conductance. *Plant Cell Environ.* **2003**, *26*, 1767–1785. [CrossRef]

36. Buckley, T.N.; Turnbull, T.L.; Adams, M.A. Simple models for stomatal conductance derived from a process model: Cross-validation against sap flux data. *Plant Cell Environ.* **2012**, *35*, 1647–1662. [CrossRef]

37. Whitley, R.; Taylor, D.; Macinnis-Ng, C.; Zeppel, M.; Yunusa, I.; O'Grady, A.; Froend, R.; Medlyn, B.; Eamus, D. Developing an empirical model of canopy water flux describing the common response of transpiration to solar radiation and VPD across five contrasting woodlands and forests. *Hydrol. Process.* **2013**, *27*, 1133–1146. [CrossRef]

38. Wang, H.; Guan, H.; Simmons, C.T. Modeling the environmental controls on tree water use at different temporal scales. *Agric. For. Meteorol.* **2016**, *225*, 24–35. [CrossRef]

39. Hong, L.; Guo, J.; Liu, Z.; Wang, Y.; Ma, J.; Wang, X.; Zhang, Z. Time-lag effect between sap flow and environmental factors of Larix principis-rupprechtii Mayr. *Forests* **2019**, *10*, 971. [CrossRef]

40. Wang, H.; Tetzlaff, D.; Soulsby, C. Hysteretic response of sap flow in Scots pine (Pinus sylvestris) to meteorological forcing in a humid low-energy headwater catchment. *Ecohydrology* **2019**, *12*. [CrossRef]

41. Zhang, R.; Xu, X.; Liu, M.; Zhang, Y.; Xu, C.; Yi, R.; Luo, W.; Soulsby, C. Hysteresis in sap flow and its controlling mechanisms for a deciduous broad-leaved tree species in a humid karst region. *Sci. China Earth Sci.* **2019**, *62*, 1744–1755. [CrossRef]

42. O'Grady, A.P.; Worledge, D.; Battaglia, M. Constraints on transpiration of Eucalyptus globulus in southern Tasmania, Australia. *Agric. For. Meteorol.* **2008**, *148*, 453–465. [CrossRef]

43. Xu, S.; Yu, Z. Environmental control on transpiration: A case study of a desert ecosystem in northwest china. *Water* **2020**, *12*, 1211. [CrossRef]

44. Verbeeck, H.; Steppe, K.; Nadezhdina, N.; Op De Beeck, M.; Deckmyn, G.; Meiresonne, L.; Lemeur, R.; Čermák, J.; Ceulemans, R.; Janssens, I.A. Stored water use and transpiration in Scots pine: A modeling analysis with ANAFORE. *Tree Physiol.* **2007**, *27*, 1671–1685. [CrossRef]

45. Chen, L.; Zhang, Z.; Li, Z.; Tang, J.; Caldwell, P.; Zhang, W. Biophysical control of whole tree transpiration under an urban environment in Northern China. *J. Hydrol.* **2011**, *402*, 388–400. [CrossRef]

46. Zeppel, M.J.B.; Murray, B.R.; Barton, C.; Eamus, D. Seasonal responses of xylem sap velocity to VPD and solar radiation during drought in a stand of native trees in temperate Australia. *Funct. Plant Biol.* **2004**, *31*, 461–470. [CrossRef]

47. Geßler, A.; Keitel, C.; Kreuzwieser, J.; Matyssek, R.; Seiler, W.; Rennenberg, H. Potential risks for European beech (*Fagus sylvatica* L.) in a changing climate. *Trees Struct. Funct.* **2007**, *21*, 1–11. [CrossRef]

48. Fritz, P. (Ed.) *Ökologischer Waldumbau in Deutschland (Ecological Reconstruction of Forests in Germany)*; Oekom Verlag: Munich, Germany, 2006; ISBN 978-3-86581-001-4.

49. Dedrick, S.; Spiecker, H.; Orazio, C.; Tomé, M.; Martinez, I. *Plantation or Conversion—The Debate! Ideas Presented and Discussed at a Joint EFI Project-Centre Conference Held 21–23 May 2006 in Freiburg, Germany*; European Forest Institute: Joensuu, Finland, 2007; ISBN 9789525453164.

50. *Green Report: Report on Forestry in the Slovak Republic per Year 2018*; Ministry of Agriculture and Rural Development of the Slovak Republic and National Forest Centre: Bratislava, Slovakia, 2019.

51. Jung, T. Beech decline in Central Europe driven by the interaction between Phytophthora infections and climatic extremes. *For. Pathol.* **2009**, *39*, 73–94. [CrossRef]

52. Mátyás, C.; Berki, I.; Czúcz, B.; Gálos, B.; Móricz, N.; Rasztovits, E. Future of beech in Southeast Europe from the perspective of evolutionary ecology. *Acta Silv. Lignaria Hung.* **2010**, *6*, 91–110.

53. Corcobado, T.; Cech, T.L.; Brandstetter, M.; Daxer, A.; Hüttler, C. Decline of European Beech in Austria: Involvement of *Phytophthora* spp. and Contributing Biotic and Abiotic Factors. *Forests* **2020**, *11*, 859. [CrossRef]

54. Střelcová, K.; Matejka, F.; Kučera, J. Beech stand transpiration assessment—Two methodical approaches. *Ekologia* **2004**, *22*, 147–162.

55. Thiel, D.; Kreyling, J.; Backhaus, S.; Beierkuhnlein, C.; Buhk, C.; Egen, K.; Huber, G.; Konnert, M.; Nagy, L.; Jentsch, A. Different reactions of central and marginal provenances of fagus sylvatica to experimental drought. *Eur. J. For. Res.* **2014**, *133*, 247–260. [CrossRef]

56. Granier, A.; Reichstein, M.; Bréda, N.; Janssens, I.A.; Falge, E.; Ciais, P.; Grunwald, T.; Aubinet, M.; Berbigier, P.; Bernhofer, C.; et al. Evidence for soil water control on carbon and water dynamics in European forests during the extremely dry year: 2003. *Agric. For. Meteorol.* **2007**, *143*, 123–145. [CrossRef]

57. van der Maaten-Theunissen, M.; Bümmerstede, H.; Iwanowski, J.; Scharnweber, T.; Wilmking, M.; van der Maaten, E. Drought sensitivity of beech on a shallow chalk soil in northeastern Germany—A comparative study. *For. Ecosyst.* **2016**, *3*, 24. [CrossRef]

58. Zimmermann, J.; Hauck, M.; Dulamsuren, C.; Leuschner, C. Climate Warming-Related Growth Decline Affects Fagus sylvatica, But Not Other Broad-Leaved Tree Species in Central European Mixed Forests. *Ecosystems* **2015**, *18*, 560–572. [CrossRef]

59. Decuyper, M.; Chávez, R.O.; Čufar, K.; Estay, S.A.; Clevers, J.G.P.W.; Prislan, P.; Gričar, J.; Črepinšek, Z.; Merela, M.; de Luis, M.; et al. Spatio-temporal assessment of beech growth in relation to climate extremes in Slovenia—An integrated approach using remote sensing and tree-ring data. *Agric. For. Meteorol.* **2020**, *287*, 107925. [CrossRef]

60. Scharnweber, T.; Manthey, M.; Criegee, C.; Bauwe, A.; Schröder, C.; Wilmking, M. Drought matters—Declining precipitation influences growth of *Fagus sylvatica* L. and *Quercus robur* L. in north-eastern Germany. *For. Ecol. Manag.* **2011**, *262*, 947–961. [CrossRef]

61. Gennaretti, F.; Ogée, J.; Sainte-Marie, J.; Cuntz, M. Mining ecophysiological responses of European beech ecosystems to drought. *Agric. For. Meteorol.* **2020**, *280*. [CrossRef]

62. Zlatník, A. Overview of groups of geobiocoene types originally wooded or shrubed. *Zprávy Geogr. ČSAV Brně* **1976**, *13*, 55–56. (In Czech)

63. FAO. *World Reference Base for Soil Resources 2014. International Soil Classification System for Naming Soils and Creating Legends for Soil Maps. World Soil Resources Reports No. 106. Update 2015*; FAO: Rome, Italy, 2015; ISBN 9789251083697.

64. Lapin, M.; Faško, P.; Melo, M.; Šťastný, P.; Tomlain, J. Klimatické oblasti [Climatic Regions]. In *Atlas Krajiny Slovenskej Republiky [Landscape Atlas of the Slovak Republic]*; Miklos, L., Ed.; Ministry of Environment of the Slovak Republic: Bratislava, Czechoslovakia, 2002; p. 334.

65. Penman, H.L. Natural evaporation from open water, bare soil, and grass. *Proc. R. Soc.* **1948**, *A193*, 120–146.

66. Granier, A.; Bréda, N.; Biron, P.; Villette, S. A lumped water balance model to evaluate duration and intensity of drought constraints in forest stands. *Ecol. Modell.* **1999**, *116*, 269–283. [CrossRef]

67. Kučera, J.; Čermák, J.; Penka, M. Improved thermal method of continual recording the transpiration flow rate dynamics. *Biol. Plant* **1977**, *19*, 413–420.

68. Jiří Kučera—Environmental Measuring Systems Sap Flow System EMS 81. User's Manual. 2nd Issue. Available online: http://www.emsbrno.cz/r.axd/pdf_v_EMS81__usermanual_u_pdf.jpg?ver= (accessed on 29 November 2020).

69. Lagergren, F.; Lindroth, A. Transpiration response to soil moisture in pine and spruce trees in Sweden. *Agric. For. Meteorol.* **2002**, *112*, 67–85. [CrossRef]

70. Rozas, V.; Camarero, J.J.; Sangüesa-Barreda, G.; Souto, M.; García-González, I. Summer drought and ENSO-related cloudiness distinctly drive Fagus sylvatica growth near the species rear-edge in northern Spain. *Agric. For. Meteorol.* **2015**, *201*, 153–164. [CrossRef]

71. Granier, A.; Biron, P.; Lemoine, D. Water balance, transpiration and canopy conductance in two beech stands. *Agric. For. Meteorol.* **2000**, *100*, 291–308. [CrossRef]

72. Gebauer, T.; Horna, V.; Leuschner, C. Canopy transpiration of pure and mixed forest stands with variable abundance of European beech. *J. Hydrol.* **2012**, *442–443*, 2–14. [CrossRef]

73. Haworth, M.; Killi, D.; Materassi, A.; Raschi, A.; Centritto, M. Impaired stomatal control is associated with reduced photosynthetic physiology in crop species grown at elevated [CO_2]. *Front. Plant. Sci.* **2016**, *7*, 1–13. [CrossRef] [PubMed]

74. Matejka, F.; Střelcová, K.; Hurtalová, T.; Gömöryová, E.; Ditmarová, L'. Seasonal changes in transpiration and soil water content in a spruce primeval forest during a dry period. In *Bioclimatology and Natural Hazards*; Střelcová, K., Mátyás, C., Kleidon, A., Lapin, M., Matejka, F., Blaženec, M., Škvarenina, J., Holécy, J., Eds.; Springer: Dordrecht, Germany, 2009; pp. 197–206.

75. Tie, Q.; Hu, H.; Tian, F.; Guan, H.; Lin, H. Environmental and physiological controls on sap flow in a subhumid mountainous catchment in North China. *Agric. For. Meteorol.* **2017**, *240–241*, 46–57. [CrossRef]

76. Bai, Y.; Zhu, G.; Su, Y.; Zhang, K.; Han, T.; Ma, J.; Wang, W.; Ma, T.; Feng, L. Hysteresis loops between canopy conductance of grapevines and meteorological variables in an oasis ecosystem. *Agric. For. Meteorol.* **2015**, *214–215*, 319–327. [CrossRef]

77. Carrasco, L.O.; Bucci, S.J.; Di Francescantonio, D.; Lezcano, O.A.; Campanello, P.I.; Scholz, F.G.; Rodríguez, S.; Madanes, N.; Cristiano, P.M.; Hao, G.Y.; et al. Water storage dynamics in the main stem of subtropical tree species differing in wood density, growth rate and life history traits. *Tree Physiol.* **2015**, *35*, 354–365. [CrossRef] [PubMed]

78. Tuzet, A.; Perrier, A.; Leuning, R. A coupled model of stomatal conductance and photosynthesis for winter wheat. *Plant Cell Environ.* **2003**, *26*, 1097–1116. [CrossRef]

79. Poyatos, R.; Llorens, P.; Piñol, J.; Rubio, C. Response of Scots pine (*Pinus sylvestris* L.) and pubescent oak (Quercus pubescens Willd.) to soil and atmospheric water deficits under Mediterranean mountain climate. *Ann. For. Sci.* **2008**, *65*, 306. [CrossRef]

80. Priwitzer, T.; Kurjak, D.; Kmeť, J.; Sitková, Z.; Leštianska, A. Photosynthetic response of European beech to atmospheric and soil drought. *For. J.* **2014**, *60*, 31–37. [CrossRef]

81. Bovard, B.D.; Curtis, P.S.; Vogel, C.S.; Su, H.B.; Schmid, H.P. Environmental controls on sap flow in a northern hardwood forest. *Tree Physiol.* **2005**, *25*, 31–38. [CrossRef]

82. Čermák, J.; Kučera, J.; Bauerle, W.L.; Phillips, N.; Hinckley, T.M. Tree water storage and its diurnal dynamics related to sap flow and changes in stem volume in old-growth Douglas-fir trees. *Tree Physiol.* **2007**, *27*, 181–198. [CrossRef]

83. Phillips, N.G.; Ryan, M.G.; Bond, B.J.; McDowell, N.G.; Hinckley, T.M.; Čermák, J. Reliance on stored water increases with tree size in three species in the Pacific Northwest. *Tree Physiol.* **2003**, *23*, 237–245. [CrossRef]

Publisher's Note: MDPI stays neutral with regard to jurisdictional claims in published maps and institutional affiliations.

Article

Carabus Population Response to Drought in Lowland Oak Hornbeam Forest

Bernard Šiška [1], Mariana Eliašová [1,*] and Ján Kollár [2]

[1] Department of Environmental Management, Faculty of European Studies and Regional Development, Slovak University of Agriculture in Nitra, Mariánska 10, 949 76 Nitra, Slovakia; bernard.siska@uniag.sk

[2] Department of Planting Design and Maintenance, Faculty of Horticulture and Landscape Engineering, Slovak university of Agriculture in Nitra, Tulipánová 7, 949 76 Nitra, Slovakia; jan.kollar@uniag.sk

* Correspondence: mariana.eliasova@uniag.sk; Tel.: +421-37-641-5614

Received: 15 September 2020; Accepted: 20 November 2020; Published: 23 November 2020

Abstract: Forest management practices and droughts affect the assemblages of carabid species, and these are the most important factors in terms of influencing short- and long-term population changes. During 2017 and 2018, the occurrences and seasonal dynamics of five carabid species (*Carabus coriaceus, C. ulrichii, C. violaceus, C. nemoralis* and *C. scheidleri*) in four oak hornbeam forest stands were evaluated using the method of pitfall trapping. The climate water balance values were cumulatively calculated here as cumulative water balance in monthly steps. The cumulative water balance was used to identify the onset and duration of drought. The number of Carabus species individuals was more than three times higher in 2018 than in 2017. Spring activity was influenced by temperature. The extremely warm April in 2018 accelerated spring population dynamics; however, low night temperatures in April in 2017 slowed the spring activity of nocturnal species. Drought negatively influenced population abundance, and the effect of a drought is likely to be expressed with a two-year delay. In our investigation, a drought in 2015 started in May and lasted eight months; however, the drought was not recorded in 2016, and 2016 was evaluated as a humid year. The meteorological conditions in the year influenced seasonal activity patterns and the timings of peaks of abundance for both spring breeding and autumn breeding Carabus species.

Keywords: Carabus; Báb; seasonal activity; abundance; recovering forest stand; climate; drought

---

1. Introduction

Ground dwelling invertebrates such as carabid beetles and spiders are often surveyed with respect of ecosystem changes. These invertebrates have strong potential as ecological indicators, as they are readily surveyed in sufficient numbers for meaningful conclusions to be drawn, have a stable taxonomy, and, at least in the case of ground beetles, are readily identified. They are, for example, good local scale indicators of ecosystem disturbance in forested landscapes at both the short- and long-term time scales, responding to clear-cut logging and fire differently [1].

Carabids belong to one of the most frequently used model groups for biological studies. The reasons for this are various, including the relatively stable taxonomy, species richness, occurrence in most terrestrial habitats and geographical areas, availability of simple collection methods, and known sensitivity to environmental changes [2]. The effect of clearcutting on Carabidae is well studied. Clearcutting has a definitive impact on Carabidae, depending on species, and notably in cases with the replacement of large forest species by the usually smaller eurytopic open field species in clearings [3]. Human impacts in forest ecosystems generally cause losses for some relict species and gains for ubiquitous species. Clearcutting leads to a drastic decrease in mean individual biomass. The regeneration of carabid fauna after clearcutting is a very slow process, and typical relict forest species are absent even after several decades of forest regeneration [4]. The impact of a drought on carabid assemblage has been studied in forests damaged by a windstorm in the High Tatra Mountains, focusing on the trends of community differentiation according to the state and management of damaged sites. The revealed trends included reversible quantitative changes recorded on stands with the fallen timber in situ, the disappearance of less tolerant forest species at sites with extracted timber, and temporal invasions of xenocoenous open-landscape species at the sites with extracted timber and additionally burned sites [5].

Phenology, which is the timing of the seasonal activities of animals and plants, is perhaps the simplest process for monitoring changes in the ecology of species in response to climate change. Species responses to climate change can disrupt their interactions with others at the same or adjacent trophic level. If closely cooperating or competing species show different responses or susceptibility to change, the outcome of their interactions may change, as suggested by long term data for terrestrial and marine organisms. Therefore, rapid climate change or extreme climatic events can be expected to change the compositions of organism communities [6]. Phenological changes caused by climate change are slower at higher trophic levels (i.e., secondary consumers) than at lower trophic levels (excluding woody plants), thus higher trophic levels are particularly sensitive to the disruption of phenological connections [7]. Carabidae represent an important food role in many terrestrial ecosystems [8,9]. They are among the most important insect predators of soil fauna, especially in temperate climate regions [10].

Climate change is expected to alter average temperature and precipitation values and increase the variability of precipitation events, which may lead to even more intense and frequent floods and droughts [11], although future projections reveal regionally different occurrences of extreme events. As projected in [12], at a pan-European scale, the regions that are most prone to a rise in flood frequency are located in northern to northeastern Europe, while the southern and southeastern European regions show significant increases in the frequencies of drought. Similarly, significant regional differences have also been projected for extreme events in China and East Africa [13,14].

As pointed out in [15], altitude and topography are strong climate-differentiating factors. The Slovak territory has been divided by [16] into nine altitudinal vegetation stages. The vegetation stages of lower elevations, i.e., oak vegetation (stage 1), are rather arid during the vegetation period (from March to September). Altitudinal analysis of the drought climatology of the Western Carpathians in Slovakia has been carried out by [15,17], and the results point out a significant trend for aridity at the lowest altitudes, which is where the oak vegetation stage occurs. The southern part of the Slovak Republic territory is the region where agricultural landscape is prevailing, and forests are scarce there and mostly occur as isolated enclaves [18]. Apart from altitudinal vegetation stage division, the Slovak Republic territory is traditionally divided into the varying agroclimatic zones, and the climatic water balance is the basis for such regionalization [19]. This practice of regionalization is still under scientific revision with respect to climate change [20] or drought occurrence analysis [21].

The aim of our research here is to determine the impact of drought on Carabus beetles populations in lowland oak hornbeam forest occurring at the driest region of the Slovakia. The influence of different types of forest stands on the Carabus beetles populations is also studied.

2. Materials and Methods

The study was carried out at the locality of Báb (Figure 1), which is situated approximately 15 km from Nitra, Slovakia, and located in the southwest of Slovakia. The forest is a rest of former climax woodland covered lowlands and hill areas of in the Nitrianska pahorkatina region. The locality belongs to the Pannonian Bio-geographical Region. The deciduous oak hornbeam forest of 66 ha is partly preserved as a natural reserve and the rest of the forest area is managed [22]. Báb research site is a part of the national LTER (Long term ecological research) network, which is part of the organizational structure of LTER-Europe (http://www.lter-europe.net/). The site was established in 1967 under the International Biological Program (IBP) framework and research started with the IBP (1967–1970) and UNESCO Man and Biosphere (1971–1974) projects. The research activities were carried out until

1974 [23]. After the complex and intensive ecosystem research in 1967–1974, research continued to a restricted extent with the work of individual researchers [24]. In November 2006, single line and shelterwood logging was carried out and, the area was than cleaned up from brushwood beginning 2007. After tree removal, four clear-cut stands were created at the locality.

Figure 1. The localization of the Báb research area (RA) and meteorological station (MS) where the meteorological data were derived. The coordinates for the RA are 17.886, 48.3033, with an average elevation of 190 m a.s.l. (minimum of 170 m a.s.l, maximum of 210 m a.s.l.). The coordinates for the MS are 17.87881, 48.30331, with an elevation of 207 m a.s.l. The data were obtained from the DEIMS-SDR (Dynamic Ecological Information Management System-Site and dataset registry) database [25].

For our study, four 20 × 20 m² study plots were chosen at the locality. Two plots were chosen to represent forest stands. The first, named L1, represented close-to-nature forest, i.e., where no forest intervention had been carried out since 1962. The characteristic features of the L1 stand are higher wood cover and lower herb cover caused by overshadowing. The characteristics of the second forest stand, named L2, were influenced by logging in the neighborhood. Higher illumination resulted in higher herb cover and lower cover of the shrub layer. The other two plots represented recovering forest after clearcutting 10 years prior. The plot R1 represented a recovering forest stand that had become overgrown with shrubs and young tree vegetation (*Acer campestre, Crataegus laevigata, Carpinus betulus, Rosa canina, Quercus cerris, Q. petrea*, etc.). The dense canopy closure of the young trees resulted in lower herb cover. A higher presence of fine woody debris was typical for the ground at the R1 study plot. The plot named R2 represented another after recovering stand after logging operations. Compared to R1, the R2 plot was characterized by the dominance of the invasive tree species *Ailanthus altissima* and the dense herb layer cover that was mostly formed by medium to tall herbs (e.g., *Mercurialis annua, Convallaria majalis, Urtica dioica, Ballota nigra*). The trees of *A. altissima* were approximately 2–3-year-old sprout shoots, as single-cut removal of this invasive species was performed in 2015. The localizations of the individual study plots are given in Figure A1.

Carabids were collected using the method of pitfall trapping (180 mL plastic cups buried in the soil and filled with vinegar to kill and preserve samples). Seven pitfall traps were installed at each research plot. Four traps were located at the corners, where the remaining three traps were placed diagonally inside the plots at regular spacings. The samples were removed weekly from the beginning of April to the end of September. The species determination was based on adult beetles morphology, where the classification method of [26] was adopted.

The occasional damage of traps is common in the field. For this reason, it was not possible to use absolute numbers of trapped individuals. The numbers of individuals for each species were evaluated as the number of individuals per trap. Weekly catches, i.e., the number of individuals per trap per week, were used for seasonal activity evaluation. Weekly evaluation allowed comparison of the dynamics of carabid populations between the two years and more precise identification of the peak activity timing. With the monthly catches, number of individuals per trap per month values were calculated for the purpose of the evaluation of the impact of drought on carabid species populations.

We used the index of dominance (D_i) to compare the occurrence levels between species and the four forest stands:

$$D_i = (n_i/N) \times 100\ (\%) \tag{1}$$

where n_i is the abundance of individual species and N is the total abundance.

The temperature and precipitation data were obtained from the automatic climatic station located in the open area next to the Báb forest (Figure 1). The meteorological conditions in the year were characterized according to the climatological normal and the period of 1961–1990 was used as normal. The monthly values of the precipitation total and the average temperature were compared to the normal values. The cumulative water balance was used for drought episode identification and the duration of the drought. Firstly, the climate water balance calculation, based on the monthly values of atmospheric precipitation and potential evapotranspiration, was derived for every month in the year. The result can be positive, representing a positive water balance in the given month, or negative, representing the water deficit. Then, the climate water balance values were cumulatively calculated as cumulative water balance in monthly steps. The criterion for identifying a drought was the occurrence of a negative cumulative value for a specific month. A polynomial fourth-order trend line described the dynamics of the drought. When the trend line cut the x-axis towards the negative values, this was considered as the onset of a drought, and vice versa, where cutting the x-axis towards positive values represented the end of a drought [20,27].

3. Results

3.1. The Occurrence of the Carabus Species

Five species of the genus *Carabus* were recorded during the two years of our research: *Carabus coriaceus*, *Carabus nemoralis*, *Carabus ulrichii*, *Carabus scheidleri*, and *Carabus violaceus*.

Carabus beetles were active during the whole period of study, i.e., from the beginning of April until the end of September. The considerable difference between the total number of caught beetles was the most obvious aspect between the two compared years, where more than triple the number of individuals were caught in 2018 than in 2017 (583 and 1789 individuals in 2017 and 2018, respectively).

The Carabid occurrence levels, evaluated as the number of individuals per trap caught through the whole research period, were not equal for the four study plots. The highest carabid occurrence was recorded at the forest stand L2, where 7.7 and 22.8 individuals per trap were caught per season in 2017 and 2018, respectively. The results for the close-to-nature forest stand (i.e., the L1 research plot) showed that Carabid occurrence there was the lowest in 2017, but the second best overall in 2018, with 3.4 and 18 individuals per trap per season, respectively. The occurrences recorded at the recovering forest stand R1 were lowest in 2018, but second best in 2017 (12.8 and 6.3 individuals caught per trap per season in 2018 and 2017, respectively). For the R2 recovering forest stand, the numbers of caught individuals per trap and season were 5.5 in 2017 and 15 in 2018.

The differences between the occurrences of the individual species were also recorded, and some specifies that related to stand preference were also recorded.

Carabus ulrichii was the most frequent species for both 2017 and 2018, with approximately 27 and 42% of collected individuals belonging to this species in 2017 and 2018, respectively (Table 1). In 2018, *C. ulrichii* dominated at all stands, and its dominance was even higher than 50% at the L2 study plot. In 2017, *C. ulrichii* dominated at the two the closest stands, which were the recovering forest plot R2 and forest plot L2 (Table 2).

Table 1. The index of dominance calculated for *Carabus* species at four different forest stands (L1 and L2 are forest plots and R1 and R2 are recovering forest plots) at the locality of Báb (southwestern Slovakia) in 2017.

Species	Index of Dominance (%)				
	Total	L1	L2	R1	R2
C. coriaceus	20.8	24.2	15.1	30.8	13.9
C. scheidleri	23.3	24.2	26.0	25.4	16.7
C. ulrichii	27.4	17.7	32.3	16.2	39.6
C. nemoralis	6.3	11.3	8.9	3.2	4.9
C. violaceus	22.1	22.6	17.7	24.3	25.0

Table 2. The index of dominance calculated for *Carabus* species at four different forest stands (L1 and L2 are forest plots and R1 and R2 are recovering forest plots) at the locality of Báb (southwestern Slovakia) in 2018.

Species	Index of Dominance (%)				
	Total	L1	L2	R1	R2
C. coriaceus	22.1	14.4	11.8	30.3	21.9
C. scheidleri	23.4	34.6	15.9	26.2	29.6
C. ulrichii	41.5	25.8	53.2	34.5	37.4
C. nemoralis	7.5	23.5	15.1	4.4	2.0
C. violaceus	5.5	1.7	4.1	4.7	9.1

The species with the lowest occurrence was *Carabus nemoralis*. The dominance of this species was slightly higher in 2018 than 2017. The dominance of this species was especially low at the two recovering forest stands (Table 2). In 2018, almost 70% of the individuals of *C. nemoralis* were caught at the L1 forest stand, and more than 90% of individuals were collected at both forest stands. In 2017, the catch percentages at the forest plots were lower, but still high with values of almost 65%.

Carabus coriaceus and *Carabus scheidleri* reached dominance percentages of approximately 20% during both survey years (Tables 1 and 2). Almost 50% of *Carabus coriaceus* individuals were caught at the R1 recovering forest stand in 2017, where it was also the dominant species (Table 1); however, the dominance of this species was still high at the same study plot in 2018 (Table 2). The same number of individuals were collected at L2. Comparing the dominance of the species *Carabus scheidleri* in the two years of study and the four studied stands, there was no clear preference for a specific stand. When comparing the absolute numbers of caught individuals, slightly fewer individuals were captured at the L1 forest stand.

The species *Carabus violaceus* was the only species with lower occurrence in 2018; however, the absolute number decrease was only 31 individuals, where several-fold increases in the numbers of individuals for other species made this species the least represented in 2018 (Table 2).

3.2. The Seasonal Activity of the Carabus Species

In 2017, Carabus species activity remained low until first ten days of May (Figure 2). Then, the activity raised gradually, reaching a peak by the end of May, followed by a sudden decline in activity in the beginning of June (22nd week). The early summer activity period reached a peak in the beginning of July. After the summer stagnation, *Carabids* performed late summer–early autumn activity period with the peak of activity recorded in the middle of August.

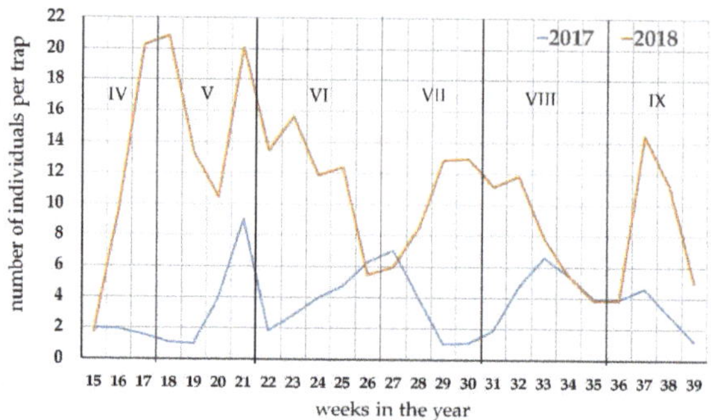

Figure 2. The seasonal activity of Carabus species in the oak hornbeam forest at the locality of Báb (southwestern Slovakia) during 2017 and 2018.

Besides the higher culmination levels, the difference in timing was another important aspect of the seasonal activity course in 2018 (Figure 2). In 2018, the beginning of Carabid activity was more dynamic when compared to the previous year. From the beginning of April, the activity rapidly increased and the spring peak of activity was recorded at the turn of April to May. After the period of decline in the middle of May, the activity raised again by the end of May and decreased in the end of June. The summer activity period culminated at the turn of July to August. The autumn period of activity culminated in the middle of September (Figure 2).

The observed activity peaks could be divided into seasonal activity for three distinct periods: (1) the spring to early summer activity period, (2) summer activity period, and (3) late summer to autumn activity period.

In 2017, the carabid activity through April and May was mostly influenced by the activity of *C. ulrichii* at all study plots (Figure 3). The peak of activity recorded by the end of May was almost exclusively influenced by this species. *C. scheidleri* started to activate from the second half of May, and the activity culmination at the turn of June to July was considerably influenced by this species, especially for both forest study plots (Figure 3); however, few individuals of the species *C. coriaceus* were recorded in April at L2 and R1. The species started its main activity at the end of June, where the maximum activity was recorded during the beginning of July. During the same period, the activity of *C. nemoralis* peaked at both forest stands (Figure 3). From the second half of July to the first half of August, summer activity stagnation was observed at all research plots. *C. violaceus* started to activate at the turn of July to August and considerably contributed to activity during middle to late August. The activities of *C. ulrichii* and *C. scheidleri* at the two forest recovering stands also significantly contributed to the August activity culmination. The autumn activity period was characterized by *C. coriaceus* and *C. violaceus* activity (Figure 3).

The very dynamic start of carabid activity in April in 2018 was affected by *C. ulrichii* activity. In contrast with the spring activity observed in 2017, the two other species, *C. scheidleri* and *C. nemoralis*, were active through April and May. *C. nemoralis* was especially active at the both forest stands (Figure 4). A sudden decline in activity in the middle of May was recorded for *C. ulrichii* at all study plots. The decline in activity was also recorded for other spring active species in 2018, with the exception of *C. scheidleri* activity in the close-to-nature forest stand (L1, Figure 4). Afterwards, the recovery of activity was recorded, and the culmination of the activity at the end of May was affected by the activity of *C. ulrichii* and *C. scheidleri* at the two forest stands (Figure 4, L1, L2). The June activity period in 2018 was affected by the three mentioned species, featuring gradual decreases in activity towards the end of June. Similar to 2017, *C. coriaceus* and *C. nemoralis* exhibited their typical summer activity period in July. The high activity rate in late summer (July and August) is the most obvious difference in the activity patterns between the two years (Figure 2). Two species were active through this period (*C. ulrichii* and *C. scheidleri*), in contrast with the same period in 2017, especially in terms of increased activity in the two forest stands. The late summer to early autumn activity of *C. coriaceus* and *C. violaceus* showed the same pattern as in 2017, with much higher occurrence levels in 2018, as in 2017 in the case of *C. coriaceus*.

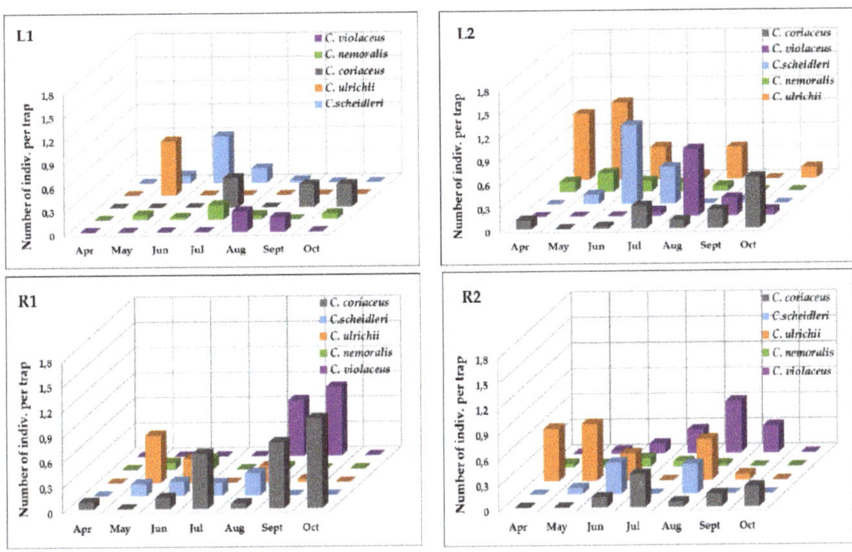

Figure 3. The monthly activity of five carabid species at Báb forest stands with four different study plots in 2017. Two forest stands: L1—close-to-nature forest plot; L2—managed forest plot. Two stands after logging operations 10 years ago: R1—recovering indigenous tree species plot; R2—alien *Ailanthus altissima* dominance.

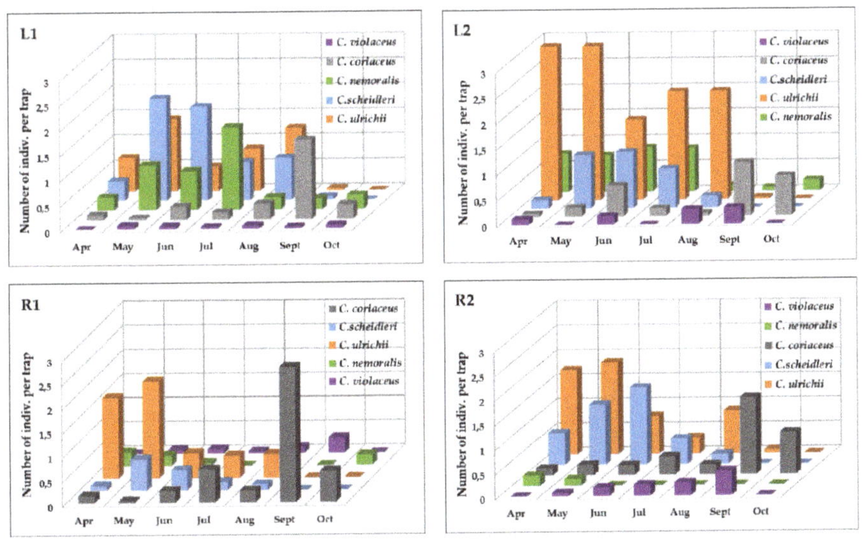

Figure 4. The monthly activity of five carabid species at Báb forest stands with four different study plots in 2018. Two forest stands: L1—close-to-nature forest plot; L2—managed forest plot. Two stands after logging operations 10 years ago: R1—recovering indigenous tree species plot; R2—alien *Ailanthus altissima* dominance.

3.3. Meteorological Conditions in 2017 and 2018 for the Locality of Báb

The year of 2017 was characterized by the onset of higher air temperatures compared to the climate normal, especially during the months of February and March. The middle of the growing season was very warm in 2017 and the end of the year was normal in terms of the temperature (Table 3). Overall, 2017 was characterized as warm with an average annual air temperature of 10.3 °C. The beginning of the growing season in 2018 was characterized by a sudden transition from cold March to extremely warm April (Table 4). Until the end of the year, the temperature was above the long-term normal. During the growing season, there were up to three extremely warm months (Table 4). Overall, 2018, with an average annual air temperature of 11.5 °C, was assessed as extremely warm in terms of the deviation of the annual air temperature from the climatic normal ΔT of +1.8 °C.

In 2017, dry conditions prevailed in the first half of the year, where two extremely dry months were recorded and May fell into the growing season (Table 3). The second half of the year was slightly richer in terms of precipitation and September was evaluated as extremely wet. Overall, 2017 was extremely dry and seven of the months were evaluated as dry (either very dry or extremely dry), and the annual precipitation total was 69% compared to the normal period 1961–1990. As in the case of temperature, the beginning of the season in 2018 was characterized by a sudden transition from normal precipitation in March to an extremely dry April (Table 4). Four of the months were evaluated as wet, and, as in the previous year, September was assessed as extremely wet (Table 4). With an annual precipitation total of 540 mm, 2018 was evaluated as a year with normal precipitation.

Table 3. The comparison of monthly precipitation totals and the average monthly temperature in 2017 with the climatological normal (period 1961–1990) for the locality of Báb (southwestern Slovakia).

Month	Average Monthly Temperature (T, °C)		Thermal Characteristic	Monthly Precipitation (P, mm)		Precipitation Characteristics
	2017	1961–1990	ΔT (°C)	2017	1961–1990	ΔP (%)
I.	−7.1	−1.7	−5.4	Extremely cold	0	31	0	Extremely dry
II.	1.9	0.5	1.4	Warm	13.2	32	41	Very dry
III.	8.4	4.7	3.7	Extremely warm	13.8	33	42	Very dry
IV.	9.5	10.1	−0.6	Normal	36.2	43	84	Normal
V.	16.2	14.8	1.4	Warm	13.8	55	25	Extremely dry
VI.	20.8	18.3	2.5	Very warm	24.6	70	35	Very dry
VII.	21.3	19.7	1.6	Warm	63.6	64	99	Normal
VIII.	22.1	19.2	2.9	Very warm	22.6	58	39	Very dry
IX.	14.3	15.4	−1.1	Cold	83.4	37	225	Extremely wet
X.	10.5	10.1	0.4	Normal	48.4	41	118	Normal
XI.	4.4	4.9	−0.5	Normal	25.8	54	48	Very dry
XII.	0.8	0.5	0.3	Normal	40.0	43	93	Normal

Table 4. The comparison of monthly precipitation totals and the average monthly temperature in 2018 with the climatological normal (period 1961–1990) for the locality of Báb (southwestern Slovakia).

Month	Average Monthly Temperature (T, °C)		Thermal Characteristic	Monthly Precipitation (P, mm)		Precipitation Characteristics
	2018	1961–1990	ΔT (°C)	2018	1961–1990	ΔP (%)
I.	1.6	−1.7	3.3	Very warm	19	31	61	Dry
II.	−1.5	0.5	−2.0	Cold	33	32	103	Normal
III.	2.9	4.7	−1.8	Cold	33	33	100	Normal
IV.	15.2	10.1	5.1	Extremely warm	5	43	13	Extremely dry
V.	18.6	14.8	3.8	Extremely warm	15	55	27	Very dry
VI.	20.0	18.3	1.7	Warm	136	70	195	Very wet
VII.	21.6	19.7	1.9	Warm	34	64	53	Dry
VIII.	22.8	19.2	3.6	Extremely warm	78	58	134	Wet
IX.	17.0	15.4	1.6	Warm	99	37	267	Extremely wet
X.	12.9	10.1	2.8	Very warm	10	41	25	Very dry
XI.	6.5	4.9	1.6	Warm	24	54	44	Very dry
XII.	0.4	0.5	−0.1	Normal	54	43	125	Wet

3.4. The Drought Occurrence at the Locality of Báb between 2015 and 2018

Figure 5 documents the occurrence of drought episodes during the years 2015 to 2018 at the locality of Báb (southwestern Slovakia). According to the cumulative water balance, drought occurrences were recorded in 2015 and 2017. Drought lasted longer in 2015 and was more pronounced in 2017. For both 2016 and 2018, positive water balance was recorded throughout the year and there was only the occurrence of a mild drought in May in 2018 (Figure 5). Comparing the trend line trajectories and the points of x-axis cutting in 2015 and 2017, the onset of drought started earlier in 2017, i.e., by the end of April, and lasted seven months. The drought episode ended in the turn of November to December in 2017; however, the drought onset started a little later in 2015, where the drought lasted until the end of the year.

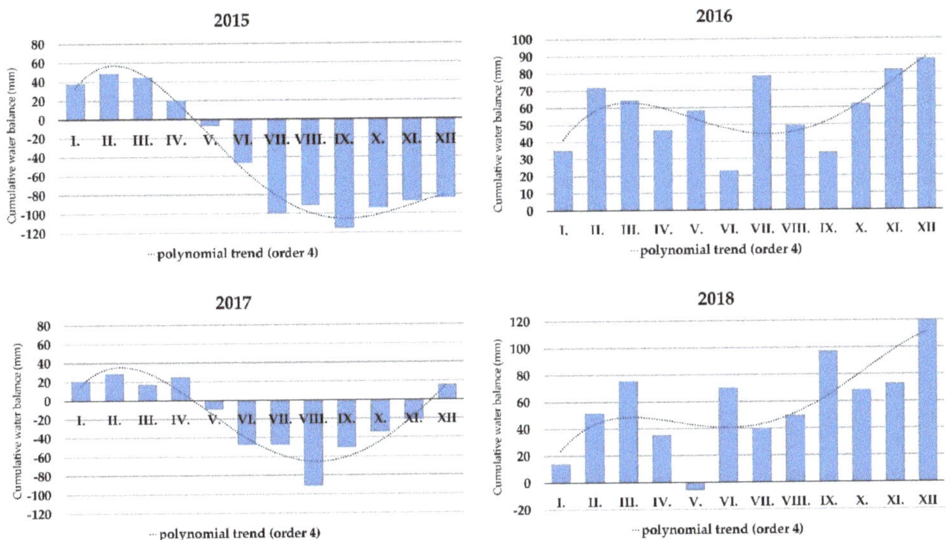

Figure 5. The cumulative water balance for the locality of Báb (southwestern Slovakia) in 2015, 2016, 2017 and 2018.

4. Discussion

In [28], changes in the occurrence of beetle families in the locality of Báb were evaluated in terms of research during the periods of 1968–1969 and 2002–2004. The results were evaluated from the point of view of changing climate parameters in the research area; however, there were no significant changes recorded for the proportions of beetle families, but some aspects of desertification were observed in the research area. The occurrence of six hygrophilous carabid species was not recorded and new thermophilous species were recorded when compared to the results of a survey thirty years ago.

The five monitored carabid species have different phenological characteristics, which has an impact on seasonal activity but also seasonal changes in carabid assemblage compositions.

In Europe and elsewhere within temperate zone, most Carabidae are single-generation [8], although adults may survive for more than one season. They can overwinter either as a larva or as an adult beetle.

Carabus coriaceus has a two-year cycle, where in Central Europe it is mostly reported to reproduce in autumn. The copulation and laying of eggs takes place from the second half of August to November, where individuals overwinter as larva of the first to second instar. The larva of the last instar require a long time to mature, so they pupate only in the following year during June to July. Newly hatched individuals appear in summer and early autumn, where after hibernation they are active in spring but do not reproduce until autumn after aestivation [29]. The work in [30] showed females with developed eggs since the beginning of August. Before this date, the female ovaries were undeveloped. [29] shows that *Carabus nemoralis* is an active species in early spring, laying eggs from March to May. Adults remain active until about July, then start aestivation during August and September. The work in [31] states that *C. nemoralis* is a species that reproduces in spring and that its seasonal activity features two peaks. The second maximum of the activity of *C. nemoralis* occurs in October, which is outside the period of our study. The course of the second maximum is influenced by temperature, where, in warmer years, newly hatched individuals are more active in autumn [31]. This situation occurred in the autumn of 2018, where the species was active at three study plots (both forest stands and the R1 recovering forest stand, Figure 4).

Carabus ulrichii overwinters as an adult [30] and therefore this species showed the earliest activity in both years (Figures 3 and 4). Wintering individuals lay eggs from the second half of April to June. Newly hatched individuals appear during the second half of August, rarely from the end of July, and are active until hibernation [30]. The spring activity of this species is affected by the activity of the overwintering population and the summer activity is affected by the population of newly hatched individuals (Figures 3 and 4). *Carabus violaceus*, like *C. coriaceus*, overwinter as larvae [30] and therefore their activity peaks later in the summer months (Figures 3 and 4). *C. violaceus* lay eggs from the second half of July to early September. The third instar larvae usually overwinter, but part of the adult population also overwinters and reproduces again during the following year [30]. The peak activity of this species was recorded during July to September, and this was the activity of the reproducing population (Figures 3 and 4).

Carabid spring activity is also influenced by the wintering conditions. For most temperate invertebrates, the length of the day is a stimulus for the onset of hibernation, but the end of winter is already controlled by other environmental factors, often in the temperate zone of the frost [32,33]. After the diapause is terminated by a frost, the synchronization of spring development is passively achieved based on the cumulative effect of rising temperatures [34]. The warm February and extremely warm March in 2017 (Table 3) could have promoted the spring activity of carabids. In reality, beetle activity declined throughout April and the first ten days of May (Figure 2). The answer is instead found in the behavioral pattern of the studied carabid species, where the species feature nocturnal activity, with the exception of *C. ulrichii*. If we take a closer look at the night temperatures (Figure A2), after approximately the 10th of April until the end of April, the night temperatures were mostly below 5 °C, including a few nights with frost. Even during the first ten days of May, the night temperatures did not rise above 10 °C. Warm nights with a temperature above 10 °C occurred regularly since the middle of May in 2017. The only species active in April in 2017, was *C. ulrichii*, the species having both nocturnal and diurnal activity. Warmer days with a temperature above 15 °C occurred regularly from the middle of March and activated this species early in the season, especially at the L2 and R2 plots, where this species was the most abundant the previous year when compared with its occurrence at the two other stands (Figure 3).

Ref [35] analyzed drought occurrence in the period of 1966–2013 and identified drought trends for the Danubian lowland area. Both a significant difference and trend were observed in April, where they confirmed a significant increase ($\alpha = 0.05$) in the number of dry periods, i.e., an arid trend. The authors evaluated the significant arid trend in April as the main warning signal. Drought occurrence analysis and future projection within the region where our research locality occurs, is vital because the region is important from an agricultural production point of view. The Danubian lowland is classified as a very dry and hot region, as projected by the climate change scenario in [21]. Cumulative water balance was used for the projection of drought occurrence alongside climate change. According to the RCP 4.5 scenario for time slice 2070–2100, the onset of drought could occur from the 10th of March in the warmest regions, and balanced cumulative water balance are projected to not occur until the end of the year [27].

Humidity significantly affects the activity of the ground beetles. Carabidae are not adapted to low humidity and Central European species are strictly hydrotactic [36]. The work in [37] studied the impact of drought on ground beetle assemblages in Norway spruce forests with different management after windstorm damage. Comparison of the changes in the numbers of individuals and species as a result of drought periods showed that changes in both the numbers of species and individuals mostly occur with an approximate delay of 1–2 years after an incidence of extreme drought or rainy years. However, only five species occurrences were evaluated during our research, similar effect was observed if we take into account the climatic characteristics of the two previous years. The year of 2015 was characterized as dry, where five months were evaluated as extremely dry and three of them (April, June, and July) fall within the main growing season. Moreover, May was evaluated as very dry as well. According to the cumulative water balance, the drought started during the turn of April to May and lasted until the end of the year. In contrast with 2015, the drought did not occur throughout whole year of 2016. Three months of 2016 were evaluated as extremely wet and two of them (July and September) fall within the main growing season. Moreover, none of the months in 2016 were evaluated as extremely warm, which contributed to the positive water balance in 2016. As in the findings of [37], we can suppose, that the two-year delay effect of the meteorological conditions on carabid populations was manifested in our study as well. The cause of such a delay stems from the relationship between the predator and the prey population dynamics. The main food sources for carabids are snails, slugs, earthworms, and other soil arthropods [29], where all are hygrophilous organisms. According to an analysis of carabid gut content, the prey group most frequently detected was earthworms, followed by slugs, especially in the case of large carabids such as *Carabus* species [38]. The activity of land gastropods is restricted to humid conditions in both time and space [39]. The effect of dry conditions on the terrestrial slug species *A. biplicata* was delayed reproduction, and a drought that lasted two weeks resulted in lower fertility [40]. A drought could also affect snail mortality where during a drought that lasted 1 month, immediately after arousal from hibernation when snails had to recover from winter fasting, mortality reached up to 70% for *H. pomatia* in Germany [41]. Regarding earthworms, [42] reported that the monthly abundance of earthworms is significantly related to the amount of rainfall. Earthworms are only active if free water is available in the soil [43]. Epigeic earthworm species living at the soil surface are strongly affected by dry and hot conditions during summer, as these species dwell in and feed on the litter layer and have a limited ability to move down into deeper soil [44]. The work in [45] also revealed strong relationship between carabid and slug population dynamics. The change in the beetle population from year to year was strongly related to both the slug population numbers in the soil and the crop mass of the beetles. This indicated that the slugs influenced the nutritional status of the beetles, and hence their reproductive success. The beetle crop mass was positively correlated with growth in the beetle population, suggesting that the greater the quantity of prey consumed one year, the greater the growth in the beetle population between years. [46] revealed that the egg production of the females of *Anchomenus dorsalis* (Carabidae) was positively affected by the amount of food, influencing the number of mature eggs in female ovaries, as well as total egg production. Starved females had no mature eggs in their ovaries.

However [47] studied the effect of temperature on the activity of selected Carabidae species. It was found that in addition to temperature, humidity also has a significant effect. The study highlighted the same temperature recorded for higher activity on a moist substrate, e.g., after rain. Both the soil moisture and humidity of the air layer near the ground are extremely important factors for limiting the distribution of the ground beetles, as well as the water balance [8]. However we did not measured the humidity conditions at the individual study plots, two plots, recovering forest stand R2 and forest stand L2, were more humid than the two other study plots. The terrain at the research site gradually decreases towards the Bábsky potok stream, and the forest stand L2 plot was situated at the foot of the slope and study plot R2 was situated within the near vicinity. The slope features a northeastern orientation, and thus study plot R2 was located in the shade of the neighboring forest vegetation for almost half of the day, causing dew to remain longer there; however, this observation is only based on subjective personal experience during field work This aspect is likely to be reflected by a higher abundance of carabids than with the two other study plots in 2017 (Figure 4).

5. Conclusions

The Carabus species populations observed in the oak-hornbeam forest considered here over two years showed large differences in terms of both the abundance and seasonal activity. The meteorological conditions during the year influenced the seasonal activity, and the temperature influenced the peak timings of Carabus species activity. Early spring activity was influenced by temperature, influencing the onset and the dynamics. Warmer spring conditions in 2018 caused a rapid onset of activity. Especially, the effect of warmer spring was observed with C. *ulrichii*, having both nocturnal and diurnal activity; however, very low night temperatures during April and the beginning of May in 2017 negatively influenced the spring activity of nocturnal species.

Drought negatively influenced the abundance of Carabus species population numbers, and this negative effect was reflected by lower populations with a two-year time delay. By analyzing the observations of other authors in terms of the effects of a drought on earthworm and mollusk populations, being the main food sources for Carabus species, we believe that the low population abundance in 2017 was influenced by the drought that lasted for eight months in 2015. In 2016, a drought was not recorded, and the populations of Carabus species in 2018 were more than three times more numerous than in 2017; however, a more detailed study, including, for example, carabid body mass evaluation, is needed to explain the mechanisms of the impact of drought on forest carabid populations at the studied locality.

The Carabus populations in the oak hornbeam forest, occupying forest enclaves in the intensively used agricultural landscape within Slovakia's driest region, are experiencing several environmental and anthropogenic pressures. The results of this study could contribute to finding appropriate management schemes for such forest ecosystems.

Author Contributions: Conceptualization, B.Š.; methodology, B.Š. and M.E.; formal analysis, M.E.; investigation, M.E., B.Š. and J.K.; data curation, M.E., J.K.; writing—original draft preparation, M.E.; writing—review and editing B.Š. and J.K.; supervision, B.Š.; project administration, B.Š. and M.E.; funding acquisition, B.Š. All authors have read and agreed to the published version of the manuscript.

Funding: This research was funded by the Grant Agency of the Slovak Republic: VEGA 1/0767/17: Response of ecosystem services of grape growing country to climate change regional impact—change of function to adaptation potential.

Appendix A

Figure A1. The forest research area in Báb with the localizations of the research plots. (R1, R2, L1, L2). The picture is a screenshot of an aerial map from the mapy.cz web application. The tool for adding the marks on the map was that used [48].

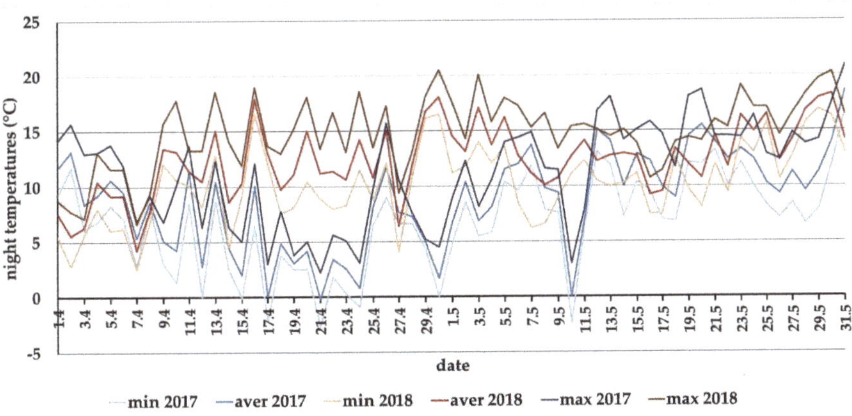

Figure A2. The night temperatures (20:00–05:00) through April and May in 2017 and 2018 at the locality of Báb (southwestern Slovakia).

References

1. Pearce, J.L.; Venier, L.A. The use of ground beetles (Coleoptera: Carabidae) and spiders (Araneae) as bioindicators of sustainable forest management: A review. *Ecol. Indic.* **2006**, *6*, 780–793. [CrossRef]
2. Kotze, D.J.; Brandmayr, P.; Casale, A.; Dauffy-Richard, E.; Dekoninck, W.; Koivula, M.J.; Lövei, G.L.; Mossakowski, D.; Noordijk, J.; Paarmann, W.; et al. Forty years of carabid beetle research in Europe—From taxonomy, biology, ecology and population studies to bioindication, habitat assessment and conservation. *ZooKeys* **2011**, *100*, 55–148. [CrossRef]
3. Niemelä, J.; Langor, D.; Spence, J.R. Effects of clear-cut harvesting on boreal ground-beetle assemblages (Coleoptera: Carabidae) in Western Canada. *Conserv. Biol.* **1993**, *7*, 551–561.
4. Skłodowski, J. Anthropogenic transformation of ground beetle assemblages (Coleoptera: Carabidae) in Białowieża Forest, Poland: From primeval forests to managed woodlands of various ages. *Entomol. Fenn.* **2006**, *11*, 296–314. [CrossRef]
5. Šustek, Z.; Vido, J. Vegetation state and extreme drought as factors determining differentiation and succession of Carabidae communities in forests damaged by a windstorm in the High Tatra Mts. *Biologia* **2013**, *68*, 1198–1210. [CrossRef]
6. Walther, G.; Post, E.; Convey, P.; Menzel, A.; Parmesan, C.; Beebee, T.J.C.; Fromentin, J.; Hoegh-Guldberg, O.; Bairlein, F. Ecological responses to recent climate change. *Nature* **2002**, *416*, 389–395. [CrossRef]
7. Thackeray, S.J.; Sparks, T.H.; Frederiksen, M.; Burthe, S.; Bacon, P.J.; Bell, J.R.; Botham, M.S.; Brereton, T.M.; Bright, P.W.; Carvalho, L.; et al. Trophic level asynchrony in rates of phenological change for marine, freshwater and terrestrial environments. *Glob. Chang. Biol.* **2010**, *16*, 3304–3313. [CrossRef]
8. Thiele, H.U. Carabid Beetles in Their Environments. In *A Study on Habitat Selection by Adaptations in Physiology and Behaviour*; Springer: Berlin/Heidelberg, Germany; New York, NY, USA, 1977; p. 369. ISBN 978-3-642-81154-8.
9. Lovei, G.L.; Sunderland, K.D. Ecology and behavior of ground beetles (Coleoptera: Carabidae). *Annu. Rev. Entomol.* **1996**, *41*, 231–256.
10. Loreau, M. Competition in a carabid beetle community: A field experiment. *Oikos* **1990**, *58*, 25–38.
11. IPCC. *Climate Change 2001: Impacts, Adaptation and Vulnerability—Contribution of Working Group II to the Third Assessment Report of the Intergovernmental Panel on Climate Change*; Cambridge University Press: Cambridge, UK, 2001.
12. Lehner, B.; Döll, P.; Alcamo, J.; Henrichs, T.; Kaspar, F. Estimating the impact of globalchange on flood and drought risk in Europe: A continental, integrated analysis. *Clim. Chang.* **2006**, *75*, 273–299. [CrossRef]
13. Duan, W.; Hanasaki, N.; Shiogama, H.; Chen, Y.; Zou, S.; Nover, D.; Zhou, B.; Wang, Y. Evaluation and Future Projection of Chinese Precipitation Extremes Using Large Ensemble High-Resolution Climate Simulations. *J. Clim.* **2017**, *32*, 2169–2183. [CrossRef]
14. Haile, G.G.; Tang, Q.; Hosseini-Moghari, S.M.; Liu, X.; Gebremicael, T.G.; Leng, G.; Kassa, A.K.; Xu, X.; Yun, X. Projected impacts of climate change on drought patterns over East Africa. *Earth's Future* **2020**, *8*, e2020EF001502. [CrossRef]
15. Škvarenina, J.; Tomlain, J.; Hrvoľ, J.; Škvareninová, J. Occurrence of dry and wet periods in altitudinal vegetation stages of West Carpathians in Slovakia: Time-series analysis 1951–2005. In *Bioclimatology and Natural Hazards*; Střelcová, K., Ed.; Springer: Dordrecht, The Netherlands, 2019; pp. 97–106.
16. Zlatník, A. *Přehled Skupin Typů Geobiocenů Původne Lesních a Křovinných v ČSSR (Overview of the Groups of the Geobiocen Types Originally Forests and Shrubs in ČSSR)*; Zprávy geografického ústavu ČSAV (Reports of the Institute of geography of the Czechoslovak academy of science), 13; The Institute of Geography of the Czechoslovak Academy of Science: Brno, Czech Republic, 1976; pp. 55–64. (In Czech)
17. Škvarenina, J.; Tomlain, J.; Hrvoľ, J.; Škvareninová, J.; Nejedlík, P. Progress in dryness and wetness parameters in altitudinal vegetation stages of West Carpathians: Time-series analysis 1951–2007. *Idojárás* **2009**, *113*, 47–54.
18. Atlas Krajiny Slovenskej Republiky (Atlas of the Slovak Republic Landscape). Available online: https://app.sazp.sk/atlassr/ (accessed on 20 October 2020).
19. Kurpelová, M.; Coufal, L.; Čulík, J. *Agroklimatické podmienky ČSSR (Agroclimatic conditions CSSR)*; Príroda: Bratislava, Slovakia, 1975; p. 270.
20. Šiška, B.; Špánik, F. Agroclimatic regionalization of slovak territory in condition of changing climate. *Meteorol. Časopis* **2008**, *11*, 61–64.

21. Šiška, B.; Takáč, J. Drought analyse of agricultural regions as influenced by climatic conditions in the Slovak Republic. *Időjárás* **2009**, *113*, 135–143.

22. Eliáš, P. Species pool of alien plants in the vicinity of Báb's Forest Research Site, south-west Slovakia. *Rosalia* **2010**, *21*, 57–74.

23. Biskupský, V. (Ed.) *Research Project Báb IBP Progress Report II. IBP Report No. 5*; Publishing House of the Slovak Academy of Sciences: Bratislava, Slovakia, 1975; p. 526.

24. Eliáš, P.; Oszlányi, J.; Matušicová, N.; Gerhátová, K.; Halada, L. Oak-Hornbeam Forest in Báb (South-Western Slovakia)—The Former Research Site of International Biological Programme in Slovakia. *Životné prostredie* **2016**, *50*, 10–17.

25. Bab—Slovakia. Available online: https://deims.org/79e10639-dd60-4f30-9c43-7b2bae0f359a (accessed on 5 November 2020).

26. Hůrka, K. *Střevlíkovití České a Slovenské Republiky, Carabid Beetles of the Czech and Slovak Republic*, 1st ed.; Kabourek: Zlín, Czech Republic, 1996; p. 565. ISBN 80-901466-2-7. (In Czech and English).

27. Šiška, B.; Žilinský, M.; Zuzulová, V. Drought and growing season on Slovakia in climate change conditions. In *Hospodaření s Vodou v Krajině, Proceedings of the Conference Hospodaření s Vodou v Krajině, Třeboň, Czech Republic, 13–14 June 2019*; Rožnovský, J., Litschmann, T., Eds.; ČHMU: Praha, Czech Republic, 2019; ISBN 978-80-87577-88-2.

28. Cunev, J.; Šiška, B. Chrobáky (Coleoptera) NPR Bábsky les pri Nitre v podmienkach meniacej sa klímy (Beetles (Coleoptera) NNR Bábsky les near Nitra in the conditions of the changing climate). *Rosalia* **2006**, *18*, 155–168, (In Slovak, Abstract in English).

29. Turin, H.; Penev, L.; Casale, A. (Eds.) *The Genus Carabus in Europe. A Synthesis*; Pensoft Publishers, Sofia-Moscow & European Invertebrate Survey: Leiden, The Netherlands, 2003; p. 512.

30. Kádár, F.; Fazekas, J.P.; Sárospataki, M.; LÖvei, G. Seasonal dynamics, age structure and reproduction of four Carabus species (Coleoptera: Carabidae) living in forested landscapes in Hungary. *Acta Zool. Acad. Sci. Hung.* **2015**, *61*, 57–72. [CrossRef]

31. Elek, Z.; Howe, A.G.; Enggaard, M.K.; Lovei, G.L. Seasonal dynamics of common ground beetles (Coleoptera: Carabidae) along urbanization gradient near Soro, Zealand, Danmark. *Entomol. Fenn.* **2017**, *28*, 27–40. [CrossRef]

32. Tauber, M.J.; Tauber, C.A.; Masaki, S. *Seasonal Adaptations of Insects*; Oxford University Press: New York, NY, USA, 1986; ISBN 9780195036350.

33. Leather, S.R.; Walters, K.F.A.; Bale, J.S. *The Ecology of Insect Overwintering*; Cambridge University Press: Cambridge, UK, 1993; p. 268.

34. Danks, H.V. *Insect Dormancy: An Ecological Perspective*; Biological Survey of Canada (Terrestrial Arthropods): Ottawa, QC, Canada, 1987; p. 439.

35. Vido, J.; Nalevanková, P.; Valach, J.; Šustek, Z.; Tadesse, T. Drought analyses of the Horné Požitavie region (Slovakia) in the period 1966–2013. *Adv. Meteorol.* **2019**. [CrossRef]

36. Weber, F. Vergleichende Untersuchungen über das Verhalten von Carabus-Arten in Luftfeuchtigkeitsgefallen. *Z. Morph. Okol.* **1965**, *54*, 551–565.

37. Šustek, Z.; Vido, J.; Škvareninová, J.; Škvarenina, J.; Šurda, P. Drought impact on ground beetle assemblages (Coleoptera, Carabidae) in Norway spruce forests with different management after windstorm damage—A case study from Tatra Mts. (Slovakia). *J. Hydrol. Hydromech.* **2017**, *65*, 333–342. [CrossRef]

38. Šerić Jelaska, L.; Franjević, D.; Jelaska, S.D.; Symondson, W.O.C. Prey detection in carabid beetles (Coleoptera: Carabidae) in woodland ecosystems by PCR analysis of gut contents. *Eur. J. Entomol.* **2014**, *111*. [CrossRef]

39. Cook, A. Behavioural ecology: On doing the right thing, in the right place, at the right time. In *The Biology of Terrestrial Molluscs*, 1st ed.; Barker, G.M., Ed.; CABI Publishing: Wallingford, CT, USA, 2001; pp. 447–487.

40. Sulikowska-Drozd, A.; Maltz, T.K. Experimental drought affects the reproduction of the brooding clausiliid Alinda biplicata (Montagu, 1803). *J. Molluscan Stud.* **2014**, *80*, 265–271. [CrossRef]

41. Nicolai, A.; Filser, J.; Lenz, R.; Bertrand, C.; Charrier, M. Adjustment of metabolite composition in the haemolymph to seasonal variations in the land snail Helix pomatia. *J. Comp. Physiol. B* **2011**, *181*, 457–466.

42. Tondoh, J.E. Seasonal changes in earthworm diversity and community structure in Central Côte d'Ivoire. *Eur. J. Soil Biol.* **2006**, *42*, S334–S340. [CrossRef]

43. Lee, K.E. *Earthworm Their Ecology and Relationship with Soils and Land Use*; Academic Press: Sydney, Australia, 1985; p. 411.

44. Eggleton, P.; Inward, K.; Smith, J.; Jones, D.T.; Sherlock, M. A six year study of earthworm (Lumbricidae) populations in pasture woodland in southern England shows their responses to soil temperature and soil moisture. *Soil Biol. Biochem.* **2009**, *41*, 1857–1865.
45. Symondson, W.O.C.; Glen, D.; Ives, A.R.; Langdon, C.J. Dynamics of the relationship between a generalist predator and slugs over five years. *Ecology* **2002**, *83*, 137–147.
46. Knapp, M.; Uhnavá, K. Body Size and Nutrition Intake Effects on Fecundity and Overwintering Success in *Anchomenus dorsalis* (Coleoptera: Carabidae). *J. Insect Sci.* **2014**, *14*, 1–6. [CrossRef]
47. Honěk, A. The effect of temperature on the activity of Carabidae (Coleoptera) in a fallow field. *Eur. J. Entomol.* **1997**, *94*, 97–104.
48. Mapy.cz. Available online: https://sk.mapy.cz/zakladni?x=17.8924454&y=48.3030312&z=18&l=0&base=ophoto (accessed on 4 November 2020).

Publisher's Note: MDPI stays neutral with regard to jurisdictional claims in published maps and institutional affiliations.

Article

Drought or Severe Drought? Hemiparasitic Yellow Mistletoe (*Loranthus europaeus*) Amplifies Drought Stress in Sessile Oak Trees (*Quercus petraea*) by Altering Water Status and Physiological Responses

Martin Kubov [1,2], Peter Fleischer, Jr. [1,2,*], Jozef Rozkošný [1], Daniel Kurjak [1], Alena Konôpková [1], Juraj Galko [3], Hana Húdoková [2,4], Michal Lalík [3,5], Slavomír Rell [3], Ján Pittner [1] and Peter Fleischer [1]

[1] Department of Integrated Forest and Landscape Protection, Faculty of Forestry, Technical University in Zvolen, T. G. Masaryka 24, 960 01 Zvolen, Slovakia; xkubovm@is.tuzvo.sk (M.K.); xrozkosny@is.tuzvo.sk (J.R.); kurjak@tuzvo.sk (D.K.); alena.konopkova@tuzvo.sk (A.K.); pittner@is.tuzvo.sk (J.P.); p.fleischersr@gmail.com (P.F.)

[2] Institute of Forest Ecology, Slovak Academy of Sciences, Ľ. Štúra 2, 960 01 Zvolen, Slovakia; hudokova@ife.sk

[3] National Forest Centre, Forest Research Institute Zvolen, T.G. Masaryka 22, 960 01 Zvolen, Slovakia; juraj.galko@nlcsk.org (J.G.); lalik@nlcsk.org (M.L.); rell@nlcsk.org (S.R.)

[4] Faculty of Ecology and Environmental Science, Technical University in Zvolen, T. G. Masaryka 24, 960 53 Zvolen, Slovakia

[5] Department of Forest Protection and Entomology, Faculty of Forestry and Wood Sciences, Czech University of Life Sciences Prague, Kamycka 129, 16500 Prague, Czech Republic

* Correspondence: p.fleischerjr@gmail.com; Tel.: +421-45-520-60-43

Received: 15 September 2020; Accepted: 22 October 2020; Published: 24 October 2020

Abstract: European oak species have long been considered relatively resistant to different disturbances, including drought. However, several recent studies have reported their decline initiated by complex changes. Therefore, we compared mature sessile oak trees (*Quercus petraea* (Matt.), Liebl.) infested versus non-infested by hemiparasitic yellow mistletoe (*Loranthus europaeus Jacq.*) during the relatively dry vegetation season of 2019. We used broad arrays of ecophysiological (maximal assimilation rate A_{sat}, chlorophyll *a* fluorescence, stomatal conductance g_S, leaf morphological traits, mineral nutrition), growth (tree diameter, height, stem increment), and water status indicators (leaf water potential Ψ, leaf transpiration T, water-use efficiency WUE) to identify processes underlying vast oak decline. The presence of mistletoe significantly reduced the Ψ by 1 MPa, and the WUE by 14%. The T and g_S of infested oaks were lower by 34% and 38%, respectively, compared to the non-infested oaks, whereas the A_{sat} dropped to 55%. Less pronounced but significant changes were also observed at the level of photosystem II (PSII) photochemistry. Moreover, we identified the differences in C content, which probably reduced stem increment and leaf size of the infested trees. Generally, we can conclude that mistletoe could be a serious threat that jeopardizes the water status and growth of oak stands.

Keywords: hemiparasite; oak stands dieback; water deficit; mineral nutrition; photosynthesis; growth response

---

1. Introduction

European oak species are dominant species in temperate European hardwood forests [1] and among the main components of the European forest economy [2]. These woody plants are, therefore, of great environmental, economic and cultural importance [3]. There is high genetic, physiological and morphological differentiation within the genus, but European oaks may be generally considered as one of the most drought-tolerant tree species in comparison with other deciduous trees of European forests [3–5]. With average air temperatures significantly rising and significant changes in precipitation regime in Europe over the last several decades [6], conditions at higher altitudes have become more suitable for these species [7]. However, they have been more often negatively affected and reports about the decline of oak stands suggest that multifactorial processes are responsible for the increased damage [8]. Inappropriate oak forest management [9], cold winters [10] and soil nutrient imbalance [11] may lead to a strong decrease in oak resistance potential. The interaction of various factors occurring on the background of climate changes leads to the more frequent extreme weather events, namely, heat stress, and a lack of precipitation during the beginning of the vegetation season and during the summer, which may affect the growth and survival of trees [12,13]. There is evidence of frequent crown and stand declines, especially at dry sites [14]. Weakened oak forests are more affected by biotic factors which may include infection by pathogenic fungi and microorganisms [15–17], and insect attacks [8,10,18]. Most of the aforementioned factors that negatively affect the vitality of trees can be amplified by an often neglected biotic agent with an increasing significance: hemiparasitic plants [19–21]. Hemiparasitic plants and their hosts form integrated systems that are stable for many years [22]. They occur more often in less dense stands [23] and even though they can assimilate CO_2, their exclusive source of water and minerals is the host plants. Thus, hemiparasites need to keep their stomata open whenever possible to sustain the stream of transpiration and the amount of water lost by hemiparasitic plants may be enormous [24]. By withdrawing water from the host trees even during drought periods, hemiparasites cause the drought effects to become more severe [23,25]. The cumulative effect of these factors leads to a more pronounced decrease in the water potential [26], stomatal conductivity and assimilation rate [23,27,28], which alter the nutrient and carbon status of host trees compared to non-infested trees growing in the same conditions [29]. Moreover, transpiration-mediated cooling of host trees is restricted, and heat stress may also impact light-dependent photosynthetic reactions. The probability of yellow mistletoe spreading is higher with rising temperatures [30], resulting in a feedback loop.

We focused on examining how yellow mistletoe (*Loranthus europaeus*) impacts the physiology of sessile oak trees (*Q. petraea* s.l.). This hemiparasitic plant is widespread in Europe, and its northern distribution border crosses Slovakia [31]. It grows mainly in oak tree crowns and occurs less frequently on stems at the places where the outer bark is cracked, and it mainly infests oaks that are more than 60 years old [19].

Lower branches of infested trees become more massive resulting in changes of overall tree shape. The accelerated growth of lower branches is a rebalancing effort of the tree to maintain the physiological performance as the upper part of the crown is more severely affected by the presence of yellow mistletoes. However, this defence mechanism is effective only when the infection is not spread and has a small extent [32]. Currently, a high population density of mistletoe can be observed and, therefore, the defence mechanism of oak trees is often ineffective. Severely infested trees respond by a premature senescence of the crown, which may eventually result in the death of the whole individual [32,33].

As results from the aforementioned, yellow mistletoe negatively affects the physiological status of host trees in many ways [23,25,26,34]. The goal of this study was to analyse a broad array of ecophysiological properties and processes in oak trees infested by yellow mistletoe during the drought period. We compared indicators of water status, photosynthetic performance, drought-induced stress nutrient status, growth and leaf traits on the infested, and non-infested mature oak trees and yellow mistletoe. We hypothesize that (i) the water potential of infested trees is significantly lower than that of non-infested trees and even lower for mistletoe, seriously influencing photosynthesis. The drought enhanced by the presence of yellow mistletoe leads to drops in both the light-dependent and light-independent stages of photosynthesis. We expect the lowest water-use efficiency (WUE) for yellow mistletoe and the highest for non-infested trees. Another hypothesis is that (ii) infested trees have significantly lower content of minerals due to losses to mistletoe and due to restricted upward transpiration stream, limiting the growth of leaves and the stem increment. Finally, we expect that (iii) higher and thicker individuals are more infested than those of lower diameters.

2. Materials and Methods

2.1. Study Site and Design Description

The research was performed in Slovakia, in the central part of the Považský Inovec Mts. (48°40′02″ N; 18°04′09″ E, 360–400 m a.s.l.). The site is located on a southwest-oriented slope with an inclination of 10–30°. The soils are acid cambisol podzol. The area belongs to a warm, dry region, hilly land and highlands.

For the research, a stand with an age of 95 years consisting exclusively of sessile oak trees (*Q. petraea* (Matt.) Liebl s.l.), was chosen. The stand belongs to managed forests with an area of 12.95 ha, stand density of 0.8 (8) and full canopy. Trees have originated from generative regeneration of 100% sessile oak seeds. The standing volume per hectare is 264 m^3 and the total standing volume is represented by 3419 m^3.

The long-term climatic data (1961–1990) for the study area were taken from the nearest meteorological station Topoľčany. The air temperature and precipitation during 2019 were measured on an open field close to the study site using our own meteorological station with Minikin Tie and Eri sensors (EMS, Brno, Czech Republic) and a built-in datalogger. The annual mean air temperature was higher in 2019 compared to the long-term mean (10.4 °C and 9.4 °C, respectively) and the average air temperature during the vegetation period (from April to September) was higher in 2019 (16.6 °C and 15.9 °C, respectively). The long-term mean annual rainfall was 549 mm, while the rainfall in 2019 was 782 mm. The amount of precipitation during the vegetation season in 2019 was 409 mm and the long-term mean was 326 mm. Even though the total precipitation was quite high compared to the long-term averages, the distribution was uneven (Figure 1) and only 23 mm fell during the 30-day period prior to the physiological measurements.

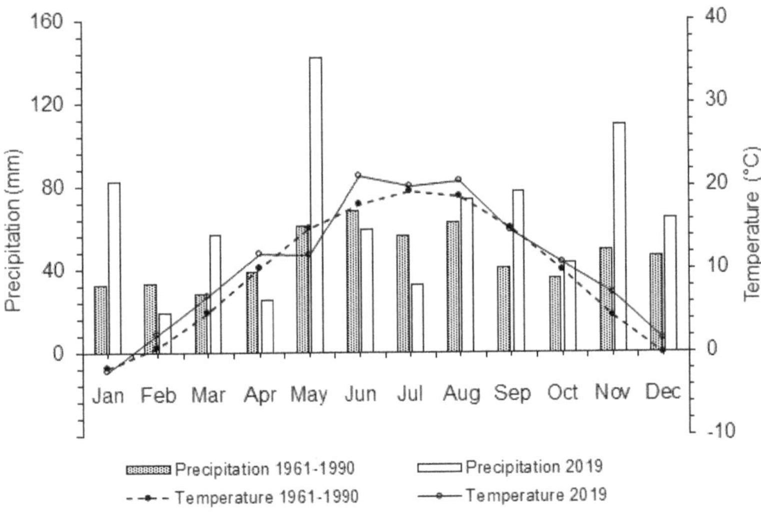

Figure 1. Precipitation and mean air temperatures for years 1961–1990 and 2019.

We identified 100 adult dominant oak trees with canopies comparably exposed to sun radiation (excluding extreme phenological forms): 50 individuals of non-infested oaks and 50 individuals infested by the hemiparasitic yellow mistletoe (*Loranthus europaeus Jacq.*), which were comparable for both, the height and diameter. All chosen trees were situated inside the stand and did not show any other visible damage, except the presence of yellow mistletoe. The diameter at breast height (DBH), the tree height, and the height of the first branch were assessed for each of 100 oak trees. The heights were measured using the ultrasonic hypsometer for the measuring of tree height and distance—Vertex IV (Haglöf, Långsele, Sweden) at late autumn of 2018. The DBH was measured in early spring and late autumn (before and after leaf unfolding) of 2018 using the steel manual band dendrometers with an accuracy of 1 mm, which were mounted at a height of 1.3 m. The difference between the two measurements represents the stem increment per year. Moreover, we visually checked each infested oak tree using binoculars, and mistletoes were counted during the winter of 2018/2019. The mean number of mistletoes per infested tree was 8.30. A total of 67% of the infested oak trees had fewer than 10 mistletoes. The maximum number of mistletoes per oak was 19, while the smallest number of mistletoes per infested oak tree was 2 individuals.

For further physiological measurements, the subset of 14 infested and 14 non-infested trees were selected, whereas we chose infested trees with between 8 and 11 mistletoes. The sampling of leaves and the in situ measurements were carried out during two sequential days on 6 and 7 August 2019 from 8:30 to 12:00 and from 15:00 to 18:00. The 2–4 m long branches from the upper, unshaded part of the crown of selected trees were cut in turns during the day by professional climbers. Immediately after cutting, the branches were put into water and measured to prevent the rapid desiccation of tissues.

The more detailed information about the individual measurements and their sample sizes are displayed in Table 1.

Table 1. The description of sample sizes for conducted measurements.

Sampling and Measurements	Oak Trees Per Variant/Repetitions	Yellow Mistletoes/Repetitions
Diameter at breast height	50/1	-
Height of tree	50/1	-
Height of the first branch	50/1	-
Stem increment	50/1	-
Leaf water potential	13/2–4	10/1
Gas exchange	14/8	14/6
Chlorophyll a florescence	14/5	-
Content of nutrients in leaves	14/1 (ca. 30 leaves)	14/1 (ca. 100 leaves)
Leaf morphological traits	14/1 (ca. 20 leaves)	14/1 (ca. 100 leaves)

2.2. Leaf Water Potential

The water potential (Ψ) was assessed for a total of 10 to 13 individuals per group. A Scholander-type pressure chamber SAPS II (Soil Moisture Equipment Corp., Goleta, CA, USA) was used and 2–4 repetitions for each tree were measured. Just one value was assessed for the mistletoe, as the variability found for one individual was low and measurements were more demanding regarding the pressure in the portable gas bottle.

2.3. Gas Exchange Measurements

Parameters related to photosynthesis and water-use efficiency were compared among infested trees, non-infested trees and yellow mistletoes. Eight and six leaves were recorded per individual oak and mistletoe, respectively. Measurements were carried out using an Li-6400XT gas exchange system connected to a standard 6 cm^2 chamber fitted with a 6400-02B light-emitting diode (LED) light source (LI-COR Biosciences, Lincoln, NE, USA). Inside the chamber, the reference CO_2 concentration and photosynthetically active radiation were maintained at 400 $\mu mol\ mol^{-1}$ and 1400 μmol photons $m^{-2}\ s^{-1}$, respectively. The air temperature inside the chamber was set to 23 °C and the relative humidity of the air was approximately 70%.

Values of the CO_2 assimilation rate (A_{sat}), transpiration rate (T), intercellular concentration of CO_2 (Ci) and stomatal conductance to water vapour (g_s) were recorded after the adaptation of leaves inside the chamber, when the values of CO_2 assimilation rate remained stable (1–2 min). The intrinsic water use efficiency WUEi was calculated as the ratio of the CO_2 fixation rate to the stomatal conductance (A_{sat}/g_S).

2.4. Chlorophyll a Fluorescence Measurements

Chlorophyll *a* fluorescence was assessed using a PAM-2500 fluorimeter (Walz, Effeltrich, Germany). Five leaves per oak tree (infested and non-infested) were measured. The leaves were kept in darkness for 30 min and then illuminated by a saturation pulse with an intensity of approximately 6580 $\mu mol\ m^{-2}\ s^{-1}$ for a duration of 1 s to assess the minimum fluorescence of dark-adapted leaves (Fo) and the maximal fluorescence of dark-adapted leaves (Fm). Then, the leaves were illuminated with an actinic light of an intensity 350 $\mu mol\ m^{-2}\ s^{-1}$ for 1.5 min to obtain the steady-state fluorescence (Fs). The second saturation pulse was used to measure the maximal fluorescence of light-adapted leaves (F'm). The maximal efficiency of the photosystem II (PSII) photochemistry (Fv/Fm), the actual efficiency of the PSII photochemistry (Φ_{PSII}), the electron transport rate (ETR) and the non-photochemical quenching (NPQ) were calculated as follows [35]:

$$\frac{Fv}{Fm} = \frac{Fm - Fo}{Fm} \tag{1}$$

$$\Phi_{PSII} = \frac{F'm - Fs}{F'm} \tag{2}$$

$$ETR = \Phi_{PSII} \times PAR \times 0.84 \times 0.5 \tag{3}$$

$$NPQ = \frac{Fm}{F'm} - 1 \tag{4}$$

2.5. Content of Nutrients in Plants and Soil

Biomass samples consisting of approximately 30 leaves per tree (infested and non-infested oaks) and 100 leaves per mistletoe individual were collected. Subsequently, the leaves were dried for 48 h in a drying oven at a temperature of 105 °C. The dry matter was then milled into a powder in a Fritsch Planetary Micro Mill (Fritsch, Markt Einersheim, Germany). The total nitrogen and sulphur content were determined with a FLASH 1112 Nitrogen, Carbon and Sulphur Analyser (Thermo Fisher Scientific Inc, Hanau, Germany). The Ca, Mg, K, and P contents were determined after mineralization of the samples in concentrated HNO_3 using microwave decomposition (UniClever type, Plazmatronika, Wrocław, Poland). The content of P was measured with an atomic emission spectrometer (AES-ICP, type LECO ICP-3000, LECO, St. Joseph, MI, USA), while the Ca, Mg and K contents were analysed using a SensAA atomic absorption spectrometer (GBC, Dandenong, Victoria, Australia).

The biological absorption coefficient (BAC) defined in 1969 by Kovalevsky [36] was calculated as the ratio between the nutrient content in the plants and the same nutrient in the soil. We utilized this relationship to determine the ability of the species to accumulate nutrients from a soil subsystem into the plant biomass or from the host plant into the parasitic plant during the growing season. Within the research plot, the infested and non-infested trees were randomly integrated, and therefore the calculation of BAC was based on the same soil samples. Sampling of the soil was carried out in August 2019. Ten samples of mineral topsoil (5–10 cm deep) were taken along two line transects (contour and fall line with lengths of 10 and 15 m, respectively) at regular distances of 5 m. These 10 samples from research plot were mixed and analysed. The collected soil sample was dried at 105 °C to a constant weight. Subsequently, the samples were milled into a fine dust using the Planetary Micro Mill. We determined the total N and S content with the FLASH 1112 CNS analyser. Available forms of the other macronutrients (Ca, Mg, K, P) were extracted according to Mehlich II, and then the samples were analysed using an atomic emission spectrometer.

2.6. Leaf Morphological Traits

Leaves from non-infested trees, infested trees, and yellow mistletoe plants were sampled and transported to the laboratory to measure of the mean leaf size (area, cm^2), using ImageJ 1.51 k software (National Institute of Health, Bethesda, MD, USA), and the leaf mass area (LMA, $g\ cm^{-2}$). LMA was calculated as the ratio between dry weight and area of leaves. We sampled approximately 20 leaves per oak tree and 100 leaves per mistletoe.

2.7. Statistical Analysis

Hierarchical analysis of variance was performed to reveal the differences among the studied parameters whenever more repetitions for one individual were measured (more leaves/twigs per oak tree/yellow mistletoe). This was used for the measurements of leaf water potential, gas exchange parameters, and chlorophyll *a* fluorescence parameters (Table 1). The analysis was performed in R (R Core Team 2017) using the lme function from the nlme package. For the measurements, where just one repetition per individual oak tree/yellow mistletoe was conducted (e.g., tree height, DBH, LMA, content of nutrient; Table 1), a one-way analysis of variance was used. Then, the Tukey tests (with significance level $\alpha = 0.05$) were performed for multiple comparisons using the glht function from the multcomp package.

Moreover, relationships between the number of mistletoes and the growth parameters of the infested trees were tested using the linear regression models.

3. Results

3.1. Leaf Water Potential and Photosynthesis

The water potential values were negative, with the lowest value in the yellow mistletoe, higher values in the infested sessile oaks and the highest value in the non-infested sessile oaks. The steps between respective variants were nearly 1 MPa, and the differences among all groups were significant (Figure 2a). Strikingly low variability was found for mistletoe.

There were no significant differences in the assimilation rate between the infested trees and the mistletoes (Figure 2b), but these two groups differed statistically from the best-performing non-infested trees. The average value of the assimilation rate for the non-infested trees was almost twice as high as that of the infested trees or mistletoes.

The yellow mistletoe showed the highest average values of transpiration rate and stomatal conductance which were 3.3 and 4 times higher than values of the infested oaks and 2.2 and 2.5 times higher than the values of the non-infested trees (Figure 2c,d). Even though the differences between the infested and non-infested trees were not as pronounced for the assimilation rate, they were statistically significant between all three groups for both, T and g_S. Despite the lower stomatal conductance in the infested trees, their intercellular concentration of CO_2 was significantly higher than that of the non-infested trees (Figure 2e). The Ci values of the yellow mistletoe was close to the ambient concentration of CO_2, and again, the variability was extremely low.

The differences in the intrinsic water-use efficiency were significant between all three groups. The lowest WUEi value was recorded for mistletoe and the highest value was recorded for the non-infested trees (19.2 and 84.5 $\mu mol\ mol^{-1}$ respectively). The WUEi values of the non-infested trees was roughly 4 times higher than that of the mistletoe (Figure 2f). The differences between the groups of oaks were significant, with higher WUEi values confirmed for the non-infested trees, in particular due to much higher assimilation rates.

Based on the maximal (Fv/Fm) and the actual photochemical efficiency of PSII (Φ_{PSII}), we did not confirm significant damage to the PSII photochemistry induced by the activity of yellow mistletoe (Figure 3a,c). However, the non-infested sessile oak trees showed a slightly (but significantly) higher efficiency of the electron transport rate (Figure 3b). The dissipation of excess energy as heat (non-photochemical quenching, Figure 3d) differed greatly between the groups of trees, with average values of 1.03 and 1.31 for non-infested and infested oaks, respectively.

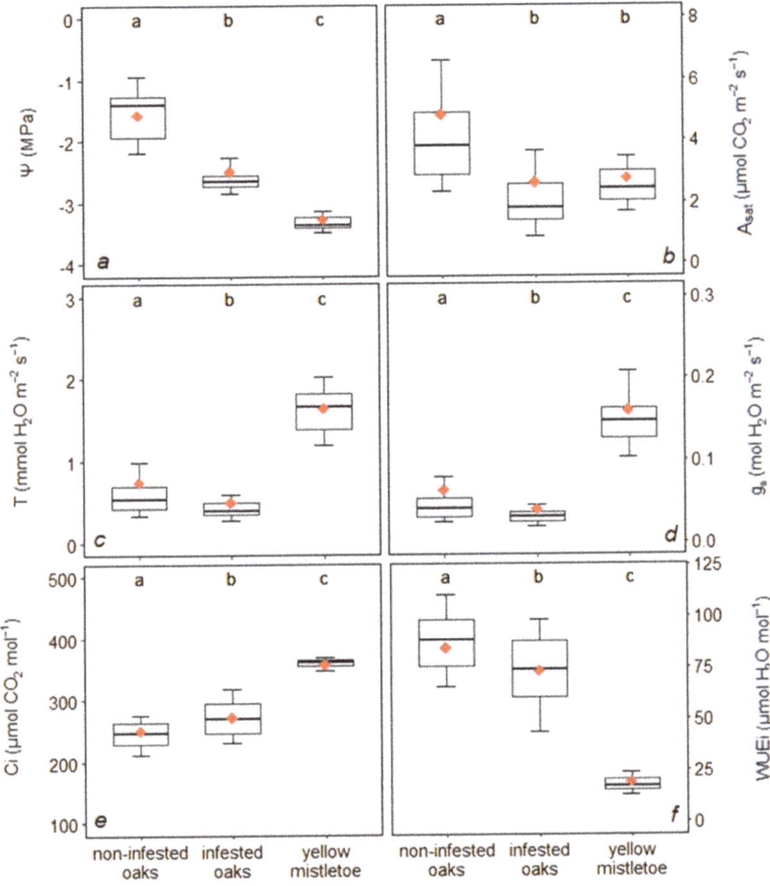

Figure 2. Water potential (Ψ, (**a**)), assimilation rate (A_{sat}, (**b**)), transpiration rate (T, (**c**)) stomatal conductivity (g_S, (**d**)), intercellular concentration of CO_2 (Ci, (**e**)) and intrinsic water-use efficiency (WUEi, (**f**)) of non-infested oaks, infested oaks and yellow mistletoes. The median (black line), 25% quantile (box; 37.5–62.5%), 50% quantile (whiskers; 25–75%) and mean (red diamond) are displayed. Lowercase letters represent statistical significance at the level of $p = 0.05$.

3.2. Content of Nutrients in Plants and Their Absorption

By comparing mistletoe and both groups of sessile oak trees we found that content of Ca, Mg, K and P in the leaves of mistletoe were significantly higher than those of both groups of oaks (Figure 4a–f). In contrast, the contents of C and N were significantly lower in the leaves of mistletoe. Most importantly, the only significant difference between the two groups of oaks was confirmed for C. Non-infested trees had a lower concentration of C in their leaves compare to the infested trees (479.2 and 498.4 g kg^{-1}, respectively).

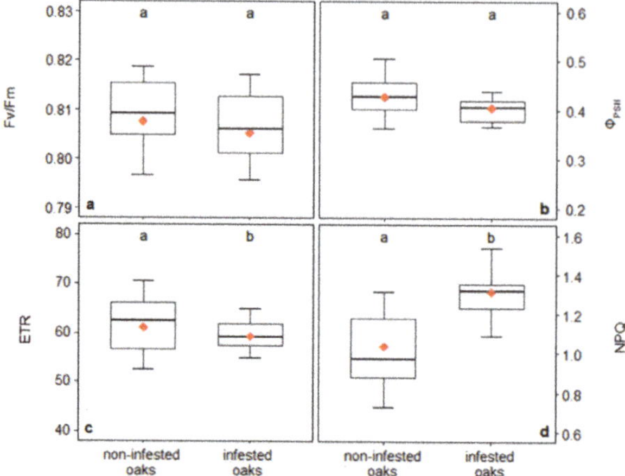

Figure 3. The maximal efficiency of the PSII photochemistry (Fv/Fm, (**a**)), actual efficiency of the PSII photochemistry (ΦPSII, (**b**)), electron transport rate (ETR, (**c**)) and non-photochemical quenching (NPQ, (**d**)) of non-infested oaks and infested oaks. The median (black line), 25% quantile (box; 37.5–62.5%), 50% quantile (whiskers; 25–75%) and mean (red diamond) are displayed. Lowercase letters represent statistical significance at the level of $p = 0.05$.

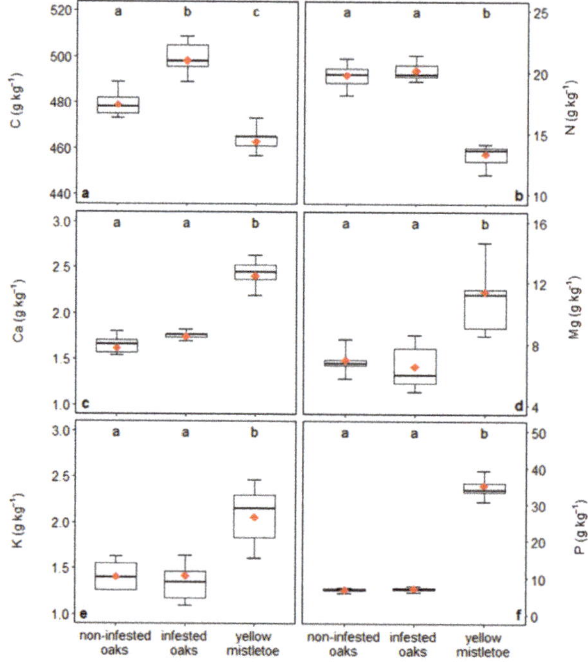

Figure 4. Contents of carbon (**a**), nitrogen (**b**), calcium (**c**), magnesium (**d**), potassium (**e**) and phosphorus (**f**) of non-infested oaks, infested oaks and yellow mistletoes. The median (black line), 25% quantile (box; 37.5–62.5%), 50% quantile (whiskers; 25–75%) and mean (red diamond) are displayed. Lowercase letters represent statistical significance at the level of $p = 0.05$.

Analyses of the mineral topsoil (5–10 cm) at the study site showed a low base saturation. On the poorest solid rocks we detected acid cambic podzol (pH 4.8). The content of C showed the highest value (22.1 g kg^{-1}) among all nutrient. On the other hand, the contents of other nutrients (N, Ca, Mg, K and P) did not exceed 1.2 g kg^{-1}. Biogeochemical flows of selected nutrients between the soil and the oaks were analysed using the biological absorption coefficient (BAC, Table 2).

Table 2. Biological absorption coefficient of nutrient (in direction from soil to oaks and from oaks to mistletoe). Mean ± standard deviations are displayed. Superscript letters (a,b) represent the statistical significance at the level of $p = 0.05$.

	Carbon	**Nitrogen**	**Calcium**	**Magnesium**	**Potassium**	**Phosphorus**
Non-infested tree	21.7 ± 0.92 [b]	17.3 ± 2.31 [a]	12.3 ± 1.96 [a]	190.9 ± 37.11 [a]	18.4 ± 4.77 [a]	55.7 ± 9.57 [a]
Infested tree	22.6 ± 0.60 [a]	17.7 ± 1.77 [a]	13.2 ± 1.28 [a]	179.8 ± 69.83 [a]	18.5 ± 5.46 [a]	59.2 ± 12.71 [a]
Mistletoe	0.9 ± 0.02	0.7 ± 0.13	1.4 ± 0.23	1.9 ± 0.58	1.5 ± 0.27	5.2 ± 1.44

We confirmed that there were no significant differences for nutrient accumulation between groups of trees (except C). The nutrients with the highest accumulation were Mg and P, while Ca had the lowest accumulation in the leaves of trees. The accumulation of Mg was 14 times higher than the accumulation of Ca for infested oak trees and almost 16 times higher for the non-infested oak trees. We found the following BACs for the nutrients in the leaves of the tested trees: Mg > P > C > K > N > Ca.

Hemiparasitic yellow mistletoe gains nutrients exclusively from its host (oak). For this reason, we calculated the BAC of the leaf nutrients ratio (mistletoe/infested oak tree). The highest values of the BAC among all the nutrients were detected for P (5.16 ± 1.44), while the lowest values were found for N (0.66 ± 0.13). By comparing the ability of mistletoes to accumulate nutrients with that of oaks, we found much lower values (Table 1) and a markedly different order: P > Mg > K > Ca > C > N. We assume that the values of the coefficients reflect the bioaccumulation potential of mistletoe and the oak–mistletoe interaction.

3.3. Growth and Rate of Infestation

We confirmed marginally significant differences in leaf area between infested and non-infested oaks. Infested oaks had significantly smaller and slightly (but not significantly) thicker leaves (Figure 5a,b).

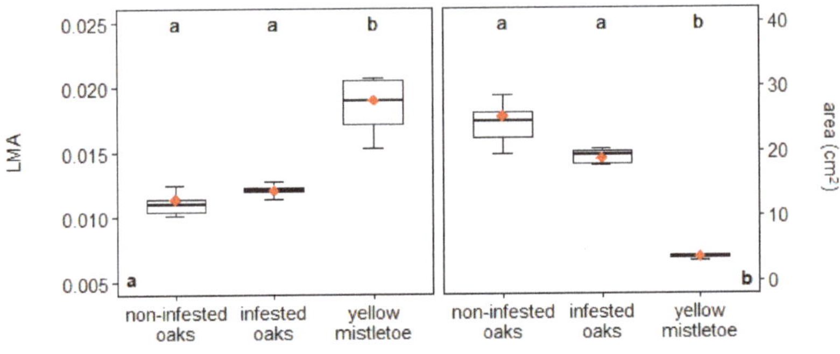

Figure 5. Leaf mass area (**a**) and leaf area (**b**) of non-infested oaks, infested oaks and yellow mistletoes. The median (black line), 25% quantile (box; 37.5–62.5%), 50% quantile (whiskers; 25–75%) and mean (red diamond) are displayed. Lowercase letters represent statistical significance at the level of $p = 0.05$.

From the point of view of tree growth, we found significant differences in all studied traits, except diameter at the breast height (Table 3). The mean height of the infested sessile oak trees was 2 m lower compared to non-infested trees with the lower-lying crown as well. Although, the differences in DBH were not significant, analysis of the average stem increment per year showed that the stem increment of the infested oaks was approximately 34% lower than of non-infested trees.

Table 3. Growth traits of non-infested and infested oak trees. Mean ± standard deviations are displayed. Superscript letters (a,b) represent the statistical significance at the level of $p = 0.05$.

	DBH (cm)	Height (m)	Height of the First Branch (m)	Stem Increment (mm)
Non-infested trees	36.6 ± 3.5 [a]	25.2 ± 1.7 [a]	9.1 ± 3.3 [a]	1.75 ± 1.4 [a]
Infested trees	39.2 ± 5.1 [a]	23.2 ± 1.9 [b]	6.7 ± 2.6 [b]	2.38 ± 1.7 [b]

Moreover, we calculated the correlation coefficient between the height of oak trees and the frequency of mistletoe in the infested oaks and found no significant relationship. On the other hand, we found a significant positive relationship between the diameter at breast height of the infested oak trees and the frequency of mistletoe ($r^2 = 0.234$, Figure 6a,b).

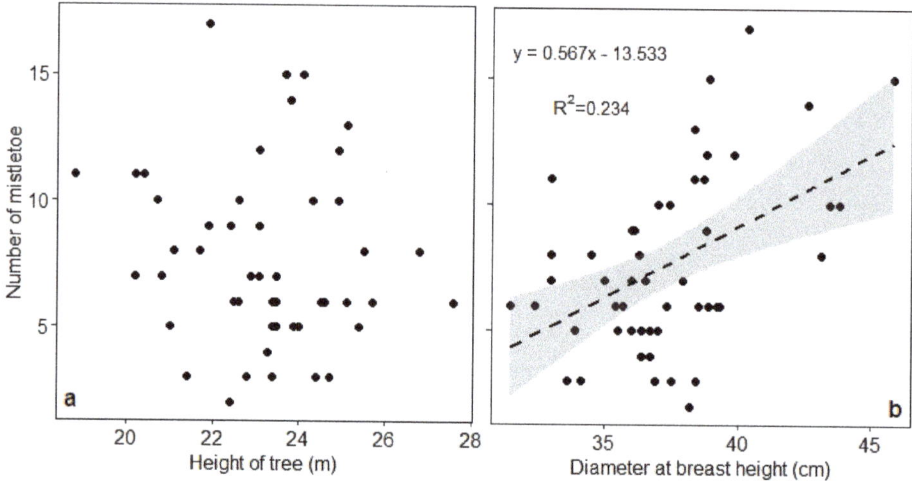

Figure 6. Relationship between the number of mistletoes and the tree height (**a**) and the tree diameter at breast height (**b**). For statistically significant relationships (**b**), the regression line (dashed line) and the 95% confidence interval (grey area) are displayed.

4. Discussion

4.1. Impact of Mistletoe on Photosynthesis and Water Relationships

Our results showed that yellow mistletoe infestation alters the water status of its host plant. Infested trees showed a more negative leaf water potential Ψ (−2.5 MPa) than did non-infected trees (−1.58 MPa). This result was expected because part of the water is drained by mistletoe, amplifying the drought stress of the host trees in our experiment. Moreover, plants try to avoid water shortage and wilting caused by drought and hemiparasite by lowering the osmotic potential via production of osmoprotective compounds, in addition to closing their stomata [24,37]. There is a close relationship between the severity of drought stress and the leaf water potential. Breda et al. [4] reported a predawn leaf water potential of −2 MPa and midday leaf water potential of −3.3 MPa as a threshold values of *Q. petraea* when the number of embolised vessels significantly increase. Cochard et al. [38] estimated the threshold value for midday water potential of −3 MPa in a 30-year-old *Q. petraea* stand. While the non-infested trees in our study were, during the observations (from 8:30 to 12:00 and from 15:00 to

18:00), in the "safe zone", the infested ones were threatened in the long term by water shortage. The high water demands of mistletoe are guaranteed by a greater negative leaf water potential (-3.5 MPa), which causes the hydrostatic pressure in the cells between the host and the hemiparasite to favour the hemiparasite and the xylem sap to flow from the host to the mistletoe [39]. The average differences in Ψ during observation hours between the infested oaks and the mistletoes was 0.75 MPa, which is a smaller value than the published values for yellow mistletoe parasitizing *Q. robur* (from -1.0 to -1.5 MPa, [40]). Davidson et al. [41] pointed out that the Ψ difference between a host and mistletoe depends on the air temperature and vapour pressure deficit (VPD). Higher air temperatures and VPDs lead to larger differences. During our experiment, the VPD ranged from 1.54 to 2.48 kPa, and the air temperature ranged from 22.0 to 26.0 °C indicating rather mild conditions. However, drought was present, and the rainless period lasted for nearly one month prior to the experiment; therefore, we assume that under more severe climate conditions (warmer and drier), the difference in Ψ between a hemiparasite and its host might be higher.

As European oaks are considered as an anisohydric species in general, which are characterized by relatively low stomatal control and under drought conditions they try to retain the CO_2 assimilation rate at the costs of water loses [42,43], we expected that the presence of yellow mistletoe would enhance the anisohydric behaviour of the oak–mistletoe system. However, based on our results as we observed the decrease in stomatal conductance and transpiration rate, we now suppose that behaviour of infested oak trees is more isohydric compared to the non-infested trees. Transpiration in the infested trees was 66% of that in the non-infested ones. Reduced transpiration of the infested trees was connected to an increased stomatal closure. In our case, the closing of stomata is a water-saving tool that prevent hydraulic failure in host trees suffering from the presence of hemiparasites. The activity of mistletoe leads to a reduced water supply in the host tree, a lower transpiration rate due to preventive stomatal closure and an eventual reduction in the cooling ability at high air temperatures [44]. Unlike host trees, mistletoe controls its own transpiration rate at very low levels and keeps stomata open allowing it to transpire large amounts of water [26]. However, several studies have described the coupled regulation of mistletoe stomata based on the host tree water status [28,40,45], light and air humidity [46] to some extent. In our study, we did not observe any of these regulatory mechanisms. Ullmann et al. [45] reported that the ratio between the transpiration rate of most mistletoe species and their hosts usually ranges between 1.5 and 7.9, but only a few studies have dealt with yellow mistletoe and *Quercus* sp. relationships. The ratio of transpiration between the mistletoe and its host in our study was 3.3, which is in good accordance with the findings published by Schultze et al. [40] and [28]. Urban et al. [47] reported that the transpiration rate of *Loranthus europaeus* was 5 times higher than that of oak. The differences in the transpiration rates might be caused by a different duration of observations and the presence of drought in our study.

In addition to the reduced transpiration in the infested oaks, the stomatal conductance was also significantly lower compared to the non-infested trees (0.039 mmol m^{-2} s^{-1} and 0.062 mmol m^{-2} s^{-1}, respectively). Low stomatal conductance on infested trees prevents water loses, but on the other hand, it causes limited CO_2 assimilation. Hence, the presence of yellow mistletoe increases the risk of carbon starvation. Under a sufficient water supply, this is not an issue for a host [24], but carbon starvation remains a challenge under drought conditions. The activity of mistletoe reduced the stomatal conductance to the same extent as that of the transpiration rate, as these processes are tidily coupled. This is in accordance with Orcutt and Nilsen [48], who demonstrated a decline in the conductance in tree leaves infested by yellow mistletoe to 68% when compared with non-infested oaks. The stomatal conductance of mistletoe (0.159 mmol m^{-2} s^{-1}) was significantly higher than in both groups of oaks confirming the huge water withdrawal from the host tree xylem sap. A high rate of water loss due to stomata openness is the most common feature among mistletoes either from temperate, tropical or arid regions [40,41,45,46,49,50].

As is apparent from the aforementioned results, we found a difference in the photosynthetic performance between the infested and non-infested trees. The close relationship between the severity of drought stress and the decline in photosynthetic activity was earlier described by Wang et al. [51], who mainly attributed the decline of photosynthesis during mild and severe drought stress to the stomatal limitation and PSII photoinhibition with accumulation of reactive oxygen species (ROS), respectively. Queijeiro-Bolaños et al. [52] stated that the presence of mistletoes impacts the photosynthetic performance of host according to the severity of infestation. The maximal rate of CO_2 assimilation (A_{sat}) in our study dropped to 55%. A similar decline (drop to 68%) was found by Johnson and Choinski [44]. For oaks infested by mistletoe, Schulze et al. [40] found A_{sat} values between 4 and 7 μmol CO_2 m^{-2} s^{-1}. Generally, the A_{sat} values of non-infested oaks are much higher. Epron et al. [53] found A_{sat} values for *Q. petraea* ranging from 12 to 14 μmol CO_2 m^{-2} s^{-1} and Morecroft and Roberts [54] published a value of 10.4 μmol CO_2 m^{-2} s^{-1} for *Q. rubra*. Osuna et al. [55] published values from 7 to 15 μmol CO_2 m^{-2} s^{-1} for *Q. douglasii* in dry years, but the A_{sat} values shifted to 15–23 μmol CO_2 m^{-2} s^{-1} during the peak of the growing season for precipitation rich years. Differentiation between our results and those mentioned above suggests pronounced drought stress during the measurements. Typically, most mistletoes have lower rates of CO_2 assimilation rate than the trees they infest, but this is not an absolute rule [56–58]. Davidson et al. [41], as in our case, observed that the A_{sat} values of a hemiparasite and its host were nearly equal. In addition to reducing the CO_2 assimilation rate of the host tree, yellow mistletoe can gain carbon from host xylem sap, as there is no phloem bridge between the host tree and hemiparasite [59]. This heterotrophic carbon gain by mistletoe species could range from 5% to 62% [57,60,61], which might lead to different impacts on host trees according to the abundance of the parasite. To date, no such data are available for yellow mistletoe.

Coupled processes of CO_2 assimilation and water transpiration on the leaf level are described by the intrinsic water use efficiency (WUEi), which is the ratio of carbon assimilation to the rate of stomatal conductance. Lower WUEi values generally means that more water is required for carbon fixation. In our study, the WUEi values of infested trees were significantly lower (73 μmol H_2O mol^{-1}) than that of non-infested trees (84 μmol H_2O mol^{-1}). This corresponds well with the results of Sangüesa-Barreda et al. [62], who reported the substantial decrease in WUEi of Scots pine (*Pinus sylvestris*) severely infested by European mistletoe (*Viscum album*) based on long-term observation. Similar results were recorded by Queijeiro-Bolaños et al. [52] for the same species. The mistletoes in our study had WUEi values 4 times lower than those of non-infested oaks, which also support the strongly anisohydric behaviour of this hemiparasitic plant.

In addition to the above, drought stress could be identified with the intercellular concentration of CO_2 (Ci). Cornic and Massachi [63] identified drought stress by declining CO_2 concentration inside the leaf. We found significantly higher Ci values accompanied by lower stomatal conductance in infested oaks. A situation in which Ci increases with declining stomatal conductance was described by Flexas and Medrano [64], who identified an inflexion point beyond which Ci starts to increase with declining conductance. The inflexion point usually lies close to the value of 50 mmol H_2O m^{-2} s^{-1}. In general, stomatal conductance below 150 causes metabolic limitation; for g_S values below 100 mmol H_2O m^{-2} s^{-1}, photochemical activity is reduced; and for g_S values below 50 mmol H_2O m^{-2} s^{-1}, permanent photoinhibition frequently occurs [65]. The stomatal conductance values in our study were notably low, at 60 mmol H_2O m^{-2} s^{-1} in non-infested tree leaves and 39 mmol H_2O m^{-2} s^{-1} in infested tree leaves. According to the association of higher Ci values with lower conductance values in infested trees, we assume that the Ci values were beyond the inflection point of the Ci curve, and photoinhibition is present. In contrast, non-infested trees show lower Ci values and higher g_S values. This points out that the inflection point of the Ci curve lies between these two values and that infested individuals undergo severe heat-light stress, which could lead to permanent photoinhibition.

The method of chlorophyll *a* fluorescence can be a sensitive indicator of plant photochemical performance under a wide range of stress factors and it is suitable for assessing photoinhibition [66]. Even though PSII has been considered relatively resistant to drought and is more suitable as an indicator of heat stress than drought [67,68], several studies have used this method and confirmed significant reduction in the photochemical performance under drought conditions [35,68,69]. Moreover, oaks in field conditions may suffer from additive effects, such as heat stress or a hindered ability to regenerate photosystems II [70]. Therefore, we expected that the unsuitable conditions intensified by yellow mistletoe ($\Psi = -1.58$ MPa in non-infested trees and $\Psi = -2.5$ MPa in infested oaks) will alter the PSII photochemistry yield in both the dark-adapted and light-adapted states. However, the Fv/Fm and Φ_{PSII} values (ranging in the interval 0.8–0.81 and 0.4–0.45, respectively) showed no significant differences between non-infested and infested oak trees. Φ_{PSII} values in the range of 0.3–0.6 were described for two oaks species by [71], whereas the values were the same during wet and dry summers, suggesting a relatively high resistance of PSII photochemistry to drought. However, the electron transport rate (ETR), which is closely coupled with the maximum velocity of carboxylation ($V_{c,max}$), can be used as a surrogate for leaf biochemistry and reflects gross photosynthesis [72,73], was significantly reduced in infested trees. This suggests that non-infested oaks utilize light energy in carboxylation more efficiently than infested trees, which corresponds well with the carbon assimilation rate results. On the other hand, we observed substantially upregulated non-photochemical quenching NPQ in infested oak trees. The higher values of NPQ means that the infested trees probably tried to avoid possible damage caused by the ROS overproduction by dissipating excess energy through the xanthophyll cycle, as they were under more severe drought stress [74,75]. Such a protective behaviour to maintain normal photochemistry has been described in several studies dealing with drought stress [51,76]. However, Wang et al. [51] described the upregulated NPQ under moderate drought stress caused by reversible photosynthetic adjustment (e.g., limited stomatal conductance, Ribulose-1-5-bisphosphate (RuBP) regeneration, adenosine triphosphate (ATP) synthesis, and/or decreases in Rubisco activity). On contrast, under a severe water deficit (with a Ψ values of approximately -3 MPa in apple tree leaves), the sharp drop of NPQ was observed, which suggests the irreversible impairment of PSII units. As we observed the higher ability of NPQ in the infested trees, we can conclude that the changes in photosynthesis of infested oaks were caused by reversible photosynthetic adjustment rather than irreversible damages of PSII.

4.2. Relationships between Oaks and Mistletoes: Growth and Mineral Nutrition

Long-term reduced photosynthesis may alter growth as well. The results of our study show significant differences in the growth parameters between non-infested and infested oak trees. We found differences in the stem increment (ca. 0.63 mm) and height (2.6 m) of non-infested oaks. Infested trees were significantly lower compare to the non-infested ones as well as their radial growth was slower.

We found that the height of trees has no significant effect on the number of mistletoes, while the relationship between the diameter at breast height and the number of mistletoes on the oaks was significant ($r^2 = 0.234$). Similar to Aukema and Martinez del Rio [77], we suppose that the tree height, crown density and position of oaks in the forest stand play important roles for infestation of the oak by mistletoe. However, the frequency and density of mistletoe are not related just to a character of the oaks and the forest stand. In general, the distribution of mistletoes is affected by climate conditions such as winter and spring frosts [31] and the presence and frequency of common bird species that feed on mistletoe fruits [78–80]. Birds usually prefer higher oak trees with a high crown density to smaller oak trees with crown lift in the forest stand [77]. According to Overton [81], larger trees are older and have more branches (space) for birds. This could be a reason why they have a higher probability of being colonized by mistletoe [82]. Matula et al. [23] found that the probability of mistletoe infection increased significantly as the stem diameter of oak increased (>27 cm). Our results confirm this observation, because mistletoe infestation occurred (Figure 6b) on trees with a stem diameter above 31 cm. If mistletoe attacks an individual oak tree, the probability of mistletoe frequency will increase due to its own reproduction and to the behaviour of birds, which disperse mistletoe seeds. This means that the most infested oaks could be those that are the nearest to other infested trees. If the distance to an infested oak tree is greater than 5 m, the likelihood of infection is much smaller [23].

Growth of oak trees has focused not only on the carbon and water cycle mechanisms, but also on the critical role of soil-available nutrients [83]. Our results show that the soil under oak forests is an important reservoir of organic C. In the soil of our study area, we found 22.1 g kg^{-1} of C. Soils from European oak forests show similar carbon concentrations [84]. On the other hand, we detected a relatively low content of Mg. According to Nilsson [85], this is not especially surprising because acidic, respectively cambic podzols are characteristic of a low availability of nutrients (particularly Mg). The ability to accumulate nutrients (other than Mg) was slightly higher in infested oaks than in non-infested oaks (Table 2), but this difference was not significant suggesting that mistletoe does not reduce the nutrient content of host trees. The reason for the low differences between infested and non-infested oaks could be the stock of nutrients that oaks have, similar to other long-living plants. We suppose that the infested trees acclimated their metabolism and mineral acquisition processes to mitigate the negative effect of the hemiparasites. On the other hand, this phenomenon may be related to the number and age of mistletoes. We expect that a low number of young mistletoes will not cause as much damage as a high number of older mistletoes. However, as there is no data about the length of time for which this particular mistletoe infestation has been in the study area, the present study can only describe one snap shot in time. Mistletoes derive nutrients from their hosts [86], so we expected lower nutrient contents in the leaves of infested trees. However, the nutrient content in leaves did not differ significantly between infested and non-infested trees in our study. Infested oaks had a relatively higher nutrient accumulation than non-infested oaks. However, the overall biomass of leaves may differ between infested and non-infested oaks. The formatting of smaller and thicker leaves and the overall reduction of growth is probably a reaction to competition for water, inorganic ions, and metabolites [87], and, at the same time, could be a mechanism to maintain a relatively high efficiency of light-dependent photosynthetic reactions. We confirmed that infested trees formed smaller leaves without significant differences in the thicknesses of the leaves between oaks. Thus, the total amounts of allocated nutrients in leaves probably differ too. The high hemiparasite to host phosphorus ratio (5× higher concentration in mistletoe) is in agreement with data published by Lamont [88]. Ca and Mg are in the plausible range, but potassium was much lower in our study. According to Hosseini et al. [86] and Glatzel and Geils [26], nutrient flow through the transpiration stream is predominantly a one-way flow (from host to mistletoe, not the opposite direction). This may be the main reason why the content of some mineral nutrients is higher in mistletoe leaves than in infested and non-infested oak trees. Many other authors have reported similar findings on various host and mistletoe species [89–91] Mistletoe parasitism is a strategy for N acquisition [61,88]. In our study, the N content was markedly lower in mistletoe than in both infested and non-infested oaks. Schulze and Ehleringer [22] and Ehleringer et al. [50] reported N availability as a limiting factor for mistletoe growth. In the study by Schultze et al. [40], the ratio of parasite/host N content was 1.08. Additionally, in the study by Bowie and Ward [92], the N content was higher in mistletoe than in host trees. In our case, the ratio is opposite (0.66), and we found higher N contents in the host tree leaves. The same findings were published Hosseini et al. [86]. However, we discarded the leaf petiole prior to determination of the elemental contents, which could cause a discrepancy with the aforementioned studies.

5. Conclusions

According to our results, we can conclude that yellow mistletoe (*Loranthus europaeus*) infestation seriously threatens the physiological and growth performance of sessile oak (*Quercus petraea* s.l.) trees. We observed a significant reduction of leaf water potential, transpiration rate and stomatal conductance in infested oak trees, which resulted in the decrease of their CO_2 assimilation rate as well. Moreover, significant changes of PSII photochemistry were observed in infested trees. Considering the nutrient status, infestation by yellow mistletoe caused a reduction in the accumulation of C, which probably led to a lower stem increment and leaf size of infested oak trees.

For this reason, we suppose that yellow mistletoe may represent a great ecological and economic problem in Europe oak forest stands, which could also reduce expected ecosystem services in these unique habitats. To determine the most effective way to combat yellow mistletoe infestation, intensive monitoring of both hemiparasite and its host is required.

Author Contributions: D.K., J.G., and P.F. conceived the design of the study, planned and coordinated the experiment; J.R., M.L., S.R., J.P. collected plant material, conducted the measurements of growth traits, and processed an overall description of stand and level of mistletoe infestation; M.K., P.F.J., J.R., D.K., A.K., H.H., and P.F. conducted the ecophysiological measurements and processed the data; P.F.J., A.K., M.K. performed the statistical analyses; M.K., P.F.J. and D.K. wrote the first version of the manuscript; all the authors contributed critically to the drafts and gave final approval for publication. All authors have read and agreed to the published version of the manuscript.

Funding: The study was financially supported by the grants of Slovak Agency for Research and Development no. APVV-17-0644, Slovak Grant Agency for Science no. VEGA 2/0120/17 as well as State enterprise Forests of the Slovak Republic (contract No. 5781/2017/LSR).

Conflicts of Interest: The authors declare no conflict of interest.

References

1. Dimopoulos, P.; Bergmeier, E.; Chytrý, M.; Rodwell, J.; Schaminée, J.H.J.; Sykora, K.V. European oak woodlands: Past, present and future. *Bot. Chron. Patra* **2005**, *18*, 1–316.

2. Rock, J.; Gockel, H.; Schulte, A. Vegetationsdiversität in Eichen-Jungwüchsen aus unterschiedlichen Pflanzschemata. *Beitr. Für Forstwirtsch. Landschaftsökologie* **2003**, *37*, 11–17.

3. Kunz, J.; Löffler, G.; Bauhus, J. Minor European broadleaved tree species are more drought-tolerant than Fagus sylvatica but not more tolerant than *Quercus petraea*. *For. Ecol. Manag.* **2018**, *414*, 15–27. [CrossRef]

4. Breda, N.; Cochard, H.; Dreyer, E.; Granier, A. Water transfer in a mature oak stand (*Quercus petraea*): Seasonal evolution and effects of a severe drought. *Can. J. Forest Res.* **1993**, *23*, 1136–1143. [CrossRef]

5. Pretzsch, H.; Schütze, G.; Uhl, E. Resistance of European tree species to drought stress in mixed versus pure forests: Evidence of stress release by inter-specific facilitation. *Plant Biol.* **2013**, *15*, 483–495. [CrossRef] [PubMed]

6. Pachauri, R.K.; Mayer, L.; Core Writing Team. (Eds.) *IPCC Climate Change 2014: Synthesis Report*; Intergovernmental Panel on Climate Change: Geneva, Switzerland, 2014.

7. Mert, A.; Özkan, K.; Şentürk, Ö.; Negiz, M.G. Changing the potential distribution of Turkey oak (IL.) under climate change in Turkey. *Pol. J. Environ. Stud.* **2016**, *25*, 1633–1638. [CrossRef]

8. Sallé, A.; Nageleisen, L.-M.; Lieutier, F. Bark and wood boring insects involved in oak declines in Europe: Current knowledge and future prospects in a context of climate change. *For. Ecol. Manag.* **2014**, *328*, 79–93. [CrossRef]

9. Olano, J.M.; Laskurain, N.A.; Escudero, A.; De La Cruz, M. Why and where do adult trees die in a young secondary temperate forest? The role of neighbourhood. *Ann. For. Sci.* **2009**, *66*, 105. [CrossRef]

10. Thomas, F.M.; Blank, R.; Hartmann, G. Abiotic and biotic factors and their interactions as causes of oak decline in Central Europe. *For. Pathol.* **2002**, *32*, 277–307. [CrossRef]

11. Bréda, N. Water shortage as a key factor in the case of the oak dieback in the Harth Forest (Alsatian plain, France) as demonstrated by dendroecological and ecophysiological study. In *Recent Advances on Oak Health in Europe. Selected Papers from a Conference Held in Warsaw, Poland, 22–24 November 1999*; Instytut Badawczy Leśnictwa (Forest Research Institute): Warsaw, Poland, 2000; pp. 157–159.

12. Doležal, J.; Mazůrek, P.; Klimešová, J. Oak decline in southern Moravia: The association between climate change and early and late wood formation in oaks. *Preslia* **2010**, *82*, 289–306.

13. Sohar, K.; Helama, S.; Läänelaid, A.; Raisio, J.; Tuomenvirta, H. Oak decline in a southern Finnish forest as affected by a drought sequence. *Geochronometria* **2014**, *41*, 92–103. [CrossRef]

14. Sonesson, K.; Drobyshev, I. Recent advances on oak decline in southern Sweden. *Ecol. Bull.* **2010**, *11*, 197–208.

15. Jung, T.; Blaschke, H.; Oßwald, W. Involvement of soilborne Phytophthora species in Central European oak decline and the effect of site factors on the disease. *Plant Pathol.* **2000**, *49*, 706–718. [CrossRef]

16. Marçis, B.; Caël, O.; Delatour, C. Relationship between presence of basidiomes, above-ground symptoms and root infection by *Collybia fusipes* in oaks. *For. Pathol.* **2000**, *30*, 7–17. [CrossRef]

17. Hajji, M.; Dreyer, E.; Marçais, B. Impact of *Erysiphe alphitoides* on transpiration and photosynthesis in *Quercus robur* leaves. *Eur. J. Plant Pathol.* **2009**, *125*, 63–72. [CrossRef]

18. Galko, J.; Nikolov, C.; Kimoto, T.; Kunca, A.; Gubka, A.; Vakula, J.; Zúbrik, M.; Ostrihoň, M. Attraction of ambrosia beetles to ethanol baited traps in a Slovakian oak forest. *Biologia (Bratisl.)* **2014**, *69*, 1376–1383. [CrossRef]

19. Millaku, F.; Sahiti, G.; Abdullahu, K.; Krasniqi, E. The spread and the infection frecuency of the Golliak (Kosovo) forest with the species hemipariastoc mistletoe (*Loranthus europaeus* L.). *Int. Multidiscip. Sci. GeoConf. SGEM* **2011**, *2011*, 1035–1040.

20. Kumbasli, M.; Keten, A.; Beskardes, V.; Makineci, E.; Özdemir, E.; Yilmaz, E.; Zengin, H.; Sevgi, O.; Yilmaz, H.C.; Caliskan, S. Hosts and distribution of yellow mistletoe (*Loranthus europaeus* Jacq. (Loranthaceae) on Northern Strandjas Oak Forests-Turkey. *Sci. Res. Essays* **2011**, *6*, 2970–2975.

21. Gebauer, R.; Volařík, D.; Urban, J. Seasonal variations of sulphur, phosphorus and magnesium in the leaves and current-year twigs of hemiparasitic mistletoe *Loranthus europaeus* Jacq. and its host *Quercus pubescens* Willd. *J. For. Sci.* **2018**, *64*, 66–73.

22. Schulze, E.D.; Ehleringer, J.R. The effect of nitrogen supply on growth and water-use efficiency of xylem-tapping mistletoes. *Planta* **1984**, *162*, 268–275. [CrossRef]

23. Matula, R.; Svátek, M.; Pálková, M.; Volařík, D.; Vrška, T. Mistletoe infection in an oak forest is influenced by competition and host size. *PLoS ONE* **2015**, *10*, e0127055. [CrossRef]

24. Zweifel, R.; Bangerter, S.; Rigling, A.; Sterck, F.J. Pine and mistletoes: How to live with a leak in the water flow and storage system? *J. Exp. Bot.* **2012**, *63*, 2565–2578. [CrossRef]

25. Rigling, A.; Eilmann, B.; Koechli, R.; Dobbertin, M. Mistletoe-induced crown degradation in Scots pine in a xeric environment. *Tree Physiol.* **2010**, *30*, 845–852. [CrossRef] [PubMed]

26. Glatzel, G.; Geils, B.W. Mistletoe ecophysiology: Host-parasite interactions. *Botany* **2009**, *87*, 10–15. [CrossRef]

27. Press, M.C.; Phoenix, G.K. Impacts of parasitic plants on natural communities. *New Phytol.* **2005**, *166*, 737–751. [CrossRef]

28. Glatzel, G. Mineral nutrition and water relations of hemiparasitic mistletoes: A question of partitioning. Experiments with *Loranthus europaeus* on *Quercus petraea* and *Quercus robur*. *Oecologia* **1983**, *56*, 193–201. [CrossRef] [PubMed]

29. Mathiasen, R.L.; Nickrent, D.L.; Shaw, D.C.; Watson, D.M. Mistletoes - ecology, systematics, ecology and management. *Plant Dis.* **2008**, *92*, 987–1006. [CrossRef]

30. Szmidla, H.; Tkaczyk, M.; Plewa, R.; Tarwacki, G.; Sierota, Z. Impact of Common Mistletoe (*Viscum album* L.) on Scots Pine Forests—A Call for Action. *Forests* **2019**, *10*, 847. [CrossRef]

31. Eliáš, P. Úhyn imelovca (*Loranthus europaeus* Jacq.) na severnej hranici rozšírenia v Európe: Slovensko (in Slovak)/Mistletoe (Loranthus europaeus Jacq.) mortality at northern geographical distribution in Europe: Slovakia. In *Dreviny v Mestskom Prostredí a v Krajine. Aktualne Trendy Dendrologického Výskumu a Praxe*; Slovak University of Agriculture: Nitra, Slovakia, 2007.

32. Kubíček, J.; Martinková, M.; Špinlerová, Z. Hemiparazité a provozní bezpečnost jimi napadených stromů (in Czech)/Hemiparasites and operating safety of infested trees. In *Provozní Bezpečnost Stromů*; Mendel University in Brno: Brno, Czech Republic, 2011; pp. 44–50.

33. McDowell, N.G. Mechanisms linking drought, hydraulics, carbon metabolism, and vegetation mortality. *Plant Physiol.* **2011**, *155*, 1051–1059. [CrossRef] [PubMed]

34. Doležal, J.; Lehečková, E.; Sohar, K.; Altman, J. Oak decline induced by mistletoe, competition and climate change: A case study from central Europe. *Preslia* **2016**, *88*, 323–346.

35. Maxwell, K.; Johnson, G.N. Chlorophyll fluorescence—A practical guide. *J. Exp. Bot.* **2000**, *51*, 659–668. [CrossRef] [PubMed]

36. Brooks, R.R. *Biological Methods of Prospecting for Minerals*; Wiley: Chichester, UK; New York, NY, USA, 1983.

37. Griebel, A.; Watson, D.; Pendall, E. Mistletoe, friend and foe: Synthesizing ecosystem implications of mistletoe infection. *Environ. Res. Lett.* **2017**, *12*, 115012. [CrossRef]

38. Cochard, H.; Bréda, N.; Granier, A. Whole tree hydraulic conductance and water loss regulation in Quercus during drought: Evidence for stomatal control of embolism? *Ann. Sci. For.* **1996**, *53*, 197–206. [CrossRef]

39. Zuber, D. Biological flora of Central Europe: *Viscum album* L. Flora—Morphol. Distrib. *Funct. Ecol. Plants* **2004**, *199*, 181–203. [CrossRef]

40. Schulze, E.D.; Turner, N.C.; Glatzel, G. Carbon, water and nutrient relations of two mistletoes and their hosts: A hypothesis. *Plant Cell Environ.* **1984**, *7*, 293–299.

41. Davidson, N.J.; True, K.C.; Pate, J.S. Water relations of the parasite: Host relationship between the mistletoe *Amyema linophyllum* (Fenzl) Tieghem and *Casuarina obesa* Miq. *Oecologia* **1989**, *80*, 321–330. [CrossRef]

42. Steckel, M.; del Río, M.; Heym, M.; Aldea, J.; Bielak, K.; Brazaitis, G.; Černý, J.; Coll, L.; Collet, C.; Ehbrecht, M.; et al. Species mixing reduces drought susceptibility of Scots pine (*Pinus sylvestris* L.) and oak (*Quercus robur* L., *Quercus petraea* (Matt.) Liebl.)—Site water supply and fertility modify the mixing effect. *For. Ecol. Manag.* **2020**, *461*, 117908. [CrossRef]

43. Sade, N.; Gebremedhin, A.; Moshelion, M. Risk-taking plants: Anisohydric behavior as a stress-resistance trait. *Plant Signal. Behav.* **2012**, *7*, 767–770. [CrossRef]

44. Johnson, J.M.; Choinski, J.S. Photosynthesis in the *Tapinanthus-Diplorhynchus* Mistletoe-Host Relationship. *Ann. Bot.* **1993**, *72*, 117–122. [CrossRef]

45. Ullmann, I.; Lange, O.L.; Ziegler, H.; Ehleringer, J.; Schulze, E.-D.; Cowan, I.R. Diurnal courses of leaf conductance and transpiration of mistletoes and their hosts in Central Australia. *Oecologia* **1985**, *67*, 577–587. [CrossRef]

46. Küppers, M.; Küppers, B.I.L.; Neales, T.F.; Swan, A.G. Leaf gas exchange characteristics, daily carbon and water balances of the host/mistletoe pair *Eucalyptus behriana* F. Muell. and *Amyema miquelii* (Lehm. ex Miq.) Tiegh. at permanently low plant water status in the field. *Trees* **1992**, *7*, 1–7.

47. Urban, J.; Gebauer, R.; Nadezhdina, N.; Čermák, J. Transpiration and stomatal conductance of mistletoe (*Loranthus europaeus*) and its host plant, downy oak (*Quercus pubescens*). *Biologia (Bratisl.)* **2012**, *67*, 917–926. [CrossRef]

48. Orcutt, D.M.; Nilsen, E.T. *Physiology of Plants under Stress: Soil and Biotic Factors*; John Wiley & Sons, Inc.: Hoboken, NJ, USA, 2000.

49. Goldstein, G.; Rada, F.; Sternberg, L.; Burguera, J.L.; Burguera, M.; Orozco, A.; Montilla, M.; Zabala, O.; Azocar, A.; Canales, M.J.; et al. Gas exchange and water balance of a mistletoe species and its mangrove hosts. *Oecologia* **1989**, *78*, 176–183. [CrossRef]

50. Ehleringer, J.R.; Schulze, E.-D.; Ziegler, H.; Lange, O.L.; Farquhar, G.D.; Cowar, I.R. Xylem-tapping mistletoes: Water or nutrient parasites? *Science* **1985**, *227*, 1479–1481. [CrossRef] [PubMed]

51. Wang, Z.; Li, G.; Sun, H.; Ma, L.; Guo, Y.; Zhao, Z.; Gao, H.; Mei, L. Effects of drought stress on photosynthesis and photosynthetic electron transport chain in young apple tree leaves. *Biol. Open* **2018**, *7*, bio035279. [CrossRef]

52. Queijeiro-Bolaños, M.E.; Malda-Barrera, G.X.; Carrillo-Angeles, I.G.; Suzán-Azpiri, H. Contrasting gas exchange effects on the interactions of two mistletoe species and their host Acacia schaffneri. *J. Arid Environ.* **2020**, *173*, 104041. [CrossRef]

53. Epron, D.; Dreyer, E.; Bréda, N. Photosynthesis of oak trees [*Quercus petraea* (Matt.) Liebl.] during drought under field conditions: Diurnal course of net CO_2 assimilation and photochemical efficiency of photosystem II. *Plant Cell Environ.* **1992**, *15*, 809–820. [CrossRef]

54. Morecroft, M.D.; Roberts, J.M. Photosynthesis and stomatal conductance of mature canopy oak (*Quercus robur*) and sycamore (*Acer pseudoplatanus*) trees throughout the growing season. *Funct. Ecol.* **1999**, *13*, 332–342. [CrossRef]

55. Osuna, J.L.; Baldocchi, D.D.; Kobayashi, H.; Dawson, T.E. Seasonal trends in photosynthesis and electron transport during the Mediterranean summer drought in leaves of deciduous oaks. *Tree Physiol.* **2015**, *35*, 485–500. [CrossRef]

56. Strong, G.L.; Bannister, P.; Burritt, D. Are mistletoes shade plants? CO_2 assimilation and chlorophyll fluorescence of temperate mistletoes and their hosts. *Ann. Bot.* **2000**, *85*, 511–519. [CrossRef]

57. Marshall, J.D.; Dawson, T.E.; Ehleringer, J.R. Integrated nitrogen, carbon, and water relations of a xylem-tapping mistletoe following nitrogen fertilization of the host. *Oecologia* **1994**, *100*, 430–438. [CrossRef] [PubMed]

58. Lüttge, U.; Haridasan, M.; Fernandes, G.W.; de Mattos, E.A.; Trimborn, P.; Franco, A.C.; Caldas, L.S.; Ziegler, H. Photosynthesis of mistletoes in relation to their hosts at various sites in tropical Brazil. *Trees* **1998**, *12*, 167–174. [CrossRef]

59. Tennakoon, K.U.; Pate, J.S. Heterotrophic gain of carbon from hosts by the xylem-tapping root hemiparasite *Olax phyllanthi* (Olacaceae). *Oecologia* **1996**, *105*, 369–376. [CrossRef] [PubMed]

60. Marshall, J.D.; Ehleringer, J.R. Are xylem-tapping mistletoes partially heterotrophic? *Oecologia* **1990**, *84*, 244–248. [CrossRef]

61. Schulze, E.D.; Gebauer, G.; Ziegler, H.; Lange, O.L. Estimates of nitrogen fixation by trees on an aridity gradient in Namibia. *Oecologia* **1991**, *88*, 451–455. [CrossRef]
62. Sangüesa-Barreda, G.; Linares, J.C.; Julio Camarero, J. Drought and mistletoe reduce growth and water-use efficiency of Scots pine. *For. Ecol. Manag.* **2013**, *296*, 64–73. [CrossRef]
63. Cornic, G.; Massacci, A. Leaf photosynthesis under drought stress. In *Photosynthesis and the Environment*; Advances in photosynthesis and respiration; Baker, N.R., Ed.; Springer: Dordrecht, The Netherlands, 1996; pp. 347–366.
64. Flexas, J.; Medrano, H. Energy dissipation in C3 plants under drought. *Funct. Plant Biol.* **2002**, *29*, 1209. [CrossRef]
65. Flexas, J.; Escalona, J.M.; Medrano, H. Down-regulation of photosynthesis by drought under field conditions in grape vine leaves. *Aust. J. Plant Physiol.* **1998**, *25*, 893–900.
66. Björkman, O.; Demmig-Adams, B. Regulation of photosynthetic light energy capture, conversion, and dissipation in leaves of higher plants. In *Ecophysiology of Photosynthesis*; Schulze, E.-D., Caldwell, M.M., Eds.; Springer Berlin Heidelberg: Berlin/Heidelberg, Germany, 1995; pp. 17–47.
67. Aziz, A.; Larher, F. Osmotic stress induced changes in lipid composition and peroxidation in leaf discs of Brassica napus L. *J. Plant Physiol.* **1998**, *153*, 754–762. [CrossRef]
68. Brestic, M.; Zivcak, M. PSII fluorescence techniques for measurement of drought and high temperature stress signal in crop plants: Protocols and applications. In *Molecular Stress Physiology of Plants*; Rout, G.R., Das, A.B., Eds.; Springer India: New Delhi, India, 2013; pp. 87–131.
69. Konôpková, A.; Húdoková, H.; Ježík, M.; Kurjak, D.; Jamnická, G.; Ditmarová, Ľ.; Gömöry, D.; Longauer, R.; Tognetti, R.; Pšidová, E. Origin rather than mild drought stress influenced chlorophyll a fluorescence in contrasting silver fir (*Abies alba* Mill.) provenances. *Photosynthetica* **2020**, *58*, 549–559. [CrossRef]
70. Awasthi, R.; Kaushal, N.; Vadez, V.; Turner, N.C.; Berger, J.; Berger, K.H.M.; Nayyar, N. Individual and combined effects of transient drought and heat stress on carbon assimilation and seed filling in chickpea. *Funct. Plant Biol. FPB* **2014**, *41*, 1148–1167. [CrossRef] [PubMed]
71. Forner, A.; Valladares, F.; Aranda, I. Mediterranean trees coping with severe drought: Avoidance might not be safe. *Environ. Exp. Bot.* **2018**, *155*, 529–540. [CrossRef]
72. Galmés, J.; Ribas-Carbó, M.; Medrano, H.; Flexas, J. Rubisco activity in Mediterranean species is regulated by the chloroplastic CO_2 concentration under water stress. *J. Exp. Bot.* **2011**, *62*, 653–665. [CrossRef] [PubMed]
73. Galmés, J.; Flexas, J.; Savé, R.; Medrano, H. Water relations and stomatal characteristics of Mediterranean plants with different growth forms and leaf habits: Responses to water stress and recovery. *Plant Soil* **2007**, *290*, 139–155. [CrossRef]
74. Demmig-Adams, B.; Adams, W.W. Xanthophyll cycle and light stress in nature: Uniform response to excess direct sunlight among higher plant species. *Planta* **1996**, *198*, 460–470. [CrossRef]
75. Jahns, P.; Holzwarth, A.R. The role of the xanthophyll cycle and of lutein in photoprotection of photosystem II. *Biochim. Biophys. Acta BBA Bioenerg.* **2012**, *1817*, 182–193. [CrossRef]
76. Kościelniak, J.; Filek, W.; Biesaga-Kościelniak, J. The effect of drought stress on chlorophyll fluorescence in *Lolium-Festuca* hybrids. *Acta Physiol. Plant.* **2006**, *28*, 149–158. [CrossRef]
77. Aukema, J.E.; Martinéz del Rio, C. Where does a fruit-eating bird deposit mistletoe seeds? Seed deposition patterns and an experiment. *Ecology* **2002**, *83*, 3489–3496. [CrossRef]
78. Monteiro, R.F.; Martins, R.P.; Yamamoto, K. Host specificity and seed dispersal of *Psittacanthus robustus* (Loranthaceae) in Southeast Brazil. *J. Trop. Ecol.* **1992**, *8*, 307–314. [CrossRef]
79. Skórka, P.; Wójcik, J.D. Winter bird communities in a managed mixed oak-pine forest (Niepolomice Forest, Southern Poland). *Acta Zool. Crac.* **2003**, *46*, 29–41.
80. Eliáš, P. Quantitative ecological analysis of a mistletoe *Loranthus europaeus* jacq. population in an oak hornbeam forest space continuum approach. *Ekologia* **1987**, *6*, 359–372.
81. Overton, J.M.C. Dispersal and infection in mistletoe metapopulations. *J. Ecol.* **1994**, *82*, 711–723. [CrossRef]
82. Arruda, R.; Fadini, R.F.; Carvalho, L.N.; Del Claro, K.; Mourão, F.A.; Jacobi, C.M.; Teodoro, G.S.; Berg, E.; Caires, C.S.; Dettke, G.A. Ecology of neotropical mistletoes: An important canopy-dwelling component of Brazilian ecosystems. *Acta Bot. Bras.* **2012**, *26*, 264–274. [CrossRef]
83. Gressler, A.; Schaub, M.; McDowell, N.G. The role of nutrients in drought-induced tree mortality and recovery. *New Phytol.* **2017**, *214*, 513–520. [CrossRef] [PubMed]

84. Michopoulos, P.; Kaoukis, K.; Karetsos, G.; Grigoratos, T.; Samara, C. Nutrients in litterfall, forest floor and mineral soils in two adjacent forest ecosystems in Greece. *J. For. Res.* **2020**, *31*, 291–301. [CrossRef]

85. Nilsson, L.-O. Forest biogeochemistry interactions among greenhouse gases and N deposition. *Water. Air. Soil Pollut.* **1995**, *85*, 1557–1562. [CrossRef]

86. Hosseini, S.M.; Kartoolinejad, D.; Mirnia, S.K.; Tabibzadeh, Z.; Akbarinia, M.; Shayanmehr, F. The effects of *Viscum album* L. on foliar weight and nutrients content of host trees in Caspian forests (Iran). *Pol. J. Ecol.* **2007**, *55*, 579–583.

87. Hosseini, S.M.; Kartoolinejad, D.; Mirnia, S.K.; Tabibzadeh, Z.; Akbarinia, M.; Shayanmehr, F. The European mistletoe effects on leaves and nutritional elements of two host species in Hyrcanian forests. *Silva Luisiana* **2008**, *16*, 9.

88. Lamont, B. Mineral nutrition of mistletoes. In *The Biology of Mistletoes*; Calder, M., Bernhardt, P., Eds.; Academic Press Australia: Sydney, Australia, 1983; pp. 185–204.

89. Kutbay, H.; Karaer, F.; Kilinç, M. The Relationships of Some Nutrients between *Cuscuta epithymum* (L.) L. var. *epithymum* and *Heliotropium europaeum* L. Available online: /paper/The-Relationships-of-Some-Nutrients-Between-Cuscuta-Kutbay-Karaer/cf9b77e40ef40542f6de9a39ef0e5a4f1d69a1ca (accessed on 1 September 2020).

90. Malicki, L.; Berbeciowa, C. Content of basic macroelements in common parasitical weeds. *Acta Agrobot.* **1986**, *39*, 123–128. [CrossRef]

91. Türe, C.; Böcük, H.; Aşan, Z. Nutritional relationships between hemi-parasitic mistletoe and some of its deciduous hosts in different habitats. *Biologia (Bratisl.)* **2010**, *65*, 859–867. [CrossRef]

92. Bowie, M.; Ward, D. Water and nutrient status of the mistletoe *Plicosepalus acaciae* parasitic on isolated Negev Desert populations of Acacia raddiana differing in level of mortality. *J. Arid Environ.* **2004**, *56*, 487–508. [CrossRef]

Publisher's Note: MDPI stays neutral with regard to jurisdictional claims in published maps and institutional affiliations.

Article

A Complex Method for Estimation of Multiple Abiotic Hazards in Forest Ecosystems

Hana Středová [1,2], Petra Fukalová [1], Filip Chuchma [2] and Tomáš Středa [2,3,*]

[1] Department of Applied and Landscape Ecology, Mendel University in Brno, Zemědělská 1, 613 00 Brno, Czech Republic; hana.stredova@mendelu.cz (H.S.); petra.fukalova@mendelu.cz (P.F.)
[2] Czech Hydrometeorological Institute, Brno—Branch Office, Kroftova 43, 616 67 Brno, Czech Republic; filip.chuchma@chmi.cz
[3] Department of Crop Science, Breeding and Plant Medicine, Mendel University in Brn, Zemědělská 1, 613 00 Brno, Czech Republic
* Correspondence: streda@mendelu.cz; Tel.: +420-545-132-408

Received: 27 August 2020; Accepted: 11 October 2020; Published: 15 October 2020

Abstract: Forest ecosystems are faced with a variety of threats, including increasingly prolonged droughts and other abiotic stresses such as extreme high temperatures, very strong wind, invasive insect outbreaks, and the rapid spread of pathogens. The aim of the study was to define crucial abiotic stressors affecting Central Europe forest ecosystems and, with regard to their possible simultaneous effect, develop a universal method of multi-hazard evaluation. The method was then applied to the particular area of interest represented by part of the Czech Republic with forest land cover (12–19 ° E, 48–51 ° N). Based on National Threat Analysis, the most significant threats of natural origin with a close relationship to forest stability were identified as drought, high temperature, and wind gusts. Using suitable indicators, a level of their risk based on occurrence and consequences was estimated. The resulting combined level of risk, divided into five categories, was then spatially expressed on a grid map. The novelty of our paper lies in: (i) all relevant climatic data were combined and evaluated simultaneously with respect to the different level of risk, (ii) the developed methodological road map enables an application of the method for various conditions, and (iii) multiple hazards were estimated for the case study area.

Keywords: forest ecosystem; climate change; abiotic stressors; drought; wind gust; high temperature

---

1. Introduction

Forest ecosystems are affected by various threats, including increasingly prolonged and severe droughts and other abiotic stresses, invasive insect outbreaks, and the rapid spread of pathogens [1]. The impact of climate change on forests is an important issue that concerns a large number of the international scientific community [2–5].

Safety and security of ecosystems, and their fundamental functions and services, is a crucial security issue. Serious damage to the environment has far-reaching implications, including endangered basic functions of states, as evidenced by environmental crises that have led to conflict escalation in the past. Environmental threats have two time horizons. Threats caused by extreme meteorological events, such as storms and floods, constitute a sudden and intense emergency, whereas others, such as droughts, are characterized by a slow and gradual development. A complex approach must address not only both time dimensions but also all phases and mutual combinations. It has been proven that environmental crises do not arise only as a consequence of particular stressors but also by their combination and interactions. Previous events clearly indicate that environmental crises are typified by synergic, cascade, and domino effects, when one stressor is forced or weakened by others. Several crises with different triggers can affect an ecosystem simultaneously and their resulting effect is much stronger. Ref. [6] developed and tested individual and collective multi-hazard risks and noted that natural hazards are often studied in isolation. However, there is a great need to examine hazards holistically to better manage the complexity of threats found in a given region. Many regions of the world have complex hazard landscapes in which risk from individual and/or multiple extreme events is omnipresent. The issue of multiple hazards is also a subject of a UN strategy for disaster risk reduction and part of the consecutive Sendai Framework for 2015–2030. The framework introduces innovative risk attitudes, ranging from risk control to risk management, thus creating the opportunity for strengthened prevention and for the application of the BBB principle—"Build Back Better"—even in the case of the ecosystems and planting of forests. Analogously, the EU Science Hub has highlighted the need for multi-hazard assessment when prioritizing security research for 2021–2030.

A wide range of national and international organizations and strategic documents exist, such as the United Nations Office for Disaster Risk Reduction (UNISDR) and the consecutive Sendai Framework for Disaster Risk Reduction 2015–2030 [7]. In addition, the European Commission's Science and Knowledge Service (EU Science Hub), the Internal Security Strategy for the European Union [8], Transforming Our World, the 2030 Agenda for Sustainable Development [9], the United Nations Framework Convention on Climate Change (UNFCCC) [10], the Strategic Concept for the Defence and Security of the Members of the North Atlantic Treaty Organisation [11], and the Conception of Environmental Security of the Czech Republic [12] each identify and address relevant security risks, including natural catastrophes and risk of natural origins. At the Czech national level, the basic security document, Security Strategy of the Czech Republic 2015 [13], includes catastrophes of natural origin with anthropological risks among the basic security issues. A detailed classification and description of particular risks was conducted in [12]. Coordinated by the Ministry of Defense, a complex Threat analysis for the Czech Republic was also carried out [14]. The final report summarizes the natural abiotic threats with an unacceptable level of risk: floods and extreme precipitation, long-term drought, extremely high temperature, and high wind speed and gusts. These risks were amplified when combined.

1.1. High Temperature and Heat Wave and Forest

The increasing trend in the number and intensity of heat waves is likely to continue throughout the 21st century. Heat waves are often accompanied by drought. Heat wave events affect a wide variety of tree functions. Photosynthesis is reduced, photooxidative stress increases, leaves abscise, and the growth rate of the remaining leaves decreases. Stomatal conductance of some species increases at high temperatures, which may be a mechanism for leaf cooling. Heat stress reduces growth and alters biomass allocation. When drought stress accompanies heat waves, the negative effects of heat stress worsen and can lead to tree mortality [15–17].

Other authors [18,19] also emphasize that common temperate European forest tree species are more vulnerable to extreme drought and heat waves. Because drought and heat events are likely to occur more frequently due to climate change, temperate European forests might approach a point at which a substantial ecological and economic transition is required. The study of [20] also highlights the urgent need for a pan-European ground-based monitoring network that is suitable for monitoring individual tree mortality.

1.2. Drought and Forest

Climate scenarios predict a rapid warming and an increase in the frequency and duration of droughts across Europe [21]. One of Central Europe's most severe summers, characterized by a long drought and heat wave, was recorded in 2018. On par with that of 2003, the summer of 2018 was classified as the most severe event in Europe of the previous 500 years. Nonetheless, the drought event in 2018 was climatically more extreme and had a larger impact on Central Europe's forest ecosystems than the 2003 drought. The most ecologically and economically important tree species in the temperate forests of Austria, Slovakia, Germany, and Switzerland showed severe symptoms of drought stress, such as exceptionally low foliar water potentials crossing the threshold for xylem hydraulic failure, widespread leaf discoloration, premature leaf shedding, and drought-induced tree mortality [22]. Moreover, in the following year the forests were highly vulnerable to secondary drought impacts, such as insect or fungal infection pressure [20,23]. Productive, hydrological, and soil protection functions are harmed in mountain forests subject to such disasters [24,25].

According to [26], drought-induced tree mortality in Israel has been observed in diffuse patterns in forests, typically affecting approximately 5% of the trees in each forest. "Hotter" droughts may have more severe impacts on terrestrial vegetation than "normal" droughts (i.e., droughts that occur at lower, more typical temperatures) [27]. Ref. [28] evaluated the consequences of the 2000 and 2012 droughts on stands of Scots pine forest in Romania. In these two years, low precipitation and warm conditions during the growing season significantly reduced Scots pine growth and productivity, and triggered mortality events in some areas. The authors concluded that it is necessary to develop adaptive forest management and mitigation strategies across drought-prone forested regions of Europe as a response to extreme climate events.

1.3. Windstorms and Wind Gusts and Forests

Windstorms and strong winds cause major disturbances and consequent economic losses in European forests. During the past century, in particular, forests in Europe have suffered from windstorms and this trend is likely to continue [18,29–31]. For example, in Finland, a total of 7 million m^3 of timber was damaged in 2001 in two separate storms (Pyry and Janika) in autumn, and four strong storms caused damage in 2010 [32].

Wind damage is influenced by forest characteristics (e.g., tree species, tree height, crown and rooting characteristics, and stand density), forest management, and the abiotic environment, particularly local wind and soil conditions [33–35]. The mean wind speed and duration, in addition to gust level, are also significant influences.

According to [34,36], the probability of wind damage increases with tree height and for certain species, e.g., Norway spruces are particularly vulnerable to wind. Forest management also influences wind damage sensitivity. Trees grown in sheltered conditions and later exposed to wind are especially sensitive to damage [34,37,38]. Areas exposed to strong wind gusts [39] or where rooting conditions are limited due to soil characteristics [40] are more predisposed to wind damage.

Ref. [41] studied combined information on windstorms and damage to forest stands from a long-term perspective (1801–2015). The concentration of extremely damaging windstorms after 1950 (9 of the 14 selected) is an indication of increasing windstorm damage to forests in the Czech Republic. This can be partly attributed to the cultivation of two tree generations of spruce monoculture in places where diverse forests once stood, and clear-felling management of those forests.

The current study identifies key abiotic stressors affecting Central Europe forests and, with regard to their combined effect, provides a universal method of multi-hazard evaluation. The method was applied to a case study (Figures 1 and 2).

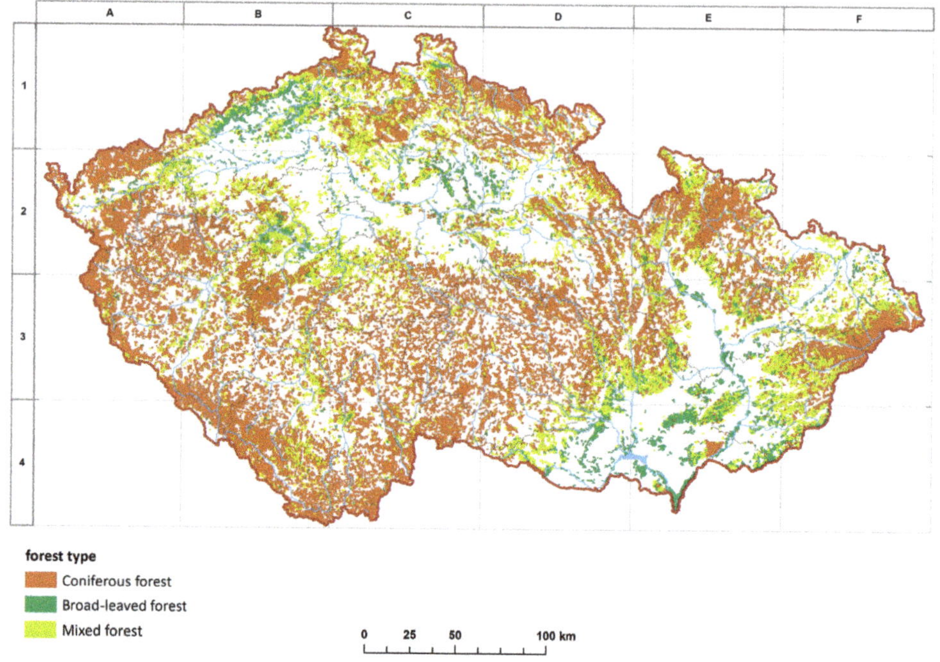

Figure 1. Area of interest.

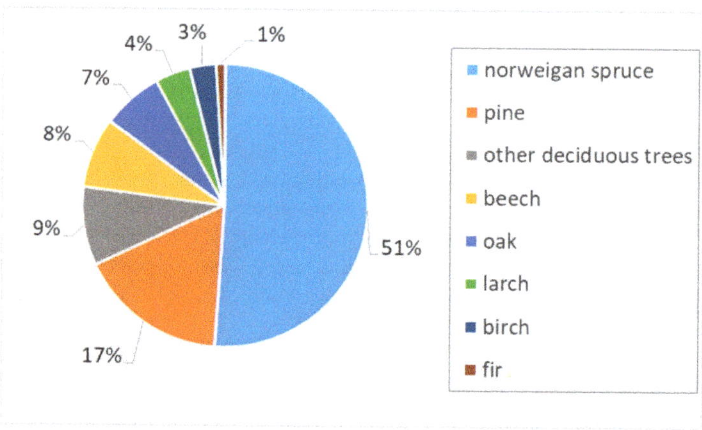

Figure 2. Species composition of forests in the Czech Republic; data source: Forest Management Institute.

2. Materials and Methods

2.1. Methodological Roadmap in Eight Steps

1. Identification of relevant threats
2. Election of suitable indicator for each threat (see Table 1)
3. Estimation of level of risk for each threat based on [14] in the Czech Republic (see Tables 2–7)
4. Calculation and spatial expression of long-term annual value of all indicators (Figures 3–5)

5. Overlapping of three risk maps
6. Final determination of combined abiotic risk according to Table 9
7. Division of each indicator into five risk categories (Table 8, Figure 6)
8. Spatial expression of combined risk (Figure 7).

Table 1. Individual abiotic threats and their indicators.

Abiotic Threat		Indicator	
High temperature	HT	long-term annual number of tropical days	TD
Drought	D	long-term annual water balance	WB
Wind gust	WG	long-term annual occurrence of wind speed above 10 m s^{-1} (% of time)	WS10

Table 2. Frequency of the threat.

Occurrence of the Threat	F
Up to half a year	10
40 to 90 years	5
1000 or more months	1

Table 3. Consequence on people coefficient.

Death Cases	Endangered People	C_{P1}; C_{P1}
1–4	1–20	1
51–100	500–1000	5
more than 1000	more than 1,000,000	10

Table 4. Consequence on the environment coefficient according to the level of environmental damage.

Affected Area	C_{En}
Water stream (up to 2 km) Water reservoirs beyond waterworks (up to 1 ha) Non-specific natural ecosystem (up to 1 ha)	1–2
Protected hydrologic areas Water stream (up to 5 km) Water reservoirs beyond waterworks (above 1 ha) Non-specific natural ecosystem (up to 3 ha)	3–5
Nature reserves, NATURA 2000 areas (up to 0.5 ha) Non-specific natural ecosystem (up to 100 ha) Protected hydrologic areas and waterworks Water stream (up to 10 km)	6–8
Nature reserves, NATURA 2000 areas (above 0.5 ha) Non-specific natural ecosystem (above 100 ha) Protected hydrologic areas and waterworks Water stream (above 10 km)	9–10

Table 5. Consequence on the economy coefficient.

Direct Economic Losses Including the Cost of Emergency and First Response	C_{Ec}
up to 4,000,000 EUR	1
4,000,000,000 to 20,000,000,000 EUR	5
more than 40,000,000,000 EUR	10

Table 6. Social consequences coefficient.

People Affected by Restrictions	Duration of the Restrictions	Level of Restrictions	C_{S1}; C_{S2}; C_{S3}
up to 1000	several h (12)	without significant restriction	1
50,000 to 125,000	several weeks	public services and basic commodities failure, blackout	5
more than 5,000,000	more than 25 years	complete destabilization of the state	10

Table 7. Estimation of level of risk for each threat; data source [14].

Threat	F	C_{P1}	C_{P2}	C_{En}	C_{Ec}	C_{S1}	C_{S2}	C_{S3}	C	R
D	7	2	6	10	5	9	6	4	5.9	41
HT	7	5	7	6	5	7	4	3	5.5	39
WG	8	2	4	8	4	5	4	4	4.5	36

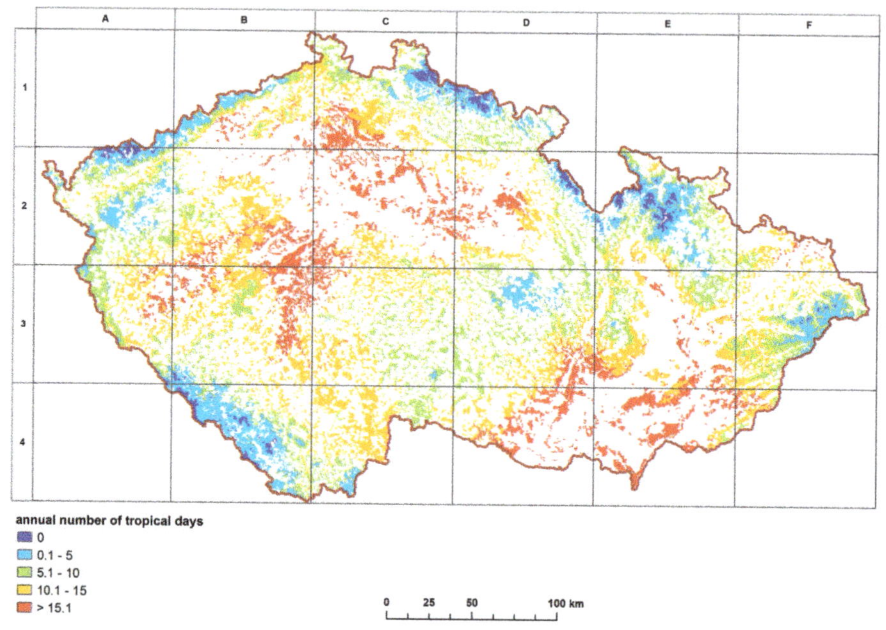

Figure 3. Categories of high temperature (HT) risk based on long-term annual number of tropical days (TDs).

Table 8. Overview of threat categories, their weights, and range of relevant indicators.

Categories of D, HT, WG	Weight of HT Categories	Range of TD (Number of Days)	Weight of D Categories	Range of WB (mm)	Weight of WG Categories	Range of WS10 (% of Time)
1	1.083	0.00	1.139	≤−150	1.000	≤3
2	2.167	<5	2.278	⟨150, 0)	2.000	⟨3, 6)
3	3.250	⟨5, 150)	3.417	(0, 150)	3.000	⟨6, 9)
4	4.333	(150, 300)	4.556	⟨150, 300)	4.000	⟨9, 12)
5	5.417	>300	5.694	>300	5.000	>12

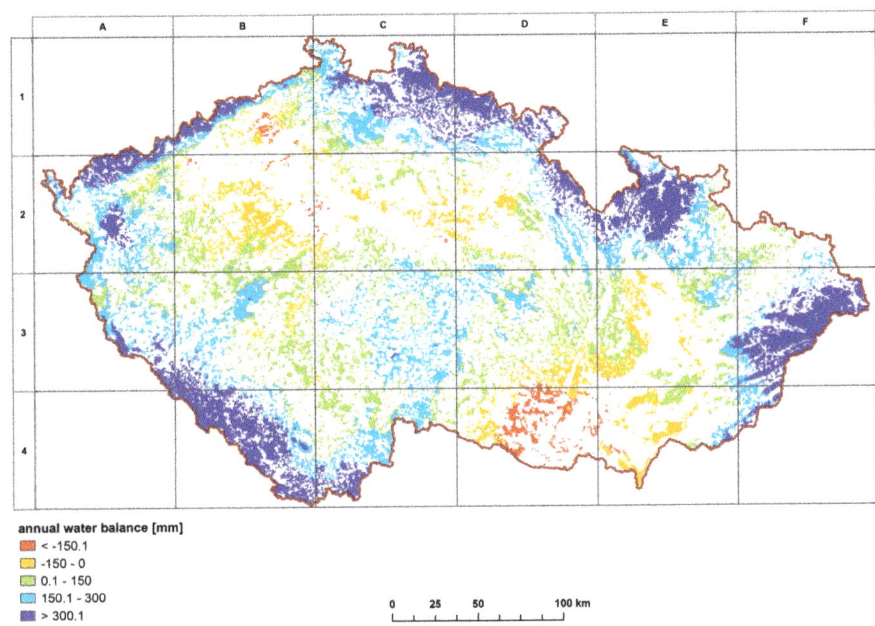

Figure 4. Categories of drought (D) risk based on long-term annual water balance (WB).

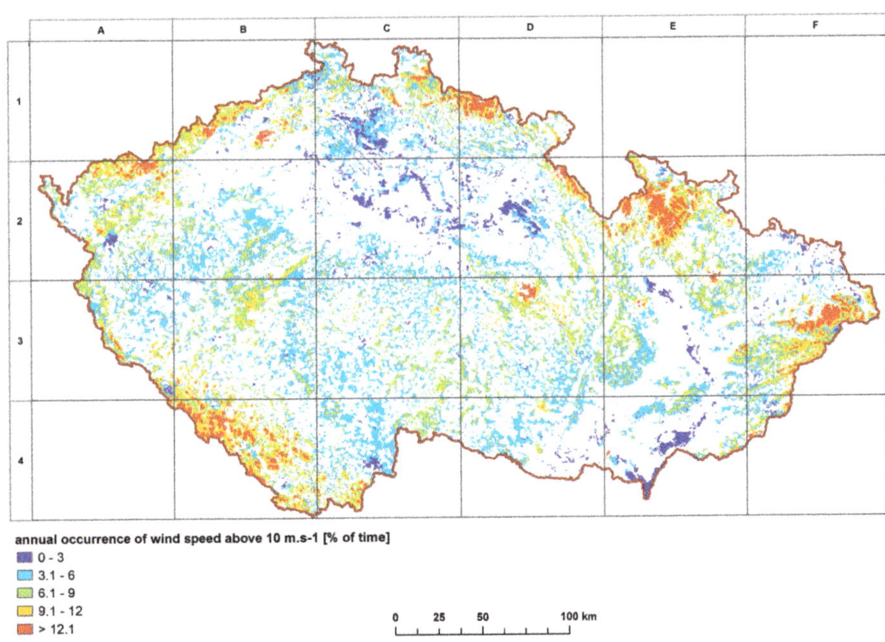

Figure 5. Categories of wind gust (WG) risk based on long-term annual occurrence of wind speeds above 10 m s^{-1} (WS10).

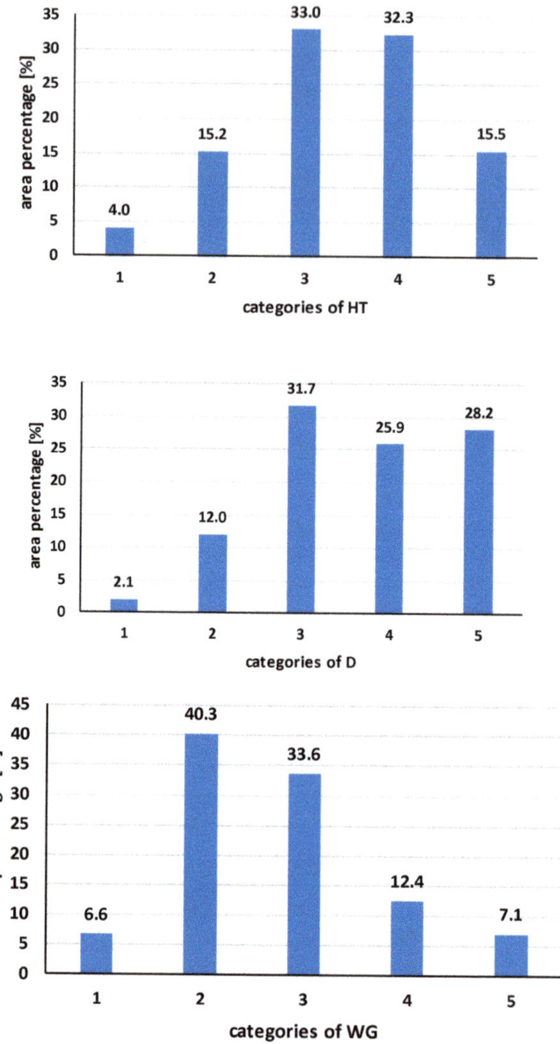

Figure 6. Percentage/area of categories of the case study area. HT = high temperature; D = drought risk; WG = wind gust.

Table 9. Matrix of degrees of combined risk for D, HT, and WG based on their weight categories.

		Weight of WG						
		1.000	**2.000**	**3.000**	**4.000**	**5.000**		
Weight of HT	**1.083**	3.222	4.222	5.222	6.222	7.222	**1.139**	
	2.167	5.445	6.445	7.445	8.445	9.445	**2.278**	
	3.250	7.667	8.667	9.667	10.667	11.667	**3.417**	**Weight of D**
	4.333	9.889	10.889	11.889	12.889	13.889	**4.556**	
	5.417	12.111	13.111	14.111	15.111	16.111	**5.694**	

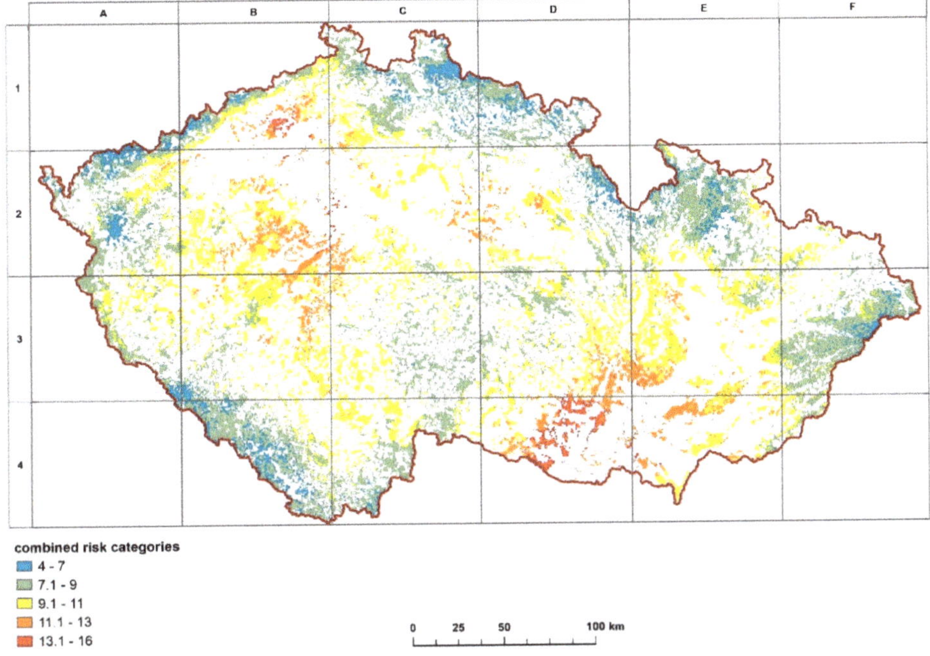

Figure 7. Spatial expression of combined risk.

2.1.1. Steps 1 and 2: Identification of Relevant Threats and Election of Suitable Indicator for Each Threat

Based on the reviews of the introduction and [14], the most significant abiotic threats of natural origin with a close relationship to woodland stability were identified as drought, high temperature, and wind gusts. Subsequently an appropriate indicator for each threat was selected (see Table 1).

2.1.2. Step 3: Estimation of Level of Risk for Each Threat

Estimation of the level of risk for each threat was conducted using the basic idea that the level of risk is related to the frequency of occurrence and the level of consequence (Equations (1) and (2)). The values of both F and C range from 0 to 10. The method of estimation of frequency and consequence is described in detail by [14].

$$R = F \times C \tag{1}$$

where R is the level of risk, F is the frequency (see Table 2), and C is the consequence.

$$C = 0.4 \times C_P + 0.2 \times C_{En} + 0.2 \times C_{Ec} + 0.2 \times C_S \tag{2}$$

where 0.4 and 0.2 are weight coefficients; C_P is the consequence on people coefficient (see Equation (3) and Table 3); C_{En} is the consequence on the environment coefficient (see Table 4); C_{Ec} is the consequence on the economy coefficient (see Table 5); and C_S is the social consequences coefficient (see Equation (4) and Table 6).

$$C_P = (C_{P1} + C_{P2})/2 \tag{3}$$

where C_{P1} is the coefficient of mortality and C_{P2} is the coefficient of endangered people.

$$C_S = (C_{S1} + C_{S2} + C_{S3})/3 \tag{4}$$

where C_{S1} is the coefficient of people affected by restrictions, C_{S2} is the duration of the restrictions coefficient, and C_{S3} is the level of restrictions coefficient.

2.1.3. Data Source

To spatially determine the degree of abiotic risk, it was necessary to prepare raster data layers with categorized values of individual selected indicators (tropical day, TD; water balance, WB; WS10—see Table 1). For the preparation of raster data layers, a data set of 268 technical climatological stations was used. A technical series of meteorological elements was created in daily intervals on the basis of measured data of the Czech Hydrometeorological Institute (CHMI) climatological station network. The process of creating a technical series of meteorological elements comprised a combination of several data quality control procedures, completion of missing values, and homogenization of time series [42–44].

2.1.4. Period of Estimation

All indicators in Table 1 were expressed as average values for the 20 year period of 2000–2019.

2.1.5. Description of Indicators

TD: A tropical day is defined as a day in which the maximum air temperature measured at a height of 2 m above the ground reaches or exceeds 30 °C. The annual number of TDs thus reflects extreme heat weather conditions of the locality.

WB: As a characteristic that indicates an increased probability of drought, the water balance is used to analyze the moisture conditions in the landscape in a straightforward manner during a certain period. The water balance is calculated as the difference between precipitation and potential evapotranspiration of grassland. Evapotranspiration in its potential form is practically identical to the maximum possible value of evaporation under optimal humidity conditions. When calculating potential evapotranspiration, the moisture conditions of the subsoil formed by the soil horizon are not taken into account. In essence, potential evapotranspiration expresses the influence of climatic conditions on the water balance (and also on evaporation) while suppressing all other factors that affect evaporation (soil moisture, etc.). This provides the scope for the concept of a climate water balance. In this study, the agrometeorological model AVISO [45–47] based on the complete Penman–Monteith combined equation (modified to the conditions of the Czech Republic) with correction for temperature of the evaporating surface and calculation of the expression of air humidity using water vapor pressure, was used to calculate grassland evapotranspiration.

WS10: To calculate the risk of wind threat, measurements of wind gusts in 15 and 10 min intervals from a set of 210 meteorological stations during the period 2000–2019 were processed. From these values, the percentage of total measurements with wind gusts above 10 m s^{-1} was calculated for the whole year. The long-term average value of each of the 3 indicators for the period 2000–2019 was then calculated individually for each station used. The preparation of climatic data and subsequent statistical analysis, including interpolation, was performed using statistical tools included in the ProClim software.

2.1.6. Software Used and Interpolation Method

The point values of long-term annual water balance, long-term annual number of tropical days, and long-term annual occurrence of wind speed above 10 m s^{-1} for the period 2000–2019 were interpolated and used to produce raster layers. The interpolation method used for raster creation was regression kriging with residuals remaining using a set of predictors (altitude, station position, slope, and terrain orientation). The resulting raster layers were processed in the ArcGIS 10.5 environment, in which raster selection was performed only on areas with forest stands. Grids were reclassified into 5 categories and were assigned a risk level value according to Table 7. The division into categories was performed according to a constructed histogram of the grid values using Sturges' rule and maintaining a partially regular range of value intervals. The result of the categorization is evident from the graphs of the percentage of raster cells in the individual classes.

The resulting raster layer of combined risk was created as the sum of reclassified raster layers of 3 indicators. Its categorization was performed in the same way as for the input indicators, when 5 categories with a range of intervals of 2 were determined. The individual layers of indicators, in addition to the combined risk layers, were processed into the final map form.

2.1.7. Area of Interest

The area of interest is located in central Europe (12–19° E, 48–51° N) and represents part of the Czech Republic, namely the area with forest land cover (see Figure 1). CHMI reports long-term (1990–2019) average annual precipitation of 614.4 mm, average annual temperature of 9 °C, annual average wind velocity in spring (March–May) of 2.5 m s^{-1}, and annual average wind velocity in autumn (September–November) of 2.2 m s^{-1}. The lowest wind velocity occurs during summer, whereas spring and autumn are windier [48].

The potential natural forest cover of the Czech Republic is 98%, whereas the present-day forested area is 34%. Roughly half of the forests in the Czech Republic are managed by Lesy CR (a state enterprise). Other major owners are private owners (25%), and towns and municipalities (17%) [49]. Of the total Czech territory (comprising 78,866 km^2), the forested area increased from 28.9% in 1845 to 33.7% in 2010 (29.0% in 1896, 30.2% in 1948, and 33.3% in 1990). The area of the main coniferous trees, i.e., spruce and pine, is declining, whereas the percentage of fir shows a slight, steady rise. The proportion of deciduous trees, particularly beech and oak, is also increasing. Figure 2 presents the species composition of forests in the Czech Republic.

3. Results

3.1. Estimation of Level of Risk for Each Threat Based on Results of the Threat Analysis for the Czech Republic (2015)

3.1.1. Step 4: Calculation of Long-Term Annual Value of All Indicators

The greatest risk of high temperatures due to the long-term number of tropical days is in areas with lower altitude, particularly South Moravia and central Bohemia (Figure 3). The risk based on the long-term annual water balance largely corresponds to the risk of high temperatures. The riskiest part of the Czech Republic is mainly South Moravia (Figure 4). The risk of wind gusts based on annual wind speeds above 10 m s^{-1} is highest in mountainous areas, i.e., the areas with the highest altitude. These mountain areas are adjacent to the borders of the Czech Republic (Figure 5).

3.1.2. Step 5: Division of Each Indicator into Five Risk Categories

According to level of risk of each threat (see Table 7), weights were assigned to indicate significance; the threat with the lowest risk level (DG: risk level of 36) was assigned a value of 1.

Table 8 provides an overall summary of all threat categories, their weights, and the range of relevant indicators.

In terms of drought risk, the majority of the case study area falls into category 3. In the case of high temperatures, most areas of the case study fall into category 3. In terms of wind gust risk, category 2 covers the majority of the area (Figure 6).

3.1.3. Steps 6–8: Overlapping of Three Risk Maps, Final Determination of Combined Abiotic Risk (According to Table 7), and Spatial Expression of Combined Risk

The combined risk of the abiotic stressors is shown in the map in Figure 7 and Table 10. The majority of the case study area in terms of combined risk falls into categories 2 (slight risk) and 3 (moderate risk) (approximately 82%) (Figure 8). It might be concluded that about 10% of the area (categories 4 and 5, i.e., high to very high risk) require special attention in terms of landscape adjustments, more detailed in situ measurement, or adaptive forest management to prevent ecological and economical losses and damage.

Table 10. Characterization of categories of combined risk in terms of their degree of combined risk (based on Table 9).

Categories of Combined Risk	Degree of Combined Risk
1–low risk	4–7
2–slight risk	7.1–9
3–moderate risk	9.1–11
4–high risk	11.1–13
5–very high risk	13.1–16

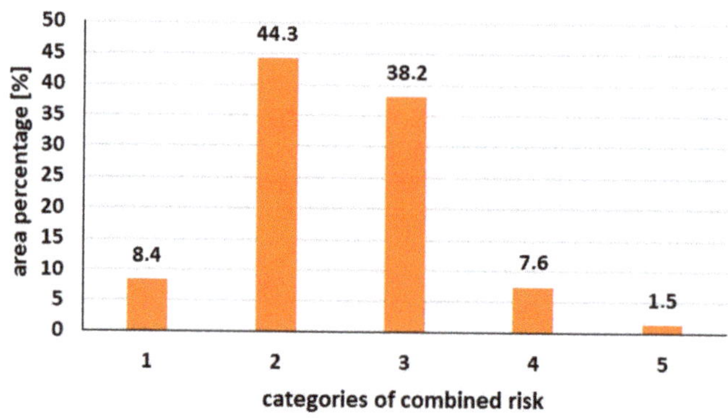

Figure 8. Categories of combined risk.

4. Discussion

Ref. [50] provided evidence of a direct link between a climate-induced increase in forest diseases and pests, and increased tree mortality. According to [17,51,52], higher temperatures also shorten the development cycles of disease and pest organisms. Many studies have found that increasing temperatures can be attributed to climate change. In their review, [53] identified six key principles for enhancing the adaptive capacity of European temperate forests in a changing climate. Climate change causes various extreme weather events, floods, and droughts, thus leading to a reduction in water quality and availability, frequent heat waves, etc. High temperature combined with severe drought and strong wind creates suitable conditions for the starting and fast spread of wildfires. Burning biomass then releases increased greenhouse gases, e.g., CO_2. Ref. [54] quantified the greenhouse gas emissions from forest fires in the European Slovak Paradise National Park, in which fire destroyed an area of 80 ha in the year 2000.

Severe and repeated droughts are currently considered one of the main factors contributing to forest dieback in Central Europe. Under Central European climatic conditions, tree species do not have sufficiently efficient defense mechanisms or strategies to survive severe drought periods without a negative impact on their physiological processes and growth. This is evidenced by the drought events in 2000, 2003, 2012, and 2018 [22,28], which had a major impact on the forest ecosystems of Central Europe. In humid temperate conditions, intense periods of drought should not be considered isolated extreme events, but events that can occur with increasing frequency in the near future [55].

Natural reserves in Central Europe, such as the Low Tatras National Park in Slovakia and the Sumava National Park in Czechia, were damaged by wind and bark beetle outbreaks, which had a major influence on the health of the forest vegetation at the end of the 20th and beginning of the 21st century.

In November 2004, forest with an area of 12,000 ha in the Tatra National Park, Slovakia, was seriously damaged by northern wind gusts exceeding 200 km h^{-1}. In July 2005, a wildfire broke out in a wind-damaged area of 220 ha [56]. The results of [57] showed that, in addition to economic losses, the fire caused significant environmental changes to the structure and properties of humic acid. However, [58] claimed that forest regeneration on sites cleared of windthrow is less intensive than in those forest ecosystems which were subjected to direct human impact, i.e., clear cuttings.

Sufficient evidence exists that individual threats, such as drought, hot weather, and extreme winds, represent a serious danger, not only for forest ecosystems, but for all natural and anthropogenic ecosystems. When these influences occur at the same time or when a given area is affected by more of them in a short time, their negative effect is considerably stronger. Although a complex analysis of forest damage was not the goal of this paper, various data on real detrimental impacts on forests provide a large amount of evidence that the health of forests largely reflects the zoning derived in the current study. Monitoring of agents that cause damage to forests (e.g., recorded damage to stands by wind; recorded damage to stands by drought; volume of spruce wood infested by bark beetles) in the Czech Republic has been undertaken by several institutions using various methods (Forestry and Game Management Research Institute [59], the Ministry of Agriculture of the Czech Republic [60], the Forest Management Institute [61]). Based on this monitoring, we can conclude that the spatial expression of combined risk in Figure 7 generally corresponds to the actual occurrence of agents causing forest damage. This is particularly true of sector D4, which was strongly affected by severe drought and an infestation of bark beetles in 1 ha of spruce stands; the area was classified as an area with significant deterioration in health status between 2016 and 2019. From this perspective, we can conclude that Figure 7 represents the potential combined risk of forest damage due to climate effects. The real impact of these risks on the forests will only be revealed in combination with actual data on stand structure, pests, and various diseases. This idea thus indicates a direction for future research i.e., validation of the findings of the current study via comparison of the "potential climate-driven combined risk" with the "real effect" of the risks on the forests.

A complex analysis of forest damage is a challenging research issue that has been addressed by a number of authors [62–64]. These previous surveys of forest damage can be significantly enhanced by a comprehensive analysis of multiple abiotic hazards which, combined with analysis of other relevant stressors, might help to identify crucial causes of forest damage.

Spatial expression of the different levels of combined risk might help optimize management and thus increase forest resilience. Theoretically, when the riskiest areas are combined with appropriate forest management, it is possible to avoid detrimental impacts on the forest landscape.

5. Conclusions

Using [14] we identified and employed key climatic elements that pose a threat to forests, namely drought, high temperature, and strong wind. When these elements are combined it is possible to evaluate the resulting level of combined risk. The developed methodological road map enables an application of the method for various conditions. We applied the method to a case study, in which we also suggested possible indicators for each element. The elements under investigation not only represent the current risks—their effect is expected to strengthen in the future due to climate change.

Although complex analysis of forest damage was not the goal of our paper, various data on real detrimental impacts on forests provide a large amount of evidence that the health of forests largely reflects the zoning derived in the current study. Biological evidence of detrimental impact is also affected by other stressors, both abiotic (such as thick and dense snow cover, winter windstorms, and heavy frosts) and biotic (e.g., the wide distribution of pests, gnawing of trees by animals), which are beyond the scope of our paper.

The potential climate-driven combined risk of forest damage might help to address the question of how information about multiple hazards in forest ecosystems can be incorporated into management decisions. Figure 7 represents one of the crucial conceptual tools to achieve a stable and resilient landscape structure. To be effective, forest management research and practice requires explicit articulation of their objectives, which should be interpreted and translated into real actions. Our results provide a useful tool for all stakeholders, particularly communities, municipalities, and government who are responsible for effective landscape planning and management. To do so, these stakeholders demand high quality data and analyses that can support their decision making. Based on our method it is possible to set threshold values that, when exceeded, trigger landscape adjustments or more detailed in situ measurements, or indicate a need of specific management.

In addition to the conceptual dimension, it is also necessary to address the issue of the real effect of abiotic threats on forest ecosystems. Follow-up research should thus focus on validation of the combined risk map using: (i) identification of the most vulnerable segments from A1 to F4, (ii) selection of appropriate data on the real occurrence of forest damage [59–61], and (iii) validation of the potential climate-driven combined risk.

Author Contributions: Conceptualization, H.S.; methodology, H.S.; software, F.C.; validation, T.S.; formal analysis, P.F.; resources, P.F.; writing—original draft preparation, P.F. and H.S.; writing—review and editing, T.S.; All authors have read and agreed to the published version of the manuscript.

Funding: This research was funded by the Ministry of Agriculture of the Czech Republic (National Agency of Agricultural Research Ministry of Agriculture of the Czech Republic) project No. QK1710197 Optimization of methods for the assessment of vulnerability to wind erosion and proposals of protective measures in intensively exploited agricultural countryside.

References

1. Mezei, P.; Jakuš, R.; Pennerstorfer, J.; Havašová, M.; Škvarenina, J.; Ferenčík, J.; Slivinský, J.; Bičárová, S.; Bilčík, D.; Blaženec, M.; et al. Storms, temperature maxima and the Eurasian spruce bark beetle Ips typographus—An infernal trio in Norway spruce forests of the Central European High Tatra Mountains. *Agric. For. Meteorol.* **2017**, *242*, 85–95. [CrossRef]

2. Andreassen, K.; Solberg, S.; Tveito, O.A.; Lystad, S.L. Regional differences in climatic responses of Norway spruce (*Picea abies* L. Karst) growth in Norway. *For. Ecol. Manag.* **2006**, *222*, 211–221. [CrossRef]

3. Bonan, G.B. Forests and Climate Change: Forcings, Feedbacks, and the Climate Benefits of Forests. *Science* **2008**, *320*, 1444–1449. [CrossRef] [PubMed]

4. Sidor, C.G.; Bosela, M.; Büntgen, U.; Vlad, R. Mixed effects of climate variation on the Scots pine forests: Age and species mixture matter. *Dendrochronologia* **2018**, *52*, 48–56. [CrossRef]

5. Chuchma, F.; Středová, H.; Středa, T. Bioindication of climate development on the basis of long-term phenological observation. In Proceedings of the International PhD Students Conference (MendelNet 2016), Brno, Czech Republic, 9–10 November 2016; Mendel University in Brno: Brno, Czech Republic, 2016; pp. 380–383.

6. Pourghasemi, H.R.; Gayen, A.; Edalat, M.; Zarafshar, M.; Tiefenbacher, J.P. Is multi-hazard mapping effective in assessing natural hazards and integrated watershed management? *Geosci. Front.* **2020**, *11*, 1203–1217. [CrossRef]

7. United Nations Office for Disaster Risk Reduction. *Sendai Framework for Disaster Risk Reduction 2015–2030*; United Nations Office for Disaster Risk Reduction: Geneva, Switzerland, 2015.

8. European Council. *Internal Security Strategy for the European Union—Towards a European Security Model*; Publications Office of the European Union: Luxembourg, 2010.

9. General Assembly Resolution A/RES/70/1. Transforming Our World, the 2030 Agenda for Sustainable Development. Available online: https://www.un.org/ga/search/view_doc.asp?symbol=A/RES/70/1&Lang=E (accessed on 22 September 2020).

10. United Nations Framework Convention on Climate Change (UNFCCC). Decision 1/CP.21: Adoption of the Paris Agreement. Available online: https://unfccc.int/files/essential_background/convention/application/pdf/english_paris_agreement.pdf (accessed on 22 September 2020).

11. North Atlantic Treaty Organization. *Active Engagement, Modern Defence, Strategic Concept for the Defence and Security of the Members of the North Atlantic Treaty Organisation Adopted by Heads of State and Government in Lisbon*; NATO Public Diplomacy Division: Brussels, Belgium, 2010.

12. Aktualizace Koncepce Environmentální Bezpečnosti, a to na Období 2016–2020 s Výhledem do Roku 2030 [in Czech]. Available online: https://www.dataplan.info/img_upload/7bdb1584e3b8a53d337518d988763f8d/koncepce-2015.pdf (accessed on 22 September 2020).

13. Security Strategy of the Czech Republic 2015. Available online: http://www.army.cz/images/id_8001_9000/8503/Security_Strategy_2015.pdf (accessed on 22 September 2020).

14. MOI (Ministry of the Interior of the Czech Republic) Threat Analysis for the Czech Republic (2015). Database of Strategies. *Portal of Strategic Documents in the Czech Republic.* Available online: https://www.databaze-strategie.cz/cz/mv/strategie/aanlyza-hrozeb-pro-ceskou-republiku-2015 (accessed on 16 September 2020).

15. Teskey, R.; Wertin, T.; Bauweraerts, I.; Ameye, M.; McGuire, M.A.; Steppe, K. Responses of tree species to heat waves and extreme heat events. *Plant Cell Environ.* **2014**, *38*, 1699–1712. [CrossRef]

16. Středa, T.; Litschmann, T.; Středová, H. Relationship between tree bark surface temperature and selected meteorological elements. *Contrib. Geophys. Geod.* **2015**, *45*, 299–311. [CrossRef]

17. Středa, T.; Středová, H.; Rožnovský, J. Orchards microclimatic specifics. In *Bioclimate: Source and Limit of Social Development. Nitra, Slovak Republic*; Slovak Agricultural University: Nitra, Slovak, 2011; pp. 132–133.

18. Seidl, R.; Thom, D.; Kautz, M.; Martin-Benito, D.; Peltoniemi, M.; Vacchiano, G.; Wild, J.; Ascoli, D.; Petr, M.; Honkaniemi, J.; et al. Forest disturbances under climate change. *Nat. Clim. Chang.* **2017**, *7*, 395–402. [CrossRef]

19. Škvarenina, J.; Tomlain, J.; Hrvol, J.; Škvareninová, J. Occurrence of Dry and Wet Periods in Altitudinal Vegetation Stages of West Carpathians in Slovakia: Time-Series Analysis 1951–2005. In *Bioclimatology and Natural Hazards*; Střelcová, K., Ed.; Springer: Dordrecht, The Netherlands, 2009; pp. 97–106. [CrossRef]

20. Schuldt, B.; Buras, A.; Arend, M.; Vitasse, Y.; Beierkuhnlein, C.; Damm, A.; Gharun, M.; Grams Thorsten, E.E.; Hauck, M.; Hajek, P.; et al. A first assessment of the impact of the extreme 2018 summer drought on Central European forests. *Basic Appl. Ecol.* **2020**, *45*, 86–103. [CrossRef]

21. Rajczak, J.; Pall, P.; Schär, C. Projections of extreme precipitation events in regional climate simulations for Europe and the Alpine Region. *J. Geophys. Res.-Atmos.* **2013**, *118*, 3610–3626. [CrossRef]

22. Vido, J.; Střelcová, K.; Nalevanková, P.; Leštianska, A.; Kandrík, R.; Pástorová, A.; Škvarenina, J.; Tadesse, T. Identifying the relationships of climate and physiological responses of a beech forest using the Standardised Precipitation Index: A case study for Slovakia. *J. Hydrol. Hydromech.* **2016**, *64*, 246–251. [CrossRef]

23. Šustek, Z.; Vido, J.; Škvareninová, J.; Škvarenina, J.; Šurda, P. Drought impact on ground beetle assemblages (Coleoptera, Carabidae) in Norway spruce forests with different management after windstorm damage–a case study from Tatra Mts. (Slovakia). *J. Hydrol. Hydromech.* **2017**, *65*, 333–342. [CrossRef]

24. Bartík, M.; Sitko, R.; Oreňák, M.; Slovik, J.; Škvarenina, J. Snow accumulation and ablation in disturbed mountain spruce forest in West Tatra Mts. *Biologia* **2014**, *69*, 1492–1501. [CrossRef]

25. Bartík, M.; Holko, L.; Jančo, M.; Škvarenina, J.; Danko, M.; Kostka, Z. Influence of mountain spruce forest dieback on snow accumulation and melt. *J. Hydrol. Hydromech.* **2019**, *67*, 59–69. [CrossRef]

26. Dorman, M.; Svoray, T.; Perevolotsky, A. Homogenization in forest performance across an environmental gradient—The interplay between rainfall and topographic aspect. *For. Ecol. Manag.* **2013**, *310*, 256–266. [CrossRef]

27. Allen, C.D.; Breshears, D.D.; McDowell, N.G. On underestimation of global vulnerability to tree mortality and forest die-off from hotter drought in the Anthropocene. *Ecosphere* **2015**, *6*, 1–55. [CrossRef]

28. Sidor, C.G.; Camarero, J.J.; Popa, I.; Badea, O.; Apostol, E.N.; Vlad, R. Forest vulnerability to extreme climatic events in Romanian Scots pine forests. *Sci. Total Environ.* **2019**, *678*, 721–727. [CrossRef]

29. Schelhaas, M.J.; Nabuurs, G.J.; Schuck, A. Natural disturbances in the European forests in the 19th and 20th centuries. *Glob. Chang. Biol.* **2003**, *9*, 1620–1633. [CrossRef]

30. Seidl, R.; Schelhaas, M.J.; Lexer, M.J. Unraveling the drivers of intensifying forest disturbance regimes in Europe. *Glob. Chang. Biol.* **2011**, *17*, 2842–2852. [CrossRef]

31. Gregow, H.; Laaksonen, A.; Alper, M.E. Increasing large scale windstorm damage in Western, Central and Northern European forests, 1951–2010. *Sci. Rep.* **2017**, *7*, 46397. [CrossRef]

32. Zubizarreta-Gerendiain, A.; Pellikka, P.; Garcia-Gonzalo, J.; Ikonen, V.P.; Peltola, H. Factors affecting wind and snow damage of individual trees in a small management unit in Finland: Assessment based on inventoried damage and mechanistic modelling. *Silva Fenn.* **2012**, *46*, 181–196. [CrossRef]

33. Mitchell, S.J. Wind as a natural disturbance agent in forests: A synthesis. *Forestry* **2013**, *86*, 147–157. [CrossRef]

34. Peltola, H.; Kellomäki, S.; Väisänen, H.; Ikonen, V.P. A mechanistic model for assessing the risk of wind and snow damage to single trees and stands of Scots pine, Norway spruce, and birch. *Can. J. For. Res.* **1999**, *29*, 647–661. [CrossRef]

35. Nykänen, M.L.; Peltola, H.; Quine, C.; Kellomäki, S.; Broadgate, M. Factors affecting snow damage of trees with particular reference to European conditions. *Silva Fenn.* **1997**, *31*, 193–213. [CrossRef]

36. Valinger, E.; Fridman, J. Factors affecting the probability of windthrow at stand level as a result of Gudrun winter storm in southern Sweden. *For. Ecol. Manag.* **2011**, *262*, 398–403. [CrossRef]

37. Lohmander, P.; Helles, F. Windthrow probability as a function of stand characteristics and shelter. *Scand. J. For. Res.* **1987**, *2*, 227–238. [CrossRef]

38. Suvanto, S.; Nöjd, P.; Henttonen, H.M.; Beuker, E.; Mäkinen, H. Geographical patterns in the radial growth response of Norway spruce provenances to climatic variation. *Agric. For. Meteorol.* **2016**, *222*, 10–20. [CrossRef]

39. Schindler, S.; Livoreil, B.; Pinto, I.S.; Araujo, R.M.; Zulka, K.P.; Pullin, A.S.; Santamaria, L.; Kropik, M.; Fernández-Méndez, P.; Wrbka, T. The network biodiversity knowledge in practice: Insights from three trial assessments. *Biodivers. Conserv.* **2016**, *25*, 1301–1318. [CrossRef]

40. Nicoll, B.C.; Gardiner, B.A.; Rayner, B.; Peace, A.J. Anchorage of coniferous trees in relation to species, soil type, and rooting depth. *Can. J. For. Res.* **2006**, *36*, 1871–1883. [CrossRef]

41. Brázdil, R.; Stucki, P.; Szabó, P.; Řezníčková, L.; Dolák, L.; Dobrovolný, P.; Tolasz, R.; Kotyza, O.; Chromá, K.; Suchánková, S. Windstorms and forest disturbances in the Czech Lands: 1801–2015. *Agric. For. Meteorol.* **2018**, *250*, 47–63. [CrossRef]

42. Squintu, A.A.; van der Schrier, G.; Štěpánek, P.; Zahradníček, P.; Tank, A.K. Comparison of homogenization methods for daily temperature series against an observation-based benchmark dataset. *Theor. Appl. Climatol.* **2020**, *140*, 285–301. [CrossRef]

43. Štěpánek, P.; Zahradníček, P.; Huth, R. Interpolation techniques used for data quality control and calculation of technical series: An example of Central European daily time series. *Időjárás* **2011**, *115*, 87–98.

44. Štěpánek, P.; Zahradníček, P.; Farda, A. Experiences with data quality control and homogenization of daily records of various meteorological elements in the Czech Republic in the period 1961–2010. *Időjárás* **2013**, *117*, 123–141.

45. Kohut, M. Water Balance of the Agricultural Landscape. Ph.D. Thesis, Mendel University in Brno, Brno, Czech Republic, 2007.

46. Kohut, M.; Roznovsky, J.; Chuchma, F. The long-term soil moisture reserve variability in the Czech Republic based on the AVISO model. In Proceedings of the Sustainable Development and Bioclimate: Reviewed Conference, Bratislava, Slovak, 5–8 October 2009; Slovak Academy of Sciences: Bratislava, Slovak, 2009; pp. 160–161.

47. Kohut, M.; Vitoslavsky, J. Agrometeorological computer and information system—The possibility of its use. In *Agro-Meteorological Forecasts and Models, Velke Bilovice, Czech Republic*; Czech Bioclimatological Society: Praha, Czech Republic, 1999; pp. 53–61.

48. Tolasz, R. *Atlas Podnebí Česka: Climate Atlas of Czechia*; Český Hydrometeorologický Ústav, Univerzita Palackého v Olomouci: Olomouc, Czech Republic, 2007.

49. MAE (Ministry of Agriculture of the Czech Republic) Forestry: Species Composition of Forest. Available online: http://eagri.cz/public/web/en/mze/forestry/species-composition-of-forests/ (accessed on 26 August 2020).

50. Sturrock, R.N.; Frankel, S.; Brown, A.V.; Hennon, P.; Kliejunas, J.T.; Lewis, K.J.; Worrall, J.J.; Woods, A.J. Climate change and forest diseases. *Plant Pathol.* **2011**, *60*, 133–149. [CrossRef]

51. Hlásny, T.; Turčáni, M. Insect Pests as Climate Change Driven Disturbances in Forest Ecosystems. In *Bioclimatology and Natural Hazards*; Střelcová, K., Ed.; Springer: Dordrecht, The Netherlands, 2009; pp. 165–177. [CrossRef]

52. Jönsson, A.M.; Harding, S.; Krokene, P.; Lange, H.; Lindelöw, Å.; Økland, B.; Ravn, H.P.; Schroeder, L.M. Modelling the potential impact of global warming on Ips typographus voltinism and reproductive diapause. *Clim. Chang.* **2011**, *109*, 695–718. [CrossRef]

53. Brang, P.; Spathelf, P.; Larsen, J.B.; Bauhus, J.; Bonččina, A.; Chauvin, C.; Drössler, L.; García-Güemes, C.; Heiri, C.; Kerr, G.; et al. Suitability of close-to-nature silviculture for adapting temperature European forest to climate change. *Forestry* **2014**, *87*, 492–503. [CrossRef]

54. Korísteková, K.; Vido, J.; Vida, T.; Vyskot, I.; Mikloš, M.; Minďáš, J.; Škvarenina, J. Evaluating the amount of potential greenhouse gas emissions from forest fires in the area of the Slovak Paradise National Park. *Biologia* **2020**, *75*, 885–898. [CrossRef]

55. Střelcová, K.; Kurjak, D.; Leštianska, A.; Kovalčíková, D.; Ditmarová, Ľ.; Škvarenina, J.; Ahmed, Y.A.R. Differences in transpiration of Norway spruce drought stressed trees and trees well supplied with water. *Biologia* **2013**, *68*, 1118–1122. [CrossRef]

56. Fleischer, P.; Koreň, M.; Škvarenina, J.; Kunca, V. Risk Assessment of the Tatra Mountains Forest. In *Bioclimatology and Natural Hazards*; Springer: Dordrecht, The Netherlands, 2009; pp. 145–154. [CrossRef]

57. Barančíková, G.; Jerzykiewicz, M.; Gömöryová, E.; Tobiašová, E.; Litavec, T. Changes in forest soil organic matter quality affected by windstorm and wildfire. *J. Soils Sediments* **2018**, *18*, 2738–2747. [CrossRef]

58. Suchockas, V.; Pliura, A.; Labokas, J.; Lygis, V.; Dobrowolska, D.; Jankauskiene, J.; Verbylaite, R. Evaluation of early stage regeneration of forest communities following natural and human-caused disturbances in the transitional zone between temperate and hemiboreal forests. *Balt. For.* **2018**, *24*, 131–147.

59. Knížek, M.; Liška, J. Occurrence of forest damaging agents in 2019 and forecast for 2020. *Zprav. Ochr. Lesa* **2020**, 1–76. (In Czech)

60. Zpráva o Stavu Lesa a Lesního Hospodářství České Republiky v Roce 2019 [In Czech]. Available online: http://eagri.cz/public/web/file/658587/Zprava_o_stavu_lesa_2019.pdf (accessed on 26 August 2020).

61. The Forest Management Institute. Available online: http://geoportal.uhul.cz/mapy/mapyzsl.html (accessed on 22 September 2020).

62. Touhami, I.; Chirino, E.; Aouinti, H.; El Khorchani, A.; Elaieb, M.T.; Khaldi, A.; Nasr, Z. Decline and dieback of cork oak (*Quercus suber* L.) forests in the Mediterranean basin: A case study of Kroumirie, Northwest Tunisia. *J. For. Res.* **2020**, *31*, 1461–1477. [CrossRef]

63. Bennett, A.C.; McDowell, N.G.; Allen, C.D.; Anderson-Teixeira, K.J. Larger trees suffer most during drought in forests worldwide. *Nat. Plants* **2015**, *1*, 15139. [CrossRef]

64. Banfield-Zanin, J.A.; Leather, S.R. Reproduction of an arboreal aphid pest, *Elatobium abietinum*, is altered under drought stress. *J. Appl. Entomol.* **2015**, *139*, 302–313. [CrossRef]

Publisher's Note: MDPI stays neutral with regard to jurisdictional claims in published maps and institutional affiliations.

MDPI

St. Alban-Anlage 66

4052 Basel

Switzerland

Tel. +41 61 683 77 34

Fax +41 61 302 89 18

www.mdpi.com

Water Editorial Office

E-mail: water@mdpi.com

www.mdpi.com/journal/water